MELANINS AND MELANOGENESIS

MELANINS AND MELANOGENESIS

GIUSEPPE PROTA
Università Degli Studi di Napoli Federico II
Dipartimento di Chimica Organica e Biologica
Napoli, Italy

Academic Press, Inc.
Harcourt Brace Jovanovich, Publishers
SAN DIEGO NEW YORK BOSTON LONDON SYDNEY TOKYO TORONTO

This book is printed on acid-free paper. ♾

Academic Press, Inc.
1250 Sixth Avenue, San Diego, California 92101-4311

United Kingdom Edition published by
Academic Press Limited
24–28 Oval Road, London NW1 7DX

Library of Congress Cataloging-in-Publication Data

Prota, Giuseppe.
Melanins and melanogenesis / Giuseppe Prota.
p. cm.
Includes bibliographical references and index.
ISBN 0-12-565970-9
1. Melanin. 2. Melanocytes. 3. Melanism. I. Title.
[DNLM: 1. Melanins–metabolism. QU 110 P967m]
QP671.M44P76 1992
612.7'927–dc20
DNLM/DLC
for Library of Congress 92-11108
CIP

PRINTED IN THE UNITED STATES OF AMERICA
92 93 94 95 96 97 BC 9 8 7 6 5 4 3 2 1

To Giovanna, Laura, Francesco, and Paolo

When the chemist in his approach to biological problems neglects the biological aspects, the work will usually remain dull and of limited general significance; when the biologist in his observations neglects the chemical aspects, the work will remain descriptive and come to a dead end.

E. B. Chain

CONTENTS

PREFACE

Human pigmentation is undoubtedly the natural phenomenon that provided the initial stimulus for melanin research. Other incentives came from interest in the physiological and pathological pigmentary changes that occur on exposure to sunlight, in aging, and in a number of genetic or acquired conditions such as albinism, vitiligo, and melanoma. Added to this is the intellectual challenge provided by the unfathomable structure of the melanin polymer, which has so far resisted investigation by any of the analytical techniques currently used to determine the chemistry of natural products. Such a broad variety of issues is well exemplified by the firm tradition of multidisciplinarity that has marked the history of melanin research since the turn of the century.

During the last two decades, an overwhelming array of work has, through numerous channels, advanced our knowledge and appreciation of the field. Considerable steps forward have been made in virtually all areas, thanks to the concerted involvement of researchers in biomimetic chemistry, enzymology, and molecular biology. Perhaps the most significant outcome of the recent studies has been the definition of a new biosynthetic pathway that has radically modified and expanded the classical notion of melanogenesis as derived from the Raper–Mason Scheme.

However, as is usually the case in periods of rapid development, most of the recent acquisitions have not yet been properly assessed or integrated into the multifaceted context of melanin pigmentation. Therefore, a need exists for a concise, yet comprehensive textbook presenting a composite picture of the new knowledge scattered in innumerable papers, reviews, and symposia proceedings.

This volume is an attempt to satisfy this quest. It is largely a reflection of my 30 years of activity in the field and continuing and frequent contact with specialists of widely diverse disciplines, which together provided the knowledge necessary to undertake the difficult task of bringing, for the first time, all aspects of melanin pigmentation into a single focus.

It is somewhat inevitable, when attempting a broad project, that selection of subject matter be controversial. Thus, the inclusion of certain topics or the omission of others might be frowned on by some expert readers from various disciplines. On the other hand, an all-inclusive treatise, besides being materially unrealizable, would have stretched the task beyond the purposes set forth above.

The core of the volume deals with the structure, physicochemical properties, and biosynthesis of the major classes of melanin pigments, including neuromelanins. Ample space is also given to the regulatory mechanisms of melanogenesis, including enzymatic, genetic, and hormonal control. Other themes include the biology of the various types of pigment-producing cells, the structure and mode of action of tyrosinase, and the chemistry of urinary melanogens and their biomedical applications as metabolic markers of melanocyte activity, especially for the follow-up of malignant melanoma. Finally, progress in the photobiology and photochemistry of melanins is considered with a special emphasis on the controversial role of these pigments in skin photoprotection.

Discussion of the various issues is made as accessible as possible to a broad range of readers, in the belief that biologists, physiologists, geneticists, and research-oriented physicians will profit from an updated coverage of the chemistry of melanogenesis, and that chemists, biochemists, and physicists will find relevant material in biological areas. Extended references to the literature, up to December 1991, have also been made, enabling anyone interested to follow up each subject in more detail. The hope is that, as structured, the volume will serve both as a basic guide for newcomers and as a handy source of specific information for practitioners in academic, medical, and industrial settings.

This effort has been made possible by the contributions of many persons. I am especially grateful to my collaborators, with whom I have share enthusiasm for research. In this regard, I am particularly indebted to Marco d'Ischia, Orlando Crescenzi, Alessandra Napolitano, Claudio Costantini, Anna Palumbo, and Giovanna Misuraca. I am also very appreciative of the contributions of my many scientific colleagues from laboratories all over the world, who have kindly provided me with some of the unpublished information and illustrative material used in this volume. Of course, I assume responsibility for all I have written and apologize in advance for any possible error or omissions.

In addition, I wish to express my gratefulness to the many publishers who gave permission to use figures and tables from their books, and to Silvana Corsani, who was invaluable in the preparation of the final manuscript. As a final remark, a word of gratitude goes to Rodolfo A. Nicolaus, who pioneered melanin research in Naples in the 1960s and guided my first steps in this intriguing field with his uncommon scientific intuition.

Giuseppe Prota

List of Abbreviations

ACTH adrenocorticotropin

AHP 4-amino-3-hydroxyphenylalanine

COMT catechol *O*-methyl transferase

5-cysdopa 5-cystein-*S*-yldopa

2-cysdopa 2-cystein-*S*-yldopa

6-cysdopa 6-cystein-*S*-yldopa

DCF dopachrome conversion factor

DCOR dopachrome oxidoreductase

DHI 5,6-dihydroxyindole

DHICA 5,6-dihydroxyindole-2-carboxylic acid

2,5-dicysdopa 2,5-dicystein-*S,S*-yldopa

2,5-diGS-dopa 2,5-diglutathion-*S,S*-yldopa

dopa 3,4-dihydroxyphenylalanine

dopac 3,4-dihydroxyphenylacetic acid

DT dopachrome tautomerase

GSH glutathione

5-GS-dopa 5-glutathion-*S*-yldopa

6-GS-dopa 6-glutathion-*S*-yldopa

2-GS-dopa 2-glutathion-*S*-yldopa

γ-GTP γ-glutamyl transpeptidase

5H6MI 5-hydroxy-6-methoxyindole

6H5MI 6-hydroxy-5-methoxyindole

6H5MICA 6-hydroxy-5-methoxyindole-2-carboxylic acid

5H6MICA 5-hydroxy-6-methoxyindole-2-carboxylic acid

HVA homovanillic acid

6-hydroxydopa 2,4,5-trihydroxyphenylalaine

5-hydroxydopa 3,4,5-trihydroxyphenylalanine

IBF indole blocking factor

ICF indole conversion factor

LH α-lipotropin

MAO monoamine oxidase

MCH melanin concentrating hormone

MFA melanin free acid

MPTP 1-methyl-4-phenyl-1,2,3,6-tetrahydropyridine

MPP^+ 1-methyl-4-phenylpyridinium ion

MSH melanocyte stimulating hormone

PDCA pyrrole-2,3-dicarboxylic acid

PTCA pyrrole-2,3,5-tricarboxylic acid

TU 2-thiouracil

CHAPTER 1

AN INTRODUCTION TO MELANIN RESEARCH

I. THE BIOLOGICAL UNIQUENESS OF MELANINS

Among the numerous pigments found in animals and plants, the melanins occupy a unique position because of their prominent appearance in mammalian tissues. In humans, almost all normal pigmentation is due to varying amounts, types, and distribution of melanins throughout the epidermis. The eventual coloration is modified by reflectance and diffraction, by subcutaneous vasculature, and by the occasional presence in the skin of other pigments, such as bile pigments, during fetal life and disease. Thus, apart from hemoglobin, melanins are the only pigments to be endogenously synthesized and found in man. In passing, it may be noted that in biomedical circles the term *pigmentation* has come to be synonymous with melanin synthesis.

The evolutionary significance of melanins has been a matter of serious consideration for many thoughtful anthropologists and biologists. A diversity of emphasis has been represented in accordance with the chief interests and viewpoints of the contributors. This is exemplified in the articles by Daniels *et al.* (1972), Riley (1980), Barr (1983), Nordlund (1985), Quevedo *et al.* (1985), Drager and Balkema (1987), and Tosti and Castelli (1990).

Stripped of myths and speculations, it appears that in nearly all mammals, and in most other vertebrates too, melanins have two important functions: They increase the optical efficiency of the eye, and they are responsible for the production of color patterns, usually of adaptive significance, in the hair and or the superficial epidermis. Their role in human skin is, however, much less clear. There is strong circumstantial evidence that melanin pigmentation in the skin has a protective value against the harmful effects of ultraviolet (UV) radiation (Quevedo *et al.,* 1974), yet the actual mechanism by which this protection is brought about is controversial, in spite of extensive investigations.

We will return to this point later in Chapter 10. Here, we anticipate that the functional significance of melanins is intimately connected with the metabolic activity of a lineage of cells *sui generis,* known as the melanocytes. These cells contain a specific enzyme, tyrosinase, which catalyzes the initial events of melanogenesis, namely the conversion of tyrosine to dopaquinone (Fig. 1.1). This is a most chemically reactive *o*-quinone and, once formed, it may give rise to a series of spontaneous transformations, including redox exchange, intramolecular cyclization, covalent binding with nucleophilic groups of cellular components, and oxidative polymerizations of the resultant products (Prota, 1980a). Viewed in these terms, melanin formation appears to be but one part of a more complex series of biochemical events which characterize tyrosine metabolism in melanocytes. The amount and type of melanins formed in melanocytes are genetically determined, but may be influenced by a variety of hormonal and environmental factors including inflammation, sex, age, and, notably, exposure to UV light, which causes the familiar phenomenon of suntanning. There are, in addition, a number of genetic or acquired developmental abnormalities which are accompanied by a decrease (hypomelanosis) or an increase (hypermelanosis) in melanin formation. Some of these pigmentary disorders are well known and include albinism, vitiligo, Mongolian spot, and melanoma (Riley, 1974a; Mosher *et al.,* 1987). The latter is a tumor with high propensity to metastasize, whose increasing incidence and resistance to conventional regimens represent a challenge to the whole scientific community. Melanoma is not limited to humans, but develops in all groups of vertebrates, including, notably, certain fish of the *Xiphophorus* genus. Studies of this animal model have greatly contributed to our understanding of the genetic mechanisms driving malignant transformations as well as to the modern concept of neoplasia (Anders, 1991).

From these brief introductory remarks, it is clear that melanin and melanogenesis, as a research tool, represent a focus where many separate disciplines converge, their range extending from basic chemistry of the pigment polymer up to the pathophysiology of the pigment-producing cells; and the list can be extended to include the role of neuromelanin in certain neurological disorders, e.g., Parkinson's disease, and the regulatory mechanisms of melanogenesis and their practical implications for controlling skin and hair pig-

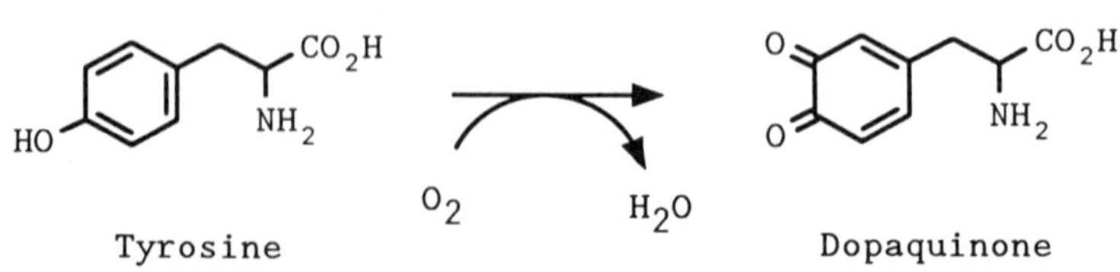

FIG 1.1 Aerobic conversion of tyrosine to dopaquinone catalyzed by tyrosinase.

mentation. Such a diversity of interests and expertise has found an excellent meeting ground in the International Pigment Cell Conferences, which were initiated in 1946 by and continued in memory of Professor Myron Gordon almost every 3 years through the most recent, held in Kobe (1990).

Besides providing a valuable permanent record of new findings in the field, these conferences have yielded another useful by-product, that is, a tradition of multidisciplinary interactions which is uncommon in research. This has been well emphasized by Fitzpatrick and his associates (see Jimbow *et al.*, 1976), who wrote, "Melanin pigmentation is an ideal subject in which to establish a dialogue between chemistry and medicine, between scientists who work exclusively on animals and clinicians who can develop new insights into basic biological concepts by studying the experiments of Nature."

II. THE INTRACTABLE NATURE OF MELANINS

The chemical investigation of natural pigments has a long and, on the whole, successful history. Coloring matters of various complexities, from the indolic historical dyes, indigo and Tyrian purple, to the tetrapyrrole prosthetic group of hemoglobin, have been isolated and their origin and structure have been established by classical and modern methods of analysis. It may therefore seem surprising that, in spite of the pressing social and biomedical motivations for studying melanins, the nature and origin of these pigments were until recently a biochemical conundrum. In an excellent, yet discouraging, review of the state of the art in the field, the late professor Blois (1978) wrote:

> *From the early commentary by Aristoteles (315 B.C.) to the most recent investigations, the study of melanins has involved virtually every applicable chemical and physical technique, and the motivation in this research has reflected the biological fashion of each era. Notwithstanding these efforts, we still do not know their chemical structure or molecular weights, their mode of synthesis is not yet agreed upon, a satisfactory nomenclature remains to be devised, and even their biological role is more a matter of surmise than proof.*

Several obstacles account for the slow progress that marks melanin research. First, natural melanins are highly insoluble materials of presumably high molecular weight and are therefore difficult to separate from the other cellular components of the organism in which they occur. Even when a melanin has been isolated, it is difficult to know whether or not it is pure, if indeed the term *pure* can rightly be applied to any melanin (Nicolaus, 1968; Swan, 1974). Moreover, with the exception of some related pigments of lower molecular weight, natural melanins are not well defined chemical entities, but mixtures of more or less similar polymers, which apparently are made up of different structural units linked through nonhydrolyzable bonds. As a result, chemical degradation is ineffective unless very drastic conditions are adopted, such as alkali fusion at 200–250°C or oxidation with permanganate, which lead to extensive breakdown of the pigment with formation of small fragments of little or no structural significance. Even more disappointing for the chemist is the fact that melanins lack well-defined physical and spectral characteristics; as a consequence, the modern structural approach by spectroscopic techniques and X-ray analysis, which has been so successful in the elucidation of complex natural products and macromolecules, has been so far defeated.

Yet, in spite of these and other difficulties, during the past two decades a considerable amount of new information about the structure and biosynthesis of melanins has been obtained by an integrated approach involving direct analysis of natural pigments and related metabolites, coupled with the study of the chemical reactivity of melanin intermediates under biomimetic conditions. There are still many things that are not clear, but the obscurity now is not so profound as before.

III. LANDMARKS IN MELANIN RESEARCH

A. The Discovery of Tyrosinase

Ethnic pigmentation, along with its complex sociopolitical implications, was the natural phenomenon that stimulated melanin research. However, while the basic questions related to the pigmentary system in man remained surrounded by confusion and controversy for a long time, the first important breakthrough to the nature of melanins came at the turn of the century, following the serendipitous discovery of tyrosinase.

In 1895 Bourquelot and Bertrand found that some colorless substance in a toadstool, *Russula nigricans,* was blackened by an enzyme present in the fungus. The melanogenic substrate was later recognized by Bertrand (1896) as the phenolic amino acid tyrosine. Examination of the substrate specificity of the *Russula* enzyme, in comparison with laccase, that does not oxidize tyrosine, led Bertrand to conclude that the fungus enzyme was a new oxidase, which then became known as tyrosinase. Within a short time, the presence of a similar enzyme was demonstrated in extracts from many other plants and from the tissues of various invertebrates and vertebrates, including cephalopods, mealworms, insects, and amphibians.

As the concentration of tyrosinase in mammalian tissues is very low, for a long time it was not clear whether this enzyme was also present in normal human skin, although it has been found in melanomas (Gessard, 1903).

Bloch (1916) attempted to settle this question by histochemical techniques. He immersed frozen sections of human skin into a dilute solution of 3,4-dihydroxyphenylalanine (dopa) at physiological pH, and after 24 hr he noted that melanin granules were deposited in the cytoplasm of the melanocytes in the basal layer of the epidermis. In subsequent studies, Bloch and Schaaf (1925) obtained evidence that the conversion of dopa to melanin was dependent upon the action of a heat-labile, cyanide-sensitive enzyme, which was apparently unable to utilize tyrosine as the substrate. Thus, the concept took form that a dopaoxidase, rather than a tyrosinase, was responsible for melanin pigmentation in mammalian tissues.

Although Bloch's hypothesis did not pass unchallenged, many years elapsed before the apparent dichotomy between melanogenesis in lower organisms and that in mammals could be reconciled. This followed the demonstration by Lerner and Fitzpatrick (1950) that when the concentration of tyrosinase is very low, as in nonirradiated human skin, the enzyme cannot be detected by Bloch's histochemical technique using tyrosine as substrate, inasmuch as the reaction is marked by a time lag before oxidation commences. This induction period can be shortened by addition of a catalytic amount of dopa, triggering the enzymic oxidation of tyrosine. By contrast, the oxidation of dopa is not characterized by an induction time, which accounts for the positive response of the melanogenic enzyme to Bloch's histochemical technique. Shortly after this work, tyrosinase was found in various mammalian tissues, and there is now little doubt that this enzyme is responsible for melanogenesis in epidermal melanocytes.

B. The Raper–Mason Scheme

In a series of classical studies carried out at Manchester in the 1920s, Raper laid the groundwork for the understanding of the mechanism whereby tyrosine is converted to melanin by the action of tyrosinase (Raper, 1928). Using an enzyme preparation from mealworms (*Tenebrio molitor*), he was able to show that in the presence of oxygen tyrosine is oxidized to dopa, and then to dopaquinone, which, as soon as it is formed, is converted to a red compound called dopachrome. Though this could not be isolated, it was correctly assigned an aminochrome-type structure, arising by ring closure of dopaquinone and subsequent oxidation of the resultant leucodopachrome (Fig. 1.2).

The proposed sequence of reactions, which is analogous to the oxidative conversion of catecholamines to aminochromes (Hawley *et. al.,* 1967), has been extensively investigated by various groups, using electroanalytical techniques (Young *et al.,* 1974), pulse radiolysis (Chedeckel *et al.,* 1984), and UV spectroscopy (Cabanes *et al.,* 1987; Rodriguez-Lopez *et. al.,* 1991). As a result

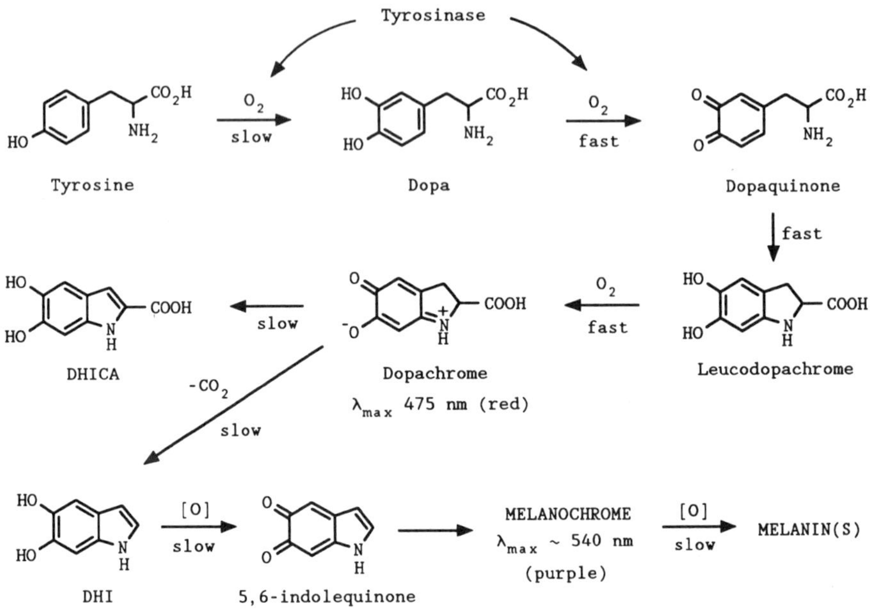

FIG 1.2 The Raper–Mason scheme of melanogenesis.

of these studies, a number of refinements have been made, but, on the whole, they all depend on the essential correctness of Raper's pioneering work.

By stopping the oxidation of tyrosine at the dopachrome stage and allowing the red solution to decolorize *in vacuo* or in the presence of sulfurous acid, Raper (1927) was able to isolate (as the dimethyl ethers) 5,6-dihydroxyindole (DHI) and a related compound in smaller amounts, identified as 5,6-dihydroxyindole-2-carboxylic acid (DHICA). On this basis, it was suggested that melanin formation resulted mainly from the oxidation of DHI and to a lesser extent from DHICA.

At that time, Raper could do no more than make suggestions about the chemistry of the melanogenic reactions following the dopachrome stage. Thus, in 1938 he wrote: "The processes taking place in the last stage of the reaction, which consists in the conversion of DHI and its 2-carboxylic acid to melanin, are obscure. Two atoms of oxygen per molecule of indole derivative are utilized, but whether they both remain in the melanin molecule or only one of them is fixed, is undecided," and he concluded: "Beyond the statement that, in the formation of this pigment (melanin) from dihydroxyindoles, quinone formation is a probable first step, it is not possible to go" (Raper, 1938).

About 10 years later Beer *et al.*, (1948, 1949) reported the synthesis of DHI and DHICA. They found that on exposure to air, especially in alkaline solution,

DHI rapidly undergoes oxidative polymerization to give a dark insoluble melanin, whereas under similar conditions autooxidation of DHICA is much slower and yields a brown solution with no precipitation of melanin. From these experiments it was inferred that of the two putative indolic precursors suggested by Raper only DHI was involved in melanogenesis. This view was later confirmed by Mason (1948), who found that under the conditions of Warburg manometry and absorption spectrophotometry, that is, utilizing dilute solutions of substrates and optimal concentration of purified tyrosinase, the major product formed in the rearrangement of dopachrome is DHI. In addition, he obtained evidence that the spectrophotometric course of the subsequent oxidation of DHI by tyrosinase is basically similar to that seen when tyrosine or dopa is oxidized by tyrosinase. These and other data were taken to suggest that melanin was a "homopolymer," that is, a regular polymer formed by self-condensation of 5,6-indolequinone (Mason, 1967).

Mason's theory of melanin structure was never satisfactory, and there was much wrangling among chemists about the actual mechanism of polymerization. Yet, for a long time, the proposed sequence of reactions up to the formation of DHI was generally recognized as the most reliable pathway for the initial events of melanogenesis *in vivo.* Thus, experimental data from studies of synthetic melanins were extended to natural melanins, regarded as being closely similar both in structure and in origin, hence the tendency to describe them collectively under the name melanin. However, the above turned out to be an oversimplification of the chemistry of melanogenesis, which will become apparent in the remainder of this chapter.

C. The Pheomelanin Pathway

As mentioned earlier, when performed *in vitro,* oxidation of tyrosine with tyrosinase results in the formation of a black, melaninlike pigment. By contrast, tyrosine metabolism in melanocytes is more colorful and variable, as seen from the various spots, stripes, and mottled appearance of many animals, as well as from the numerous shades of yellow, red, brown, and black of mammalian hair. Some colors arise from differences in amount, size, and distribution of the pigment granules rather than from qualitative differences in the chemical composition. Optical effects such as diffraction, scattering, and interference also contribute significantly to extend the palette of colors produced by melanins *in vivo* (Fox and Vevers, 1960).

However, at the chemical level, it had been known since the earliest times that the polychromy of melanin pigmentation involves at least two distinct groups of pigments: the black and brown insoluble eumelanins (ευ, good; μελας, black) and the alkali-soluble pheomelanins (φαιος, dusky), ranging from yellow to reddish-brown (Görnitz, 1923). Both types of pigments are frequently found in the same animal, for example a dark pigment in the eyes may coexist with red pheomelanic hair. Moreover, in some mammals, individual hair is not uniformly colored, but shows a terminal or subterminal yellow band, attributable to pheomelanin, while the basis is pigmented with eumelanin. This "bicolored" or agouti type of hair is of particular interest, for it provides evidence that the same melanocyte can produce both eumelanin and pheomelanin depending upon the prevailing conditions in the pigment cell environment.

Based on these and other biological observations, in 1958 Fitzpatrick *et. al.*, in a comprehensive review of early studies on the nature of hair pigments, suggested that, whereas eumelanins arise biogenetically from tyrosine, the formation of pheomelanins probably requires the intervention of another substrate. This was thought to be tryptophan or a related metabolite, for example, 3-hydroxykynurenine, which could be enzymatically oxidized to yellow pigments only in the presence of intermediates of the eumelanin pathway, such as dopaquinone. Under these conditions, dopaquinone would be reduced back to dopa as fast as it is produced, so that no black pigment could be formed (Fig. 1.3). However, efforts to demonstrate the involvement of tryptophan metabolites in pheomelanin formation by histochemical techniques were unsuccessful, and the proposed pathway was therefore abandoned.

The first indication of the nature of pheomelanins came from studies of a minor group of epidermal pigments which characterize certain types of yellow or reddish hair and feathers. These pigments were first described by Sorby (1878), who found that extraction of red human hair with hot mineral acid gave a purple pigment which could not be obtained from the hair of any other color. About 55 years later, the pigment was eventually isolated in a crude form by Flesch and Rothman (1945), who suggested a tentative formula, $(C_{15}H_{20}N_2O_9)_2Fe$, and proposed the name trichosiderin (ϑριξ, hair; σιδεροσ, iron), to emphasize the presence of iron. In a series of later studies (Flesch *et al.*, 1967; Flesch, 1968), the possibility that iron was a constitutive element of

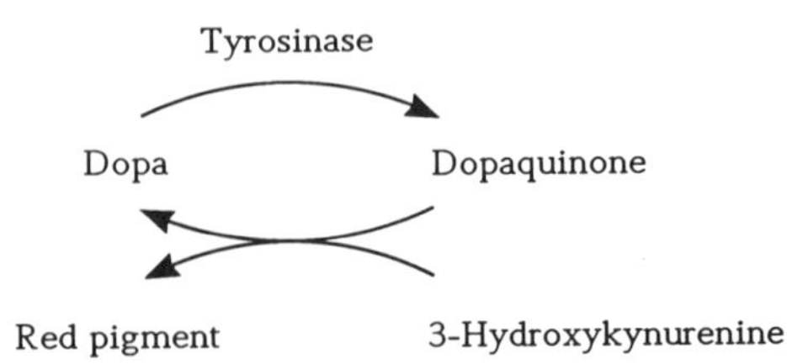

FIG 1.3 An early hypothesis of the biogenesis of pheomelanin (Fitzpatrick *et al.*, 1958).

the trichosiderin chromophore was further supported, and the pigment was redescribed as a protein complex of the general structure chromophore-Fe(III)-protein.

In 1956, however, Barnicot obtained evidence that trichosiderin was probably an artifact; extraction of red hair with cold dilute alkali yielded an orange-yellow pigment that was converted into the red trichosiderin when heated with acid. Apparently unaware of Barnicot's work, Boldt (1964) reexamined trichosiderin obtained by acid extraction and found that the crude pigment described by Flesch and Rothman was in fact a mixture of several components, some of which contained sulfur, in addition to nitrogen (Boldt and Hermestedt, 1967).

These results were confirmed and extended by parallel studies at Naples (Prota *et al.*, 1966; Prota and Nicolaus, 1967a) which also provided conclusive evidence that the acid-extractable pigments were artifacts arising from unstable yellow precursors. These were then isolated by a mild alkaline extraction procedure using as starting material red feathers from New Hampshire hens (Prota and Nicolaus, 1967b). Elemental analysis of the pure pigments thus obtained confirmed the presence of nitrogen and sulfur, but no iron was found, though it was detectable during the first stages of purification. The name trichosiderin(s) therefore appeared inappropriate and later the name trichochromes was proposed for the yellow native pigments (Prota and Thomson, 1976).

Almost at the outset of these studies, it was recognized that trichochromes were chemically related to the pheomelanins found also in red hair and feathers, both types of pigments arising from a deviation of the eumelanin pathway through the intervention of cysteine. This finding was the starting point of a new exciting era in melanin research which has witnessed a radical change of the traditional concepts of melanogenesis, far beyond the boundaries of the Raper–Mason scheme. It is surprising, however, to see how slowly these advances have been integrated into the biological and clinical context of melanin pigmentation: Even today a number of papers in the literature continue to report on experiments based on the tyrosine–tyrosinase concept of melanogenesis.

IV. THE ORIGIN AND MISUSE OF THE TERM *MELANIN*

When and how the term *melanin* came into use is controversial. According to Becker (1959), it was introduced by Robin (1873), while Nicolaus (1968) and others give credit to Berzelius (1840) for being the actual originator. It seems, however, that the term *melanin* had already been used by Vizio (Skinner, 1961) and other anatomists of the early nineteenth century, who made the first relevant observations on the origin of racial differences with crude but dedicated efforts at experimentation. Whatever the origin of the name melanin, it is worth noting that originally it was used in a rather restrictive sense to denote the pigments found in skin, hair, eye, and melanomas of man and other mammals, with the notable exception of the pigment in the ink of cephalopods. This latter was recognized at an early time by the French physician Le Cat (1765) to exhibit the same general features as the pigment of human skin and the choroid of the eye, a deduction which later proved to be correct (Nicolaus, 1968).

However, the lack of a precise chemical definition of what constitutes a melanin has led gradually over the years to a considerable broadening of the term, due largely to the tendency to classify every natural or synthetic dark-looking pigment as melanin, especially if it had defied analysis.

This state of affairs is well illustrated by a quotation from Jacobson and Millot (1953) who, in discussing the nature of a pigment from a marine invertebrate, made the following statement: "Although chemical tests specific for melanin are non-existent, it is possible to identify, at least provisionally, a naturally occurring brown or black pigment as a melanin, provided that it shows certain characteristics, namely resistance to solvents, bleaching when subjected to the action of oxidants, and the capacity to reduce directly ammoniacal solutions of silver nitrate."

Reliance on such inadequate diagnostic criteria has resulted in the widely held misconception that melanins are present in almost all forms of life, including higher plants, fungi, and bacteria. In fact, very few of the dark pigments found in plants have been investigated, and the scanty evidence which has been obtained by direct examination points mainly to nonmelanic pigments (Thomas, 1955). For example, the black phytomelanins found in the seedcoats and pericarps of many *Compositae* are oxidation products of polyphenols. The humic acids, which occur in the soil and peat, are also polymeric and phenolic in nature (Ziechmann, 1980), but clearly they are not melanins. Another case in point is the glucoside aucubin (1.**1**) which is responsible for the blackening of the leaves and fruits of the Japanese laurel, *Aucuba japonica,* as well as of the leaves and seeds of many other plants. The black material is an oxidation product of the aglycon aucubigenin, which is liberated by the action of a β-glucosidase present in the tissues (Thomson, 1962).

More cogent examples of melaninlike pigments are found in fungi or bacteria (Wheeler and Bell, 1988). Unlike animal melanins, which are formed intracellularly from tyrosine, the fungal pigments are usually wall bound or extracellular in nature and arise by enzymic oxidation of non-nitrogenous phenols and their subsequent interaction with proteins. Accordingly, they have sometimes been referred to as allomelanins (from αλλοσ, other) (Nicolaus, 1968). One example is the black pigment of the sporophores of the large ascomicete *Daldinia concentrica,* frequently found as parasite on ash

(Fraxinus excelsior). Allport and Bu'Lock (1960) showed that the main chromogen is 1,8-dihydroxynaphthalene (1.**2**), which is oxidized enzymatically to give an almost black quinone (1.**3**), 4,9-dihydroxyperylene-3,10-quinone and a polymer of the type 1.**4**.

The black pigment aspergillin, produced by *Aspergillus niger,* is probably similar in structure and origin, as suggested by analytical and degradative studies (Quilico,1937; Barbetta *et al.,* 1966; Ray and Eakin, 1975). The enzyme involved in the biosynthesis of aspergillin and related dihydroxynaphthalene pigments has not been identified, though available evidence suggests that it is probably a wall-bound laccase or a phenolase.

It appears that some fungal pigments, like those in cell walls of the basiodiospores from *A. bisporus* and *A. campestris,* are synthesized from γ-glutaminyl-4-hydroxybenzene (1.**5**), presumably by a pathway involving tyrosinase-catalyzed hydroxylation to γ-glutaminyl-3,4-dihydroxybenzene (1.**6**) and subsequent formation of the quinone (1.**7**) (Fig. 1.4). This may undergo a nonenzymatic, base-catalyzed cleavage of the γ-glutaminyl residue, to give the hydroxy-benzoquinone-imine (1.**8**), a highly reactive substance which is very prone to polymerize (Boekelheide *et al.,* 1979; Rast *et al.,* 1981).

A further example of fungal allomelanin is seen in the spore of *Ustilago maydis* D.C., a well known corn parasite. Analytical and degradative studies indicate that the *Ustilago* pigment is an irregular polymer or mixture of similar polymers made up of catechol units at varying oxidation levels, linked to one another by C–C and C–O–C bonds (Piattelli *et al.,* 1965) (Fig. 1.5). Such a structure is supported by the comparison of the properties of the natural pigment with those of a similar black polymer obtained by enzymic oxidation of catechol (Piattelli *et al.,* 1965). The biosynthetic origin of the *Ustilago* pigment and the precise nature of the enzyme(s) involved in its formation are unknown. Though catechol is a good substrate of tyrosinase, there is no evidence that this enzyme is associated with the biosynthesis of *Ustilago* melanin or of related pigments found in other fungi and in the seeds of various plants (Nicolaus *et al.,* 1964).

Another point for discussion concerns the dark pigments formed by injury to plant tissues, for example, potatoes, bananas, apples, etc. The blackening process, known as enzymic browning, is a matter of economic importance and has therefore received a good deal of attention (Hurrell and Finot, 1984). In most cases, the blackening has been shown to result from the action of tyrosinase on tyrosine or related nitrogen containing phenolic sustrates, such as dopa, hydroxytyramine, and dopamine. These reactions are akin to those that occur in the formation of melanins and are therefore very useful model systems for studies of melanogenesis. However, from the strictly physiological viewpoint, the melanins thus formed cannot be regarded as naturally occurring pigments, inasmuch as they are produced extracellularly under

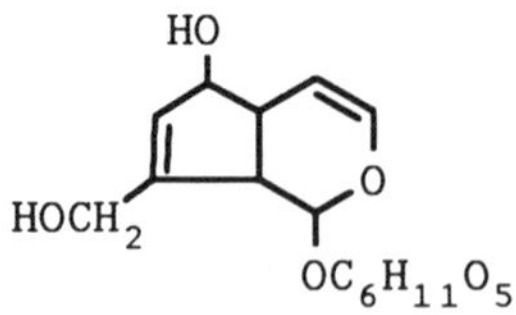

1.1

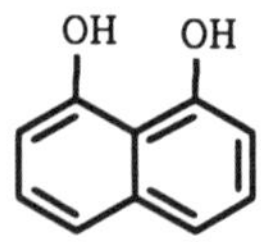

1.2

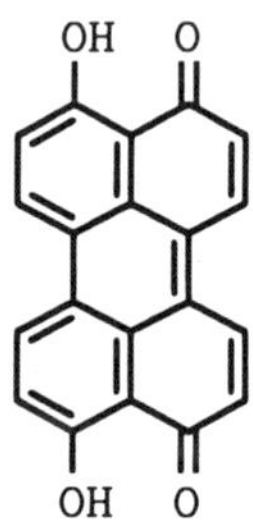

1.3

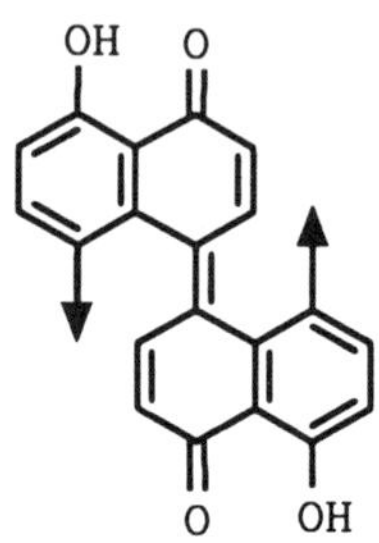

1.4

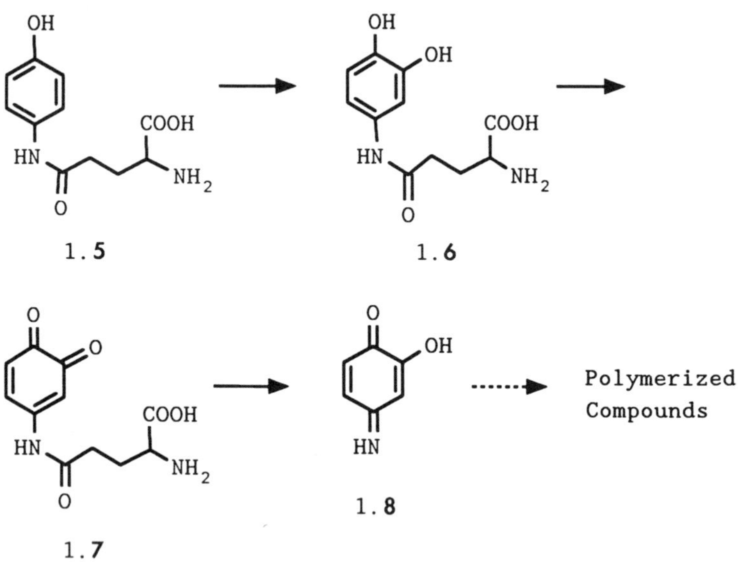

FIG 1.4 Tyrosinase-catalyzed oxidation of γ-glutaminyl-4-hydroxybenzene (Boekelheide *et al.*, 1979).

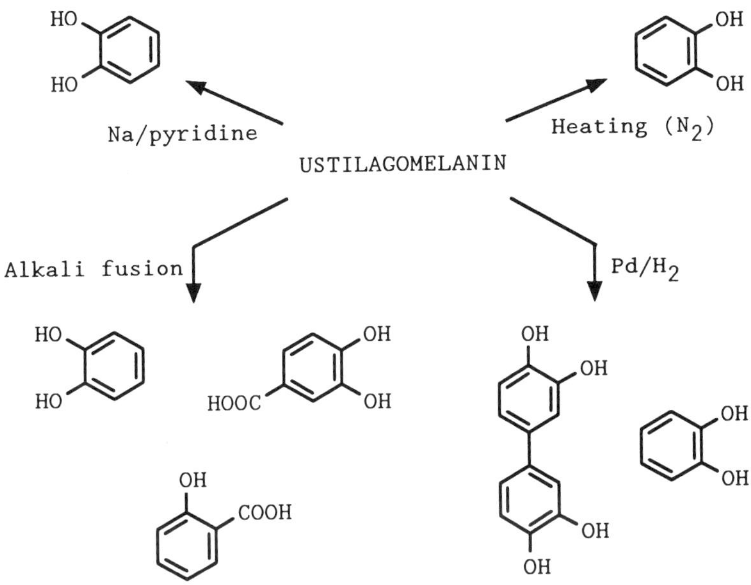

FIG 1.5 Chemical degradations of *Ustilago* melanin (Piattelli *et al.*, 1965).

circumstances where the enzyme and the substrate are allowed to interact.

To summarize, though tyrosine and tyrosinase are widely distributed in living organisms, the occurrence of authentic melanins, in the modern usage of the term, is much more restricted, for it is associated with the metabolic activity of a specific cellular system, the melanocyte, found typically in vertebrates and occasionally in invertebrates. The dark pigments found in higher plants, fungi, and bacteria include a wide range of products derived from a variety of substrates that have little structural relationship except for the presence of phenolic groups. Yet, from the standpoint of comparative biochemistry, it is noteworthy that a number of these non-nitrogenous melanins are formed as a result of the activity of an enzymatic system analogous to tyrosinase found in animal tissues. From this perspective, nitrogenous melanins may be regarded as an evolutionary diversion of a more general pathway of oxidative metabolism of phenols.

CHAPTER 2

MELANIN-PRODUCING CELLS

I. INTRODUCTORY REMARKS

In keeping with their importance in skin pigmentation, epidermal melanocytes have commanded a good deal of attention and, as is usually the case, they have become the archetype of the melanin-producing cells. There is, however, a considerable range of variation in the morphology and functional activity of melanin-producing cells from different organisms, and even within the same species various types of melanin cells are usually found. It seems appropriate, therefore, that at the onset a comparative survey be made of the biological features of melanin cells, with special reference to melanogenic activity. Additional information on the subject can be found in a number of general reviews (Billingham and Silvers, 1960; Niebauer, 1968; Quevedo *et al.*, 1987; Boissy, 1988).

II. MAIN TYPES OF MELANIN CELLS

Among vertebrates, there are at least two distinct but ontogenically related types of melanin-producing cells, which differ markedly from one another in form, function, location, and response to hypophyseal hormones. One group of these cells includes the cutaneous and extracutaneous melanocytes, typical of homotherm vertebrates such as avians and mammals. The other group is typical of cold-blooded vertebrates and is conventionally referred to as melanophores, though the use of this term is controversial (Bagnara and Hadley, 1973). Many authors consider the term *melanophore* synonymous with *melanocyte* and use the two terms indifferently, whereas others consider inappropriate the suffix *-phore* and call the poikilotherm melanin cells melanocytes. Though the proposed terminology is by no means satisfactory, it may serve, at least provisionally, to emphasize certain basic differences between the two groups of related melanin-producing cells.

Yet the problem still remains with regard to other types of melanin-producing cells, such as, for example, the Kupffer cells and, notably, the ink gland of cephalopods which, though generally regarded as melanocytes, do not originate from the neural crest. Thus, if the ontogenesis from the neural crest is to be taken as a fundamental criterion for true melanogenesis to occur in biological systems, as is commonly believed among biologists, it follows that the latter melanin-producing cells occupy an equivocal position. With these arguments in mind, in the following discussion we shall try to give an operational definition of melanin-producing cells in terms of ontogenesis, morphology, and functional activity.

III. DEVELOPMENTAL ASPECTS

Both melanophores and melanocytes found in vertebrates are embryologically derived from a stem population of melanoblasts that originates from the neural crest (Duncker, 1985), from which they migrate to populate the various body regions (Fig. 2.1). The melanoblasts that happen to reach the correct environment mature into functionally active melanocytes or melanophores.

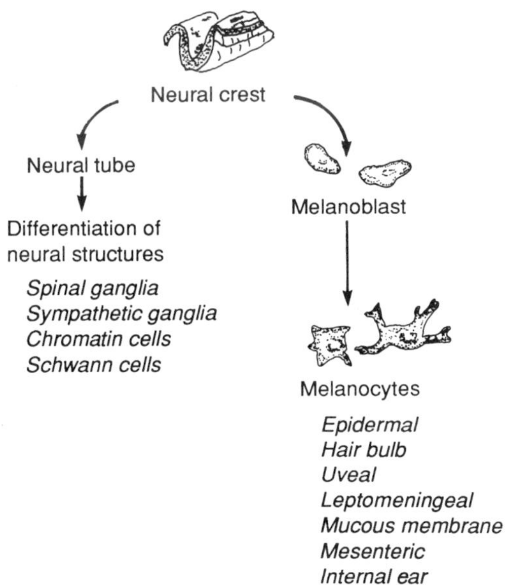

FIG 2.1 Diagram showing the common origin of melanocytes and nerve cells from the neural crest.

The ontogenesis of melanocytes was first demonstrated by DuShane (1935) using amphibian embryos. He showed that when segments of the neural crest are removed from embryos as their neural folds develop, the embryos mature into larvae which have no melanophores in the body segment belonging to regions that have been excised. These observations were later extended to avian (Dorris, 1939) and mammalian melanocytes (Rawles, 1948; Zimmermann and Becker, 1959) by properly designed experiments, confirming that only those tissues containing neural crest or migrant cells derived from it could produce pigmented melanocytes.

The migration and dispersion of melanoblasts from the neural crest was beautifully illustrated by Mintz (1967), using allophenic mice. The procedure was to take cleavage-stage eggs from mice with different coat colors and treat them with pronase in order to produce disaggregation of the cells. The cells were then mixed together at a very early stage of development, and the mixed blastocysts (chimeric) thus formed were implanted into the uterus of a female who had been made pseudopregnant by mating with a vasectomized male. In a large series of coat-color gene combinations, Mintz found an irregular stripping occurring in the coats of mice in which potentially pigmented and unpigmented genotypes were mixed. For example, from mixtures of albino and black mice, she obtained mice with a variable number of black and white stripes, that is, a zebra mouse (Fig. 2.2). From these and other experiments, it was concluded that all the melanocytes giving rise to hair color in adult mice are derived from 34 pigment stem cells. These migrate from the neural crest and subsequently undergo proliferation to produce clones of melanoblasts which will differentiate into melanocytes in particular localized skin areas.

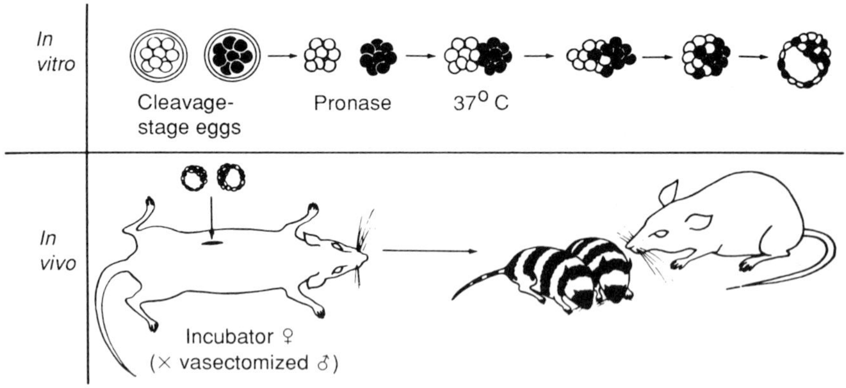

FIG 2.2 Schematic illustration of the experimental procedure for producing allophenic mice. After Mintz (1967).

There is evidence that all stages in the differentiation of melanoblasts are influenced by their intrinsic genetic machinery as well as by the nature of the embryonic environment in which they migrate. However, the mechanism of melanoblast migration, homing, and differentiation still awaits elucidation (Holbrook *et al.*, 1988). They probably move along in a segmental fashion, together with the primitive neurovascular structures supplying the skin. The final stage of this migration can often be observed in newborn infants of Negroid or Mongolian origin who show the so-called Mongolian spots, that is, brown or bluish-black macules located in the sacral region (Mosher *et al.*, 1987). Under this condition, the melanocytes remain in the dermis instead of continuing to their proper destination in the basal layer of the epidermis. The blue nevus is also the result of the occurrence of compact masses of melanocytes in the dermis and is probably similar in origin to Mongolian spots. Likewise, a defect in migration and/or the failure of neural crest melanoblasts to differentiate may account for white spotting, piebaldism, or variegation in skin and hair color.

Whimster (1979) has proposed a mechanism to explain regional differences in the behavior of melanocytes and regeneration of melanocyte patterns in the skin of vertebrates. He has emphasized that the ability of the local tissue environment to maintain cutaneous pigmentary patterns depends on its innervations. Although unambiguous evidence in support of this interpretation is lacking, Whimster's concept of a cutaneous mosaicism stemming from early development is of value in the search for an explanation of some patterns of differentiated melanocytes in mammalian skin.

IV. MELANOPHORES

Typical of cold-blooded vertebrates, for example, fishes, amphibians, and reptiles, melanophores appear as black or brown flattened cells with radially directed dendrites or, in other cases, they look basketlike with processes directed upward away from the body of the cell (Fig. 2.3). They are mainly distributed in the dermis and appear in far greater number in the dorsal integument, which partly explains the fact that the ventral surface of most poikilotherms is light colored (Bagnara and Hadley, 1973). The basis for such a difference in coloration is not known, although it could partly be due to the failure of ventral integument to support the differentiation of melanoblasts that enter in that area.

An alternate mechanism follows from the work of Fukuzawa and Ide (1988) implying the presence, in the ventral skin, of a melanization inhibiting factor (MIF). The apparent absence of such a factor in dorsal skin would suggest that while melanoblasts invaded both dorsal and ventral skin, the presence of MIF

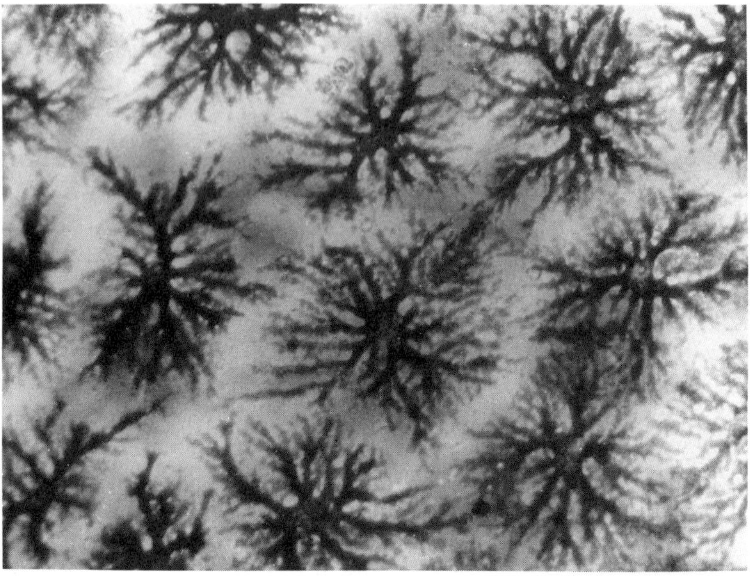

FIG 2.3 Dermal melanophores from dorsal surface of *Xenopus laevis* larva ×150. Courtesy of Professor J. T. Bagnara.

activity in the ventrum prevents melanoblast differentiation, while dorsal melanoblasts become pigmented, leading to the circumscribed dorsal–ventrum pattern. In line with this view, subsequent studies (Fukuzawa and Bagnara, 1989; Kreutzfeld *et al.*, 1989) have demonstrated the presence of MIF activity in the ventrum of other species of frogs and have found it also to be a potent inhibitor of melanization in a number of biological systems, including murine melanoma cell lines. Preliminary experiments indicate also the existence of a putative stimulatory factor of melanogenesis apparently localized in the dorsal skin of frogs and other vertebrates (Bagnara *et al.*, 1992).

Typical dermal melanophores are melanogenically active only at some stages of embryonic development and do not transfer the melanin granules into any other cell. In amphibians, for example, *Rana pipiens,* melanogenically active melanophores are also found just above the germinative layer where they form a relatively uniform network. These epidermal cells are generally thin and have a highly branched structure similar to the epidermal melanocytes of homotherm vertebrates. A major difference, however, is that the poikilotherm epidermal cells, like the dermal melanophores, are able to translocate their pigment granules, both out of and toward the cell center in response to a variety of physiological clues, including various hormones and light. These movements are usually rapid (few seconds to hours) and result in the spectacular color changes that are observed in amphibians and other cold-blooded vertebrates as well as certain invertebrates, for example, cephalopods (Bagnara and Hadley, 1973). Such "physiological" color changes are often followed by the less rapid "morphological" color changes resulting from a quantitative difference in the overall amount of melanin of the skin.

Both types of pigmentary phenomena are seen when frogs are adapted to a white or black background (Fig. 2.4). In the first case, within a few hours the melanin granules are withdrawn from the dendrites and aggregated in the body cells without variation in the absolute amount of the pigment. Over a period of a week, there is a marked decline in the amount of melanin in the epidermal layer (Hadley and Quevedo, 1966). On the other hand, when frogs are adapted to a black background, the pigment is rapidly dispersed through the highly branched dendrites of melanophores with consequent darkening of the epidermis. This rapid color change is followed by a gradual accumulation of large amounts of melanin within epidermal melanocytes and keratinocytes, further reinforcing the primary physiological adaptation to the background coloration. A third expression of morphological color change results from a remarkable increase in the total number of pigmented cells in the skin of the adapting animal. However, whether this is due to a proliferation of existing melanophores or to melanization of latent undifferentiated melanophores is not clear.

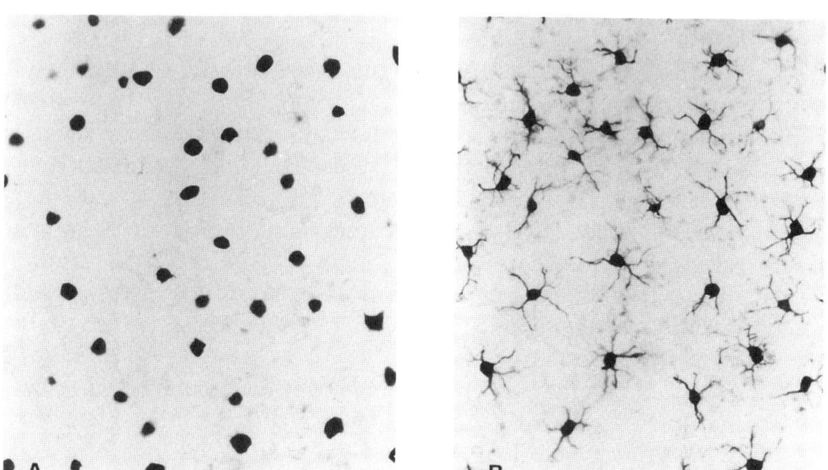

FIG 2.4 Adaptive responses of epidermal melanophores of the frog *Rana pipiens* to a white (A) and a black (B) background. After Bagnara and Hadley (1973).

In addition to melanophores, other pigment-containing organelles stemming from the neural crest, such as xanthophores and iridophores, are involved in color changes seen in lower vertebrates and invertebrates. Though the three types of chromatophores are obviously very different from one another, it has been suggested that they all originate from a primordial organelle with the potential of becoming any of the circumscribed pigment cells in response to external cues present in the milieu of their final destination (Bagnara *et al.,* 1979; Bagnara, 1983).

V. MELANOCYTES

A. CUTANEOUS

The melanocyte system of higher vertebrates, such as avians and mammals, consists of two anatomically and, at least to some extent, physiologically distinct parts: that belonging to the basal layer of the epidermis, and that belonging to the hair bulbs, which is responsible for pigmentation of hair.

The quantity of melanocytes per area of skin at various body sites in humans varies dramatically, with the whole-body average ranging from 1000 to 1500/mm^2 (Quevedo *et al.,* 1987). Some of the most densely populated areas are the sun-exposed skin and, notably, the least exposed areas, that is the genitals. Several studies, reviewed by Szabo *et al.* (1988), have shown that the color differences within the same species are due to functional, not anatomical, factors, as the number of melanocytes is the same regardless of the genetically determined color of the skin.

It has been estimated by Rosdahl and Rorsman (1983) that the overall mass of the epidermal melanocytes in the adult human would have a size larger than the parathyroid glands or the hypophysis but somewhat smaller than the adrenal glands. In subjects older than 25–30 years the number of melanogenically active melanocytes decreases with age by approximately 10–20% per decade in both sun-exposed and protected areas. Little is definitively known about the relationship between melanocyte activity and aging (Ortonne, 1990). It is noteworthy, however, that melanocytes derived from adult skin have a considerably shortened life span *in vitro,* with decreased response to specific mitogens (Gilchrest, 1985).

Epidermal melanocytes may vary in size and shape according to their population density. In an area where there are few melanocytes, they have very long dendrites and the cell body resembles the branching cells of the central nervous system. In some other areas, where a high population density of melanocytes can be found, the melanocytes have shorter dendrites and are round rather than polygonal. Functionally, epidermal melanocytes are best referred to as "incontinent" because they transfer the pigment they produce to the surrounding kerotinocytes, a process that Masson (1948) called cytocrine activity. Moreover, the melanization and transfer of pigment granules proceeds throughout their life and is triggered by various internal or external stimuli, especially sunlight.

The melanocytes in regenerating hair merit a separate description, in view of their remarkable ability to synthesize melanin only during the growing stage (anagen) of the hair cycle (Quevedo *et al.,* 1987). In this stage melanin granules are actively produced and supplied to the Malpighian cells of the upper bulb which are to become the cells of the medulla and of the cortex. On histological examination at this time, the follicular melanocytes possess tyrosinase activity. In the catagen and in the resting (teleogen) stage of the hair cycle, however, tyrosinase activity is no longer detectable. It has not yet been definitively demonstrated what happens to these melanocytes during the autolytic involutionary events of catagen and during teleogen. Electron microscopy studies of this process have provided evidence that unpigmented melanocytes do exist in the epithelial collar during catagen and in the hair germ during teleogen (Sugiyama and Kukita, 1976). This would suggest that during teleogen these melanocytes dedifferentiate, are reactivated in the subsequent anagen, and keep recycling until they are shed with a hairshaft prior to graying (Sugiyama, 1979).

The pigmentary activity of the hair bulb frequently differs both qualitatively and quantitatively from that of the superficial epidermis (Billingham and Silvers, 1960). For example, in white human beings dark hair may grow in almost colorless skin. In mice and rabbits, the melanocytes of the superficial epidermis of the trunk skin are normally nonmelanogenic, whereas those in the follicle may be very active. In red and black patched guinea pigs, red hair may issue from black skin in the transitional zone between red and black areas, and genetic studies on the guinea pig have shown that the melanocytes of the two subsystems may be affected differently by certain genes (Quevedo *et al.,* 1981). Furthermore, the melanocytes of the hair follicle may cease to function, as in graying of hair, whereas those of the superficial epidermis of the same skin retain their ability to produce melanin in response to appropriate stimuli.

B. Extracutaneous

In nearly all mammals, as well as in most other vertebrates, melanocytes are found at various internal sites, such as the uveal tract of the eye, the inner eye, the leptomeninges, and the oral mucosa (Breathnach, 1988). The retinal pigment epithelium (RPE) is often included among these extracutaneous melanin cells, although it does not originate from the neural crest but from the outer wall of the optic cup. There are, indeed, several similarities between melanogenesis in RPE and the other extracutaneous melanocytes. They all share the property of synthesizing melanin only in the early stage of their embryonic life. Accordingly, Masson (1948) aptly described these transiently melanogenic active cells as continent since, once they have elaborated their quota of pigment, they apparently retain it throughout their life. However, the demonstration of tyrosinase activity in extracts of adult bovine eye (Dryja *et al.*, 1978) and repeated clinical observations that melanocytes of the choroid or the iris may become neoplastic, giving rise to melanomas, suggest that the loss of pigmentary activity in these cells is not necessarily complete or irreversible. More recently, *de novo* synthesis of pigment granules that are ultrastructurally similar to those formed in epidermal melanocytes has been observed in porcine RPE cells *in vitro* (Dorey *et al.*, 1990). There is also evidence that the melanocytes in the human leptomeninges and inner ear may become melanogenically active under both physiological and pathological conditions (Goldgeier *et al.*, 1984; Meyer zum Gottesberge, 1988).

Cable and Steel (1991) have identified two types of melanocytes within the intermediate layer of the stria vascularis in the cochlea of normally pigmented mice: dark and light intermediate cells. The dark cells, present only in adult mice, contain few cell organelles other than melanin granules accumulated within lysosomal bodies and they do not stain after incubation in dopa, suggesting that they are melanogenically inactive. The light cells persist throughout life and are dendritic with numerous cytoplasmic vesicles that show tyrosinase activity and melanin granules in various stages of development.

Notwithstanding these and other observations, it is clear that pigment formation, if any, in adult extracutaneous melanocytes is normally negligible. The question then arises as to the functional significance of these cells. Most suggestions that have been made point to the special physical and biochemical properties of the pigment rather than to the cell itself. Thus, it has been thought that melanin pigmentation in the inner ear could serve as a sink for free-radical species, as a reservoir for trace elements, and/or as an intracellular buffering system for calcium, and that it may provide protection against ototoxic drugs (Lyttkens *et al.*, 1979; Wasterstrom, 1984; Meyer zum Gottesberge, 1988). Likewise, the vital role of the retinal pigment epithelium in the nourishment and repair of the neural retinal cells has been alternatively explained in terms of shielding of the photoreceptors from scattered light or of phagocytosis of membrane lamellae from tips of the outer segments of the rods and cones (Feeney-Burns, 1980; Kinnear *et al.*, 1985).

The more recent literature on the protective action of ocular melanins has been carefully reviewed by Sarna (1992) with special emphasis on the scavenging of oxygen radical species arising from the interaction of redox metal ions and hydrogen peroxide. On the assumption that bound-to-melanin metal ions are less damaging to the pigment cells, he postulated that photooxidation of the RPE melanin may lower its metal binding capability and reduce its antioxidant ability with adverse effects on the functional performance of the chorioretinal complex. Such a mechanism would be in keeping with the high flux of oxygen which characterizes the RPE–choroid layer as well as with a reduced zinc content in the pigment of degenerating RPE cells (Samuelson *et al.*, 1991). It may be argued however, that melanin photoaging in the RPE may be a secondary event following depletion of the cellular antioxidant systems, for example, ascorbic acid, glutathione (GSH), and vitamin E.

VI. OTHER TYPES OF MELANIN-PRODUCING CELLS

Much of the vertebrate glia tissue has a common origin with the neural tissue and the cells often are shaped similarly to the melanocytes, forming a sheath around some peripheral nerves. Certain neurons, particularly those of the dopaminergic system in the midbrain of primates, also show a grey or dark color. The pigment, known as neuromelanin, has been extensively investigated, and a survey of what is currently known on its nature and origin is given in Chapter 6. Moreover, there are several other records in the literature of melanic material in cellular systems or organs of non-neural origin (Needham, 1974). In most cases, however, there is no evidence that the pigment is actually a melanin in the modern usage of the term or that it is produced intracellularly by enzymic oxidation of tyrosine via dopa and related metabolites. Thus, for the sake of uniformity, in the following survey attention will be restricted to those few examples of cells that have been shown to exhibit melanogenic activity, such as the endothelial cells of vertebrate liver or kidney and the ink gland of cephalopods, the latter being the most well-defined example of a melanin-producing system found in invertebrates.

A. *Endothelial Tissues*

In Amphibia and Reptilia clusters of black or brown pigment cells are commonly seen in the liver, wherein they are localized mainly near the blood vessels as well as in the connective tissue of the portal canals and the sinusoidal walls (Hack and Helmy, 1983; Duncker, 1985). Unlike dermal melanophores, the pigment cells of liver are globular rather than dendritic in shape and contain, besides melanin, hemosiderin and lipofuscin.

Since these melanin cells do not originate from the neural crest, for a long time it was thought that they were simply melanophages, being therefore unable to synthesize melanin. In contrast with this view, Cicero *et al.* (1982) found that injection of ^{14}C-labeled tyrosine or dopa into the peritoneal cavity of living frogs leads to a significant incorporation of these precursors in both skin and liver melanins; in particular, incorporation occurred specifically in Kupffer cells. Moreover, a comparison of the chemical properties of the hepatic pigment with those of the melanin isolated from frog skin has shown that they are structurally similar with respect to elemental composition and degradation behavior. These results strongly suggest that Kupffer cells have the ability to synthesize their own melanin by a pathway similar to that accounting for melanogenesis in the dermal melanophores. Accordingly, they have been classified as "extracutaneous pigment cells of histocytic origin" (Sichel, 1988).

The existence of similar types of melanin-producing cells has been demonstrated by ultrastructural and biochemical techniques in the kidney of *Sparus auratus* (Zuasti *et al.*, 1989). At the morphological level, the pigment granules within these cells are remarkably similar to those found in the internal viscera, liver, spleen, or kidney of other teleost fish (Agius, 1985), implying a common functional role. This has been envisaged as that of repository centers of effete metabolic materials of host origin, such as ions or free radicals, which can be trapped by the pigment polymer. Such a role would be consistent with the observed increase in the content of visceral melanin as the fish grow older. In another study by Sichel and associates (1987), it was found that in various species of vertebrates there is a statistically significant inverse relationship between Mn–SOD (superoxide dismutase) activity and the pigment content in the liver. This finding, coupled with the ability of melanins to trap superoxide, has led these authors to suggest that the pigment in the liver cells could have, at least in part, a SOD-like activity and, as such, could play a role in scavenging free radicals transiently generated during biological oxidation processes (Geremia *et al.*, 1989; Scalia *et al.*, 1990).

What has been said so far would point to the existence of certain similarities between the pigment granules in the liver or kidney of poikilotherm vertebrates and the melanic material found in the heart and other organs of aging persons (Altschule and Hegedus, 1977; Hegedus, 1977). If proved, such a relationship would open new vistas into the physiological and pathological significance of visceral melanins.

B. The Ink of Cephalopods

The essential histological features of this peculiar melanin-producing system were first studied by the French biologist Giroud in 1882. He showed that the gland consists of cylindric cells which have a basal nucleus and an apical secretory zone containing melanin granules (Fig. 2.5). These are found densely packaged in the vacuoles which eventually empty into the lumen of the gland and are transferred to the sac, where they accumulate to form the black ink that the animal discharges when frightened.

Surprisingly enough, no biological study of any significance was carried out in subsequent years until the 1960s, when a reexamination by Szabo (1965) led to a reappraisal of the ink gland as a most perfect biological model of a melanin-producing system. Its value in studies of melanogenesis depends on the fact that nearly all the ink gland cells are actively engaged in melanin synthesis, whereas only about 10% or less of the cells in melanoma and less than 1% of the cells in the skin may produce melanin.

Ultrastructural observations (Ortonne *et al.,* 1981) have shown that the developmental stages of the melanin cells in the apical zone of the gland are quite similar to those of vertebrate melanophores (Turner *et al.,* 1975). The same authors also confirmed Giroud's view that the more these cells became pigmented, the closer to death they are. This would imply that melanogenesis is a malevolent event in the life history of these cells that correlates with their aging.

After maturation and subsequent degeneration, the whole melanophore content, including pigment granules and other cellular components, is transferred to the ink sac, acting as a reservoir of the exhausted material. The relatively large amount of pigment (1 g or more) which is normally present in the ink sac of cephalopods, for example, the cuttlefish *Sepia officinalis,* is probably the result of a long period of storage, as evidenced by labeling experiments (Nicolaus, 1968). This implies either that cephalopods use small amounts of ink for defense purposes or that a special mechanism exists for the rapid synthesis of the pigment.

The chemistry of sepiomelanin will be discussed later. It may be appropriate here to spend a few words on the role of melanin in the ink cloud discharged by a threatened cephalopod. Since time immemorial this defense mechanism has been attributed to functioning as a smoky screen, allowing the animal to escape the predator unseen. Such a view, however, does not withstand critical analysis, first because the size of the ink cloud alone is seldom enough to conceal the cephalopod or to baffle a serious predator and, second, because many cephalopods live in very deep waters, where perception of colors is almost absent for the lack of visible light. Indeed, there is reason to believe that the true function of the ink is not as a screen to cover retreat but rather as a chemical decoy to delay or prevent pursuit (MacGinitie and MacGinitie, 1968; Fox, 1974). Attempts to identify the active component of the ink have been scarce and not very informative, though some experiments by Kittredge *et al.* (1974) would suggest that dopachrome or another melanogenic intermediate may account for the chemosensory properties of cephalopod ink.

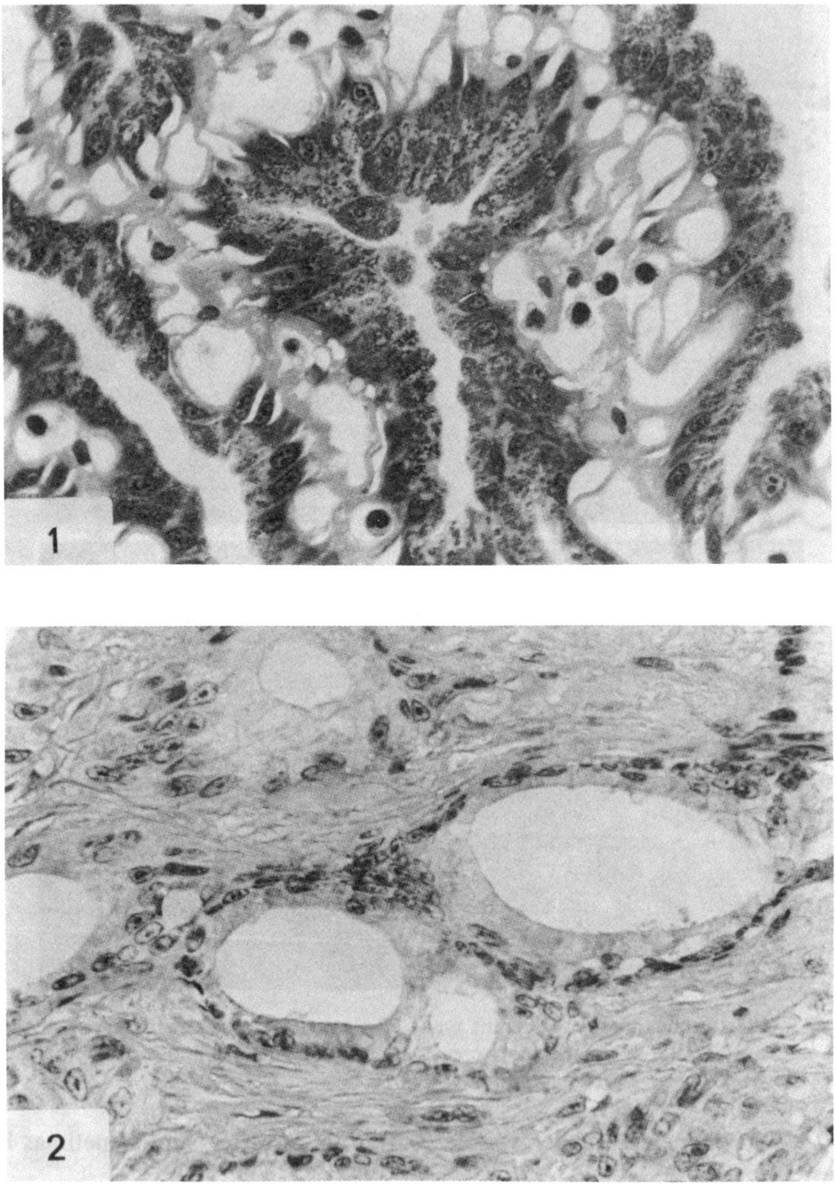

FIG 2.5 Light microscopy of the ink gland of the cuttlefish *Sepia officinalis*. In the apical portion (1) the epithelial cells arranged in orderly single layers upon a basement membrane contain melanin granules. In the caudal portion (2) the epithelial cells contain few or no pigment granules.

TABLE 2.1
TYROSINASE ACTIVITY IN CEPHALOPOD INK[a]

Species	Whole ink (units/ml)[b]	Melanin-free ink (units/ml)[b]	Protein (mg/ml)	Specific Activity (units/mg of protein)
Octopus vulgaris	1.8	1.6	5	0.32
Sepia officinalis	1.3	1.1	10.7	0.10
Loligo vulgaris	0.8	0.7	1	0.7

[a]Average values obtained from ink of eight animals for each species. From Prota *et al.*, 1981.
[b]Units expressed as μmol of tyrosine (0.5 m*M* L-tyrosine, pH 6.8) oxidized per minute under standard assay conditions; for comparison 1 mg of commercial mushroom tyrosinase corresponds to 1.1–1.3 units.

An alternate explanation follows from the finding that the melanin-free ink from *Octopus vulgaris* contains large amounts of tyrosinase in a relatively pure form (Prota *et al.,* 1981). One milliliter of freshly collected, melanin-free ink contains as much enzyme as that present in 1 mg of commercial mushroom tyrosinase. Comparable amounts of the enzyme were also found to be present in the ink from other cephalopods, for example, *Sepia officinalis* and *Loligo vulgaris* (Table 2.1).

These results would suggest that the active component in the ink of cephalopods is not melanin, as usually believed, but the enzyme tyrosinase, which could effectively catalyze the *in situ* oxidation of phenols to toxic quinones, capable of paralyzing the olfactory sense of the predator. The black color of the cephalopod secretion could probably serve as a focus of visual attention to attract the predator right through the ink cloud containing the active principle, tyrosinase, in the highest concentration. This interpretation provides a caveat to the common anthropocentric view of melanogenesis in terms of visual perception, which may be quite misleading.

VII. MELANOSOMES

A. ONTOGENESIS

Within melanin-producing cells, pigment formation appears to take place in a distinct cytoplasmatic organelle, known as the melanosome, containing structural proteins, tyrosinase, and possibly other functional proteins. Generally speaking, two basic types of melanosomes are formed in melanocytes: the black or brown eumelanosome and the yellow or reddish pheomelanosome.

The former is a rather large ellipsoidal organelle with a highly ordered protein matrix, whereas the latter looks spherical and is composed of loosely aggregated small granules (Parakkal, 1967). Usually, epidermal melanocytes will form one of the two types of melanosomes, however, in the agouti mouse the same melanocyte can apparently synthesize one type of melanosome after the other during a cycle of hair generation (Sakurai *et al.,* 1975; Jimbow and Takeuchi, 1979). Coexistence of both eumelanosomes and pheomelanosomes in melanocytes has also been observed in a rare case of human congenital pigmentary disorder (Mishima *et al.,* 1988).

The melanosome concept was developed in 1961 by Seiji and his associates following the application of techniques similar to those used by Siekevitz and Palade (1958) in the study of the zymogen granules in exocrine cells of the pancreas. These involve isolation of the subcellular fractions of the melanocyte by sucrose gradient ultracentrifugation combined with histochemical and electron microscopic analysis. The bulk of the numerous cytological studies and tracer experiments accumulated since then (reviewed by Wick *et al.,* 1987) would suggest that the structural proteins and tyrosinase are not incorporated simultaneously into the melanosomes, though both are formed on membrane-bound ribosomes. The structural proteins appear to be assembled into the smooth endoplasmic reticulum (SER) or multivesicular bodies that are connected to the rough endoplasmic reticulum (RER). On the other hand, tyrosinase is assembled in a particular area of the Golgi apparatus, to form eventually small vesicles or "coated" vesicles which transfer the enzyme into premelanosomes, where the structural proteins have already been assembled (Maul and Brumbaugh, 1971; Turner *et al.,* 1975; Mishima *et al.,* 1979).

Transcription and translational mechanisms probably interplay to affect not only the rate of melanosome synthesis but also the melanogenic potential of these organelles. However, what specifically regulates the onset of the synthesis of enzymes and structural proteins found in the various subcellular compartments of melanocytes is not clear. New evidence (Hatta *et al.,* 1988; Mishima, 1990) is accumulating suggesting that coated vesicles play more than a passive role of carriers for the translocation of tyrosinase from the Golgi apparatus into the premelanosomes, but their functional significance in the melanization process of premelanosomes is still being assessed (see Chapter 8).

B. Staging and Tyrosinase Processing

Four stages of melanosome development have been recognized by electron microscopy (Fig. 2.6). In the first of these stages, the basic structural unit consists of tyrosinase and probably a protein matrix. During stage II, these

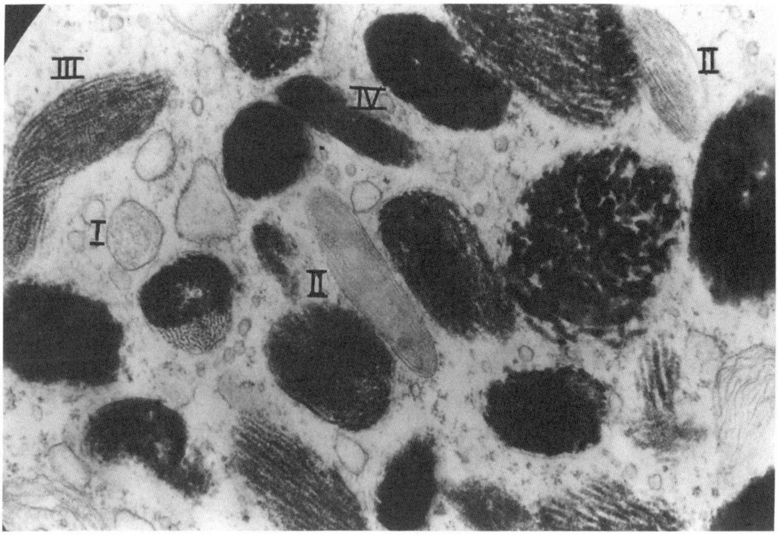

FIG 2.6 Melanosomes at different developmental stages (I–IV) in melanocyte from human fetus retinal pigment epithelium cells. The same morphology and staging are seen in epidermal melanocytes. Courtesy of Professor A. Breathnach.

units assume an ordered pattern to form the inner membranous structures of the melanosome, after which melanin biosynthesis begins on the inner membranes (stage III). As melanin gradually accumulates on the inner membranes, the melanosome is eventually transformed into a uniformly dense particle with visible structure. During this stage, the melanosome gradually loses its tyrosinase activity, and when the melanization is complete no tyrosinase can be detected (Seiji, 1981). The completely melanized melanosome corresponds to what is commonly referred to as a melanin granule (Seiji *et al.*, 1963). These fully melanized granules tend to accumulate at the periphery of the cell, particularly in relation to the dendrites.

The structural changes which occur in the enzyme during melanosome development have been best worked out on malignant melanocytes, especially Cloudman S91 and B16 lines (Hearing and Jimenez, 1987), but there are grounds to believe that tyrosinase synthesis and activation in normal human melanocytes are basically similar (Halaban *et al.*, 1983; Burchill *et al.*, 1990). Following synthesis in ribosomes, the enzyme is transferred to the Golgi apparatus, where it undergoes posttranslational processing involving addition of sialic acid and neutral sugars through *N*- and *O*-glycosidic linkages (reviewed by Mishima and Imokawa, 1985). The mature tyrosinase is then packaged into coated vesicles and delivered to the melanosome, where it becomes membrane bound and aggregates into a high-molecular-weight form that appears to be actively involved in melanogenesis.

As has been shown in lysosome systems (von Figura *et al.,* 1979), the carbohydrate moiety of tyrosinase seems to function intracellularly as a signal for its translocation from coated vesicles into melanogenic compartments of melanocytes. In accord with this view, Imokawa and Mishima (1984) found that addition of glycosylation inhibitors of protein, such as glucosamine and tunicamycin, results in a reversible loss of melanization in cultured B16 melanoma cells with a slight to moderate decrease in the growth curve. Moreover, analysis of the electrophoretic pattern of tyrosinase present in the treated cells revealed that inhibition of melanogenesis was associated with a disappearance of, or a marked decrease in, the membrane-bound enzyme and an increase in the soluble form. These observations on the subcellular dynamics of oligosaccharide synthesis have been explained in terms of a model of tyrosinase processing (Fig. 2.7) in which formation and translocation of the glycosylated form of the enzyme to premelanosomes is envisaged as an essential requirement for melanogenesis to occur.

In some melanocyte lines, the subcellular apparatus of melanogenesis is induced by the uptake of L-tyrosine, and this induction includes premelanosome synthesis and tyrosinase translocation from *trans*-Golgi reticulum to melanosomes (Slominski *et al.,* 1988; Slominski and Costantino, 1991). It has also been reported that L-dopa, while increasing the melanogenic effect, may determine which type of energy-yielding metabolism is operative in the cell by stimulating anaerobic glycolysis and the pentose phosphate pathway

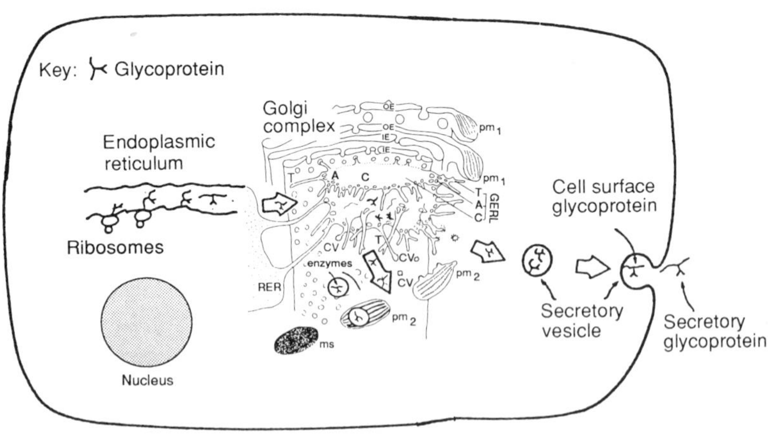

FIG 2.7 Visualization of subcellular dynamics of oligosaccharide synthesis in different melanogenic compartments. After Mishima and Imokawa (1985).

(Scislowski *et al.,* 1984). In addition, physicochemical factors, such as intramelanosomal pH, oxygen, and ionic strength, have been reported to affect tyrosinase activity and melanization of melanosomes in both normal and malignant melanocytes in culture (Laskin *et al.,* 1980; Moellmann *et al.,* 1988; Horikoshi *et al.,* 1991), but the relevance of these studies to the *in vivo* situation remains open to question.

C. Transfer and Fate of Melanin Granules

As emphasized earlier, melanization of epidermal tissues is a bipartite process involving the formation of melanin(s) within melanocytes and the subsequent redistribution of the pigment granules or melanosomes to the surrounding keratinocytes. A similar bipartite process accounts for the morphological color changes commonly seen in the skin of frogs, where melanin granules are transferred from epidermal melanocytes to keratinocytes as an adaptational response to changes in background coloration (Hadley and Quevedo, 1966). It follows, therefore, that the intensity of visible pigmentation of skin, hair, and feathers is determined not only by the pigmentary activity of melanocytes, but also by the rate of transfer and mode of distribution of melanosomes within the keratinocyte population. Once inside the epidermal cells, the melanin granules tend to shift to the upper part that faces the outside, arranging themselves in the form of a cupola above the nucleus. Such a typical distribution is often used as an argument to support the view that melanins may provide a device to protect the nuclei against sunlight.

The number, type, size, and distribution patterns of melanosomes in epidermal tissues are genetically controlled, as documented by ultrastructural observations (Szabo *et al.,* 1988). Typically, melanosomes in dark-skinned people, especially Negroids, are larger than those in light-skinned people. Moreover, the former are transferred singly inside the keratinocytes, whereas the latter are transferred in groups of two or several bound together by a limiting membrane. When inside the keratinocytes, the large melanosomes of dark-skinned people remain intact, but the clumps of melanosomes seen in white skin become attached to lysosomes and are eventually disrupted, leaving a "melanin dust." The observed differences in the shape and stability of melanosomes may account in part for racial differences in skin color and in skin photoprotection (Quevedo *et al.,* 1974). It is not clear, however, whether formation of melanin dust is due to racial differences in the nature of individual melanosomes or to racial differences in the amount of antioxidant enzymes in the epidermal cells. Yohn *et al.* (1991) found that cultured human keratinocytes have more peroxidase, catalase, glutathione peroxidase, and superoxide dismutase activity than melanocytes. However, no significant differences in the level of these enzymatic activities were observed between the two type of cells taken from black and Caucasian individuals. The individual size of pigment granules may also be a function of the rate of pigment synthesis. Thus, exposure of white skin to UV rays causes the pigment granules to change their size to a larger dimension characteristic of unstimulated Negro skin (Toda *et al.*, 1972).

Studies of the mechanism of melanin transfer and distribution suggest that the keratinocytes acquire the melanin particles by phagocytizing portions of the dendrites of melanocytes (Mottaz and Zelickson, 1967; Wolff *et al.*, 1974; Okazaki *et al.*, 1976). It is also possible that the keratinocytes play some active role in controlling the rate of melanin synthesis by the melanocytes and that the whole question of pigmentation involves complementary activities by

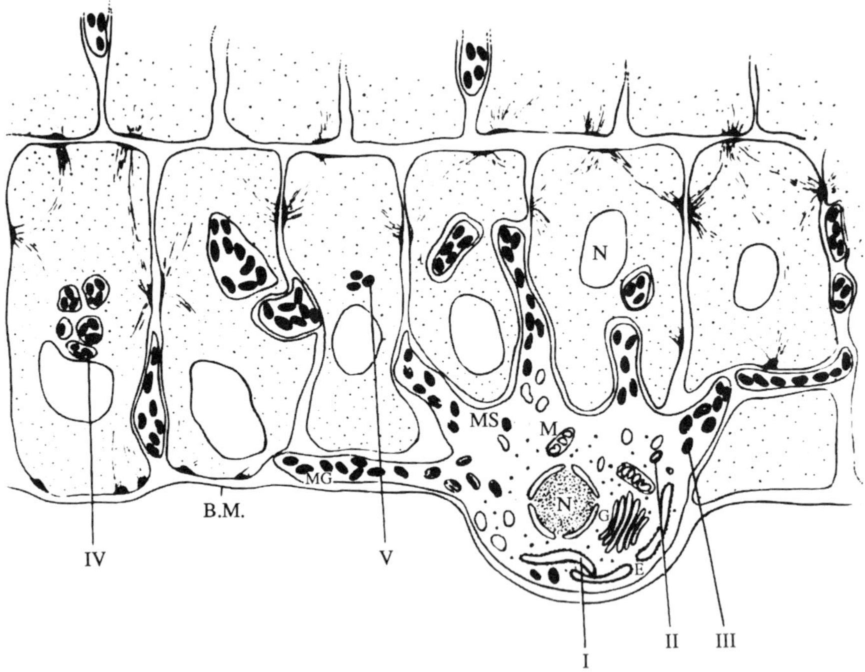

FIG 2.8 Schematic drawing of an epidermal melanin unit showing a melanocyte and the group of keratinocytes with which it maintains functional contacts. I, endoplasmic reticulum where tyrosinase is synthesized; II, premelanosome stage I; III, fully melanized melanosome; IV, passage of melanin granules from the melanocyte to keratinocytes; V, liberation of melanosomes free in the cytoplasm of keratinocytes. B.M., basal membranes; M, mitochondria; MS, melanosome; N, nucleus; E, endoplasmic reticulum; G, golgi; M.G., melanin granules. From Fitzpatrick and Breathnach, 1963.

melanocytes and keratinocytes. This has been concisely expressed in the concept of the epidermal melanin unit (Fig. 2.8), which has been proposed by Fitzpatrick and Breathnach (1963) to emphasize the existence of a structural and functional organization of melanocytes and keratinocytes at a biological level higher than that of the individual cell.

The number of these units may vary markedly between different regional sites in the human skin. It is noteworthy, however, that the ratio of keratinocytes to melanocytes within epidermal melanin units remains constant, implying the existence of a relative field of influence of the melanocyte dendrites. This is consistent with numerous observations showing that factors affecting the integrity and proliferation of keratinocytes, such as mechanical injury, heat, UV radiations, inflammation, etc., also have an indirect effect on the morphology and functional state of melanocytes (Prunieras, 1986; Nordlund *et al.*, 1988; Prota, 1989b; Tosti and Castelli, 1990). Indeed, it appears that the melanocytes behave as if they know exactly what is going on in the keratinocytes, and vice versa, and that this knowledge profoundly influences the reciprocal behavior. Though this is well documented at the biological level, the nature of the chemical signals involved in the physiological communication between melanocytes and keratinocytes is still little understood and represents a most promising area for new discoveries.

CHAPTER 3

TYROSINASE

I. INTRODUCTION

Tyrosinase is a copper-containing enzyme which can catalyze two distinct reactions, the *ortho*-hydroxylation of a monophenol (hydroxylase activity) and the conversion of an *o*-diphenol to the corresponding *o*-quinone (oxidase activity). The two enzymatic activities are intimately coupled and are also referred to as cresolase or phenolase and catecholase or diphenolase activities, respectively (Fig. 3.1).

Though there are other enzymes that can effect these transformations separately, there is apparently no enzyme that can catalyze both reactions like tyrosinase. In the IUB classification tyrosinase has therefore been given two entries: monophenol, dihydroxyphenylalanine: oxygen oxidoreductase, EC 1.14.18.1, for monophenol monooxygenase, and EC 1.10.3.1. for catechol oxidase (Robb, 1984). The current classification of tyrosinase as a monooxygenase or a mixed-function oxidase, transferring only one atom of oxygen from molecular oxygen to the substrate, was established by Mason *et al.* (1955), using ^{18}O-labeled oxygen. The two-electron donor required in the hydroxylation reaction of tyrosinase is the *o*-diphenol, which is generated internally from the monophenol substrate.

II. DISTRIBUTION AND PROTEIN STRUCTURE

A. BIOLOGICAL AND FUNCTIONAL ASPECTS

As mentioned earlier, tyrosinase is of widespread occurrence throughout the phylogenetic scale, from microorganisms to plants and animals, up to man. Representative sources have been tabulated by Dawson and Tarpley (1951) and Yasunobu (1959), and some further examples are mentioned by Lerch (1981) and Robb (1984). The isolated enzymes differ from one another with respect to molecular size, amino acid composition, and protein structure: Yet,

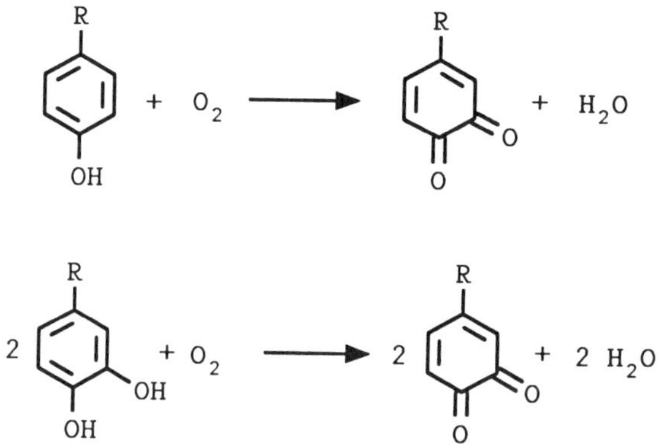

FIG 3.1 Enzymatic activities of tyrosinase.

they all seem to have evolved from a common ancestor protein containing the essential structural elements for the catalytic activity (Lerch and Germann, 1988). This enzymic system evidently appeared at an early stage of evolution, although its purpose seems to have changed with later developments. In animal tissues, the enzyme is apparently involved in biosynthesis of melanins and related metabolites, but the significance of tyrosinase in plants and lower organisms is much less clear.

Hirsch (1954) found that tyrosinase is indispensable for the perithecial development during sexual differentiation of *Neurospora crassa*. Moreover, the enzyme is completely absent in the vegetative state; however, under restrictive growth conditions, high levels of tyrosinase can be induced. Inhibitors of protein synthesis, for example, cycloheximide, D-amino acid, and amino acid analogs, also induce tyrosinase formation in liquid cultures (Horowitz *et al.*, 1970). This induction of *de novo* tyrosinase synthesis has been interpreted in terms of a model in which the rate of enzyme formation is controlled by an unstable or rapidly turning over repressor protein. According to Feldman and Thayer (1974), catabolite repression could also be an important factor for tyrosinase synthesis, since the enzyme is also induced in the presence of cyclic AMP and phosphodiesterase inhibitors. However, the functional significance of the induction of tyrosinase synthesis in *N. crassa* and other organisms, which are devoid of melanins, remains elusive. It may be that the redox potential of the tyrosinase-containing tissues of these organisms is too low to permit melanogenesis, or it may be that the enzyme may act on substrates other than tyrosine to form colorless products having some as yet unknown biological function. Unfortunately, little work in this direction has appeared so far.

B. Tyrosinases from Microorganisms and Plants

The best known example of bacterial tyrosinase is found in *Streptomyces glaucescens.* It is a monomeric, single-chain protein with a molecular weight of 30.9K, which places it as the smallest representative of tyrosinases so far studied. The structural gene for this enzyme has been identified, and the amino acid sequence of the encoded protein has been deduced from the DNA structure (Huber *et al.,* 1985). The protein structure of the related single-chain enzyme from *Streptomyces antibioticus* has been established by Bernam *et al.* (1985). Other bacterial tyrosinases that have been examined in some detail include those from *Streptomyces nigrifaciens* (Nambudiri *et al.,* 1972) and in *Vibrio tyrosinaticus* (Pomerantz and Murthy, 1974). This latter is apparently present in the form of two isozymes with molecular weights of 41K and 38.5K which, however, are poorly characterized with respect to copper content and quaternary structure assembly.

Among fungal tyrosinases, the enzyme from the edible mushroom, *Agaricus bisporus,* has attracted a good deal of attention since the early work of Keilin and Mann (1938). However, its structural characterization has been rather slow due to the occurrence of multiple forms and protein contamination. Mushroom tyrosinase was originally described as occurring in four chromatographically separated isozymic forms (α, β, γ and δ) exhibiting different monophenol/*o*-diphenol activity ratios and characterized by a concentration-dependent association–dissociation equilibrium (Jolley *et al.,* 1969). Apparent molecular weights ranged from minima of 26–32K to maxima of 120K (Kertesz and Zito, 1965).

Since the 120-kDa form was shown to contain four copper atoms, the enzyme was supposed to consist of four identical subunits, each containing one copper atom per molecular weight of 30K (Duckworth and Coleman, 1970). More recent work (Strothcamp *et al.,* 1976), however, indicated that native mushroom tyrosinase is made up of two types of polypeptide chains, a heavy subunit (H) of 43 kDa and a light subunit (L) of 13.4 kDa. Thus, the 120-kDa form is regarded as arising from a quaternary assembly of the type L_2H_2. The heterogeneity of mushroom tyrosinase toward hydroxyapatite has further been explained in terms of two types of the H subunits, designated Ha and Hb, differing in molecular weight by about 3K. The high hydroxylase enzyme (α) would be Ha_2L_2 and the high oxidase enzyme (γ) Hb_2L_2.

Another remarkable example of tyrosinase is found in the ascomicete *N. crassa.* Chemical and spectroscopic studies of this enzyme by Horowitz *et al.* (1970) and Lerch (1983, 1988) have provided important leads to the understanding of the structure and electronic and geometric arrangement of the active site of tyrosinases from both prokaryote and eukaryote organisms.

The enzyme occurs in many wild-type strains of *N. crassa* as allelic forms differing in thermostability and electrophoretic mobility. Two of these forms have been isolated and obtained in crystalline state. Both isozymes, designated TL (thermolabile) and TS (thermostable), exhibited identical sedimentation behavior, associating reversibly in a range of molecular weights from 35K to 120K.

The determination of the amino acid sequence, by chemical techniques, showed the enzyme as a single-chain protein of 407 amino acids with a molecular weight of 46K including 2 g-atoms of copper per mole. From this analysis two interesting features emerged: (1) a blocked amino terminus, identified as *N*-acetylserine by computer-assisted mass spectrometry; and (2) the presence of a unique chemically modified cysteine residue (Fig. 3.2) in which the side chain of Cys-94 is linked via a thioether bridge to the 2 position of the imidazole ring in the neighboring His-96 (Lerch, 1982).

Although the possibility of an isolation artifact cannot be ruled out, it has been suggested that this unusual cross-linking could arise from a posttranslational processing leading to the activation of the enzyme, consistent with the presence in crude extracts of both an active and an inactive (protyrosinase) form; however, conclusive proof of this mechanism is still lacking. In plants, tyrosinase can be found in gymnosperms and angiosperms; it is absent in a variety of higher plants, such as *Cucurbita* and *Brassica* species (Robb, 1984). When present, it is predominantly soluble (potato tuber) and in some species it may be associated with chloroplasts (spinach, spinach beet, green olive), but it is not yet entirely clear whether all these are true tyrosinases. This is partly due to the fact that in most sources tyrosinase is found associated with other phenol oxidases, including laccase and peroxidase, which may affect or mimic to some extent its catalytic activity. Relevant in this context is also the property of tyrosinase to loose its monophenolase activity during the purification pro-

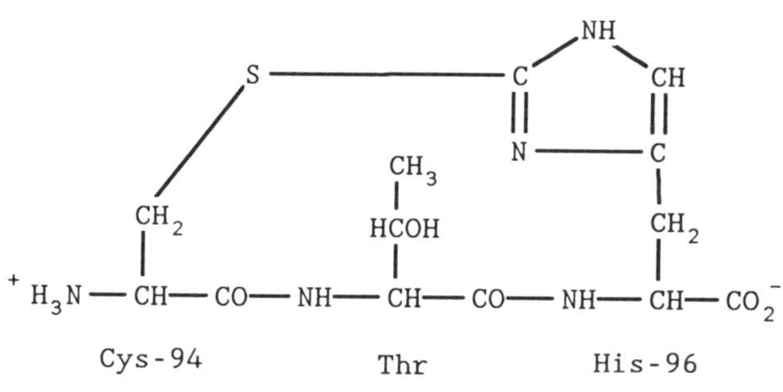

FIG 3.2 Structure of the thioether-containing tripeptide residue of *Neurospora crassa* tyrosinase.

cess. Thus, unless care is taken, a purified tyrosinase preparation may be almost devoid of cresolase activity and become a polyphenoloxidase.

C. Animal Tyrosinases

Although tyrosinase is likely to be widely distributed throughout the animal kingdom, the relevant literature is not so rich as with plants. Details are available of methods for extraction and purification of tyrosinase from various sources (Robb, 1984; Lerch, 1988). Usually, several purification steps involving the use of detergents and/or phospholipases and proteases are needed to obtain the enzyme in sufficient amounts for biochemical characterization. A remarkable exception is the ink of cephalopods, which represents the richest and most convenient source so far known in the animal kingdom for the extraction of tyrosinase in a relatively pure form (Prota *et al.,* 1981). An enzyme preparation from cuttlefish (*Sepia* sp.) has recently become commercially available.

Polyacrylamide gel electrophoresis of preparations from three closely related species (*Octopus vulgaris, Sepia officinalis, Loligo vulgaris*) revealed a single band on staining with L-tyrosine, with estimated molecular weights of about 205K for *Octopus,* 125K for *Sepia,* and 135K for *Loligo* (Prota *et al.,* 1981). Despite differences in molecular sizes and electrophoretic mobilities, the properties of cephalopod enzymes seem to parallel those of tyrosinases from other sources. A noteworthy difference is the marked preference for the D-form of the substrate amino acids tyrosine and dopa, as evidenced from the apparent Michaelis constants of $3.3 \times 10^{-4}\,M$ for D-tyrosine, $1.7 \times 10^{-4}\,M$ for L-tyrosine, $10 \times 10^{-3}\,M$ for L-dopa, and $4.3 \times 10^{-3}\,M$ for D-dopa.

A distinguishing feature of tyrosinases from insects is the frequent occurrence in the form of inactive protyrosinases that become active following a proteolytic stage (Munn and Bufton, 1973; Yamaura *et al.,* 1980). The proenzyme of the silkworm *Bombyx mori* consists of two subunits of 40 kDa with only one copper in the divalent state per subunit; these could aggregate in a dimeric form depending on the ionic strength of the medium and the protein concentration (Ashida, 1971). Upon the action of a highly specific protease, isolated from the cuticle, this protyrosinase releases a small peptide, thereby becoming active (Dohke, 1973). This activation process is apparently accompanied by a reduction of cupric copper or by a reorganization of the active site leading to an electron paramagnetic resonance (EPR) silent, binuclear copper pair. Inactive and proteolytically activated tyrosinases have also been isolated from frog (*Rana pipiens*) epidermis (Barisas and McGuire, 1974). These appear to consist of subunits of 54 kDa (proenzyme) and 50 kDa (active enzyme) with a copper content of 0.15%. A tetrameric structure with subunits of smaller molecular weight (ca. 30K) has also been reported for the active enzyme (Mikkelsen and Triplett, 1975). In another study (Penafiel *et al.,* 1982) on *Rana esculenta ridibunda,* it was found that both protyrosinase and tyrosinase are composed of a single type of subunit of about 68 kDa and 62 kDa, respectively. They, however, have different quaternary structures, the proenzyme occurring as dimer, and the enzyme as tetramer.

Among animal tyrosinases, the mammalian enzyme has naturally attracted most of the attention. Since in normal epidermal tissues it is present in exceedingly low concentrations, pigmented melanomas of laboratory animals, especially mice (B16 and Cloudman S91 lines), have been instrumental for studies aimed at characterizing the enzyme. The accumulated evidence, reviewed by Hearing and Jimenez (1987, 1989), indicates that the murine tyrosinase is an acidic glycoprotein which occurs in a number of distinct isozymic forms, both soluble and membrane bound. Polymorphism derives from post-translational processing of the newly synthesized enzyme in the Golgi area, whereby sialic acid and 4 moles of asparagine-linked carbohydrate chains containing mannose, glucosamine, galactose, and fucose are added to each mole of the enzyme. As a result, an almost continuous spectrum of tyrosinase isozymes can be analytically resolved on polyacrylamide gels (Ferrini *et al.,* 1987). In mouse melanocytes, the *de novo* synthesized enzyme is a single-chain protein with a molecular weight of ca. 55–58K and an isoelectric point of 4.3. After glycosylation, it is converted into a number of electrophoretically faster migrating forms ranging in molecular weight from 65 to 80K (Hearing and Jimenez, 1989; Yurkow and Laskin, 1989). However, the exact molecular weights of these isozymes are uncertain since the carbohydrate moiety may alter their migration in SDS polyacrylamide gels. The membrane-bound enzyme appears to be an aggregate of high molecular weight (ca. 150K) with an isoelectric point of 5.3, which can be dispersed by ionic detergent solubilization back to its monomeric form.

The amino acid composition of murine tyrosinase has been determined by five different laboratories with comparable results (Burnett, 1971; Pomerantz and Li, 1973; Hearing *et al.,* 1981; Laskin and Piccinini, 1986; Ferrini *et al.,* 1987). Assuming a molecular weight of 58K for the unglycosylated enzyme, the basic amino acids (His, Lys, and Arg) combined represent 10% of the residues in the enzyme, while the contribution of the acidic amino acids (Asp and Glu) may be as high as 18% depending upon the identity of AsX and GlX. As to the sulfur-containing amino acids, Cys and Met, they are each present at 9 mol/mol protein, the cysteine possibly being involved in internal disulfide bonds (Laskin and Piccinini, 1986).

More recently, tyrosinase has been isolated from a human melanoma cell line (IGR 1), and both the particulate and the soluble forms have been characterized (Wittbjer *et al.,* 1989, 1990a). On polyacrylamide gel electrophoresis, the molecular weight of the membrane-bound enzyme was found to be 73K. After trypsin treatment, this gave an enzyme with an apparent molecular weight of 62K, somewhat larger than that determined for the natural soluble form, that is, 53K. This contrasts with the report by Nishioka (1978) that trypsination of the membrane-bound tyrosinase gives a product with the same molecular weight as the soluble form.

In another study by the Swedish group (Wittbjer *et al.,* 1990b) it was shown that tyrosinase from bovine eyes occurs mainly in a soluble form having molecular weight and isoelectric behavior similar to those of soluble human tyrosinase. The N-terminal amino acid sequence, determined by automated Edman degradation, proved identical to that of the N-terminal region of mouse melanoma tyrosinase predicted from a cDNA clone by Kwon *et al.* (1988) and showed homology to that of human tyrosinase with the exception of three amino acids. Moreover, histidine was identified as the N-terminal amino acid.

Efforts to determine the protein structure of mammalian tyrosinases, be they from mouse melanomas, bovine eyes, or human malignant melanocytes, by conventional chemical sequencing techniques have been so far unsuccessful. Reasons for this are multiple and include, besides the occurrence of the enzyme in low concentration (less than 0.01% of the protein content of melanocytes), the presence of a blocked amino terminus and the heavily glycosylated nature of the enzyme, accounting for its marked resistance to proteolytic enzymes (Imokawa and Mishima, 1984; Yurkow and Laskin, 1989).

Given these difficulties, information on the structure of mammalian tyrosinases has mainly been sought through investigation of melanocyte-specific cDNA clones which may encode the enzyme (see Chapter 9). The original sequence reported by Shibahara *et al.* (1986) for the murine gene was a prematurely truncated version of the full-length transcript, the complete sequence of which was subsequently established by various groups (Kwon *et al.,* 1988; Yamamoto *et al.,* 1989; Terao *et al.,* 1989).

The open reading frame of the mouse tyrosinase cDNA clone encodes a protein 533 amino acids in length with a putative leader peptide of 18 amino acid residues and 6 potential glycosylation sites (Asn/X/Ser or Asn/X/Thr), as indicated in Fig. 3.3. Cleavage of the leading peptide would then lead to a protein of 515 amino acid residues with a molecular weight of 58,547, which is close to the estimated molecular weight of the unglycosylated T3 isoform of mouse tyrosinase of 55–58K (Burnett, 1971; Hearing *et al.,* 1981). The deduced amino acid sequence of the human cDNA clone, established by Kwon *et al.* (1987), is similar in structure and molecular weight, consisting of 529 amino acid residues.

```
-180  GCTTAGTGTA AAACAGGGTG AGAGTATTTG ATGTAAGAAG OCCAGTGGTT ATATAGOTCT TAGCCAAAAC ATGTGATACT
      Hind III
-80   CACTCCAGOG GTTOCTOCAA AAGAACTCTG TGACACTCAT TAACCTATTG CTGCAGATTT TGTATGATCT AAACCAGAAA
1     ATG TTC TTG GCT GTT TTG TAT TGC CTT CTG TOG ACT TTC CAG ATC TCT GAT GGC CAT TTT     60
1     Met Phe Leu Ala Val Leu Tyr Cys Leu Leu Trp Ser Phe Gin Ils Set Asp Gly Mis Phe     20
61    GCT CGA OCC TOT OCC TCC TCT AAG AAC TTG TTG GCA AAA GAA TOC TGC CCA CCA TGG ATC    120
21    Pro Arg Ala Cys Ala Ser Ser Lys Aso Leu Leu Ala Lys Glu Cys Cys Pro Pro Trp Met     40
121   GCT GAT GOG AOT CCC TGC COC CAG CTT TCA GOC AGA GGT TCC TGC CAG GAT ATC CTT CTG    180
41    Gly Asp Gly Ser Pro Cys Gly Gln Leu Ser Gly Arg Gly Ser Cys Glm Asp Ile Leu Leu     60
101   TCC AGT GCA CCA TCT GCA CCT CAG TTC CCC TTC AAA GOG GTG CAT GAC CGT GAC TCC TGG    240
61    Ser Ser Ala Pro Ser Gly Pro Gls Phe Pro Phe Lys Gly Val Asp Asp Arg Glu Ser Trp     80
241   CCC TCT GTG ITT TAT AAT AGG ACC TGG CAG TOC TCA OOC AAC TTC ATG CGT ITC AAC TGC    300
81    Pro Ser Val Phe Tys Agp Arg Thr Cys Gls Cys Ser Gly Asp Phe Mat Gly Phe Aso Cys    100
301   OGA AAC TCT AAG TTT GCA TTT OGG CCC CCA AAT TCT AGA GAG AAG COA OTC TTC ATT AGA    360
101   Gly Asp Ser Lys Phe Gly Phe Gly Gly Pro Asn Cys Thr Glu Lys Arg Val Leu Tle Arg    120
361   AGA AAC ATT TTI GAT ITG AGT CTC TCC GAA AAG AAT AAG TTC ITI ICI IAC CTC ACT TTA    420
121   Arg Asn Ile Phe Asp Leu Ser Val Ser Glu Lys Asn Lys Phe Phe Ser Tyr Leu Thr Leu    140
421   GCA AAA CAT ACT ATC AOC TCA GTC TAT OTC ATC CCC ACA GGG ACC TAT OOC CAA ATC AAC    480
141   Ala Lys Ile Thr Ils Ser Ser Val Tyr Val Ile Pro Thr Gly Thr Tyr Gly Gln Mat Aso    160
481   AAT GGG TCA ACA CCC ATG TTT AAT GAT ATC AAC ATC TAC GAG CTC ITI OTA TGG ATO CAT    540
161   Asn Gly Ser Thr Pro Mat Phe Asn Asp Ile Asp Ile Tyr Asp Leu Phe Val Trp Mat His    180
541   TAC TAT GTG TCA ACG GAC ACA CTG CTT GOG GCC TCT GAA ATA TCG AGG GAC ATT CAT TTT    600
181   Tyr Tyr Val Ser Arg Asp Thr Leu Leu Gly Gly Ser Glu Ile Trp Arg Asp Ile Asp Phe    200
601   GCC CAT CAA CCA CCA GOG TTT CTO CCT TGG CAC AGA CTT TTC TTG TTA TTG TGG GAA Caa    660
201   Ala His Glu Als Pro Gly Phe Leu Pro Trp His Arg Leu Phe Leu Leu Leu Trp Glu Cln    220
661   CAA ATT CGA GAA CTA ACT GGG GAT GAG AAC TTC ACT GTT CCA TAC TCG GAT TOG AGA CAT    720
221   Glu Ils Arg Glu Leu Thr Gly Asp Glu Asn Phe Thr Val Pro Tyr Trp Asp Trp Arg Asp    240
721   GCA GAA AAC TGT GAC ATT TCC ACA GAT GAG TAC TTC GGA GGT CGT CAC CCT GAA AAT CCT    780
241   Als Ulu Asp Cys Asp Ils Cys Thr Asp Glu Tyr Leu Gly Gly Arg His Pro Glu Asp Pro    260
781   AAC TTA CTC AGC CCA GGA TCC TTC TTC TCC TCC TOU GAG ATC ATT TCT AGC AGA TCA GAA    840
261   Asp Leu Leu Ser Pro Als Ser Phe Phe Ser Ser Trp Gln Ils Ils Cys Ser Arg Ser Glu    280
841   CAG TAT AAT AGC CAT CAG GTT TTA TGC GAT GGA AGA CCT GAG GGA CCA CTA TTA GOT AAT    900
281   Glu Tyr Asp Ser His Gln Val Leu Cys Asp Gly Thr Pro Clu Gly Pro Leu Leu Arg Asn    300
901   CCT GGA AAC CAT GAC AAA GCC AAA ACC CCC AGG CTC CCA TCT TCA GCA GAT GTG GAA TTT    940
301   Pro Gly Asn His Asp Lys Ala Lys Thr Pro Asp Leu Pro Ser Ser Ala Asp Val Glu Phe    320
961   TCT GTG AGT TTG ACC CAG TAT GAA TCT GGA TCA ATC GAT AGA ACT GCC AAT TTC AGG TTT   1020
321   Cys Leu Ser Leu Thr Cln Tyr Glu Ser Gly Ser Mal Asp Arg Thr Ala Agn Phe Ser Phe    340
1021  AGA AAC ACA CTC GAA GTA TTT GCC AGT CCA CTC AGA GGG ATA OCA GAT CCT TCT CAA AGT   1080
341   Arg Asn Thr Leu Glu Val Phe Ala Ser Pro Leu Thr Gly Ile Ala Asp Pro Ser Cln Ser    360
1081  AGC ATG CAC AAT GOC TTA CAT ATC ITT ATO AAT GGA ACA ATG TCC CAA CTA CAG GGA TCG   1140
361   Ser Mat His Asn Ala Leu His Ile Phe Het Asg Gly Thr Mat Ser Cln Val Cln Gly Ser    380
1141  GCC AAC GAT CCC ATT TTT CTT CTT CAC CAT CCT TTY GTC GAC AGT ATT TTT GAA CAA TOG   1200
381   Ala Asn Asp Pro Ile Phe Leu Leu His His Ala Phe Val Asp Ser Ile Phe Glu Cln Trp    400
1201  CTG CGA AGG CAC CGC CCT GTT TTG GAA GTT TAC CCA GAA CCC AAT GCA CCT ATC GOC CAT   1260
401   Leu Arg Arg His Arg Pro Leu Leu Glu Val Tyr Pro Glu Ala Asn Ala Pro Ile Gly His    420
1261  AAC AGA GAC TCT TAC ATG GTT CCT TTC ATA CCC CTC TAT AGA AAT GGT GAT TTC TTC ATA   1320
421   Asn Arg Asp Ser Tyr Mat Val Pro Phe Ile Pro Leu Tyr Arg Asn Gly Asp Phe Phe Ile    440
1321  ACA TCC AAG GAT CTG GGA TAT GAC TAC AGC TAC CTC CAA GAG TCA GAT CCA GGC TTT TAC   1380
441   Thr Ser Lys Asp Leu Gly Tyr Asp Tyr Ser Tyr Leu Gln Glu Ser Asp Pro Gly Phe Tyr    460
1361  AGA AAT TAT ATT CAG CCT TAC TTG GAA CAA GCC AGT CGT ATC TOO CCA TOO CTT CTT GGo   1440
461   Arg Asn Tyr Ile Glu Pro Tyr Leu Glu Gln Ala Ser Arg Ile Trp Pro Trp Leu Leu Gly    480
1441  GCA GCA CTC CTC CGA OCT GTT ATT CCT OCA GCT CTC TCT GOG CTT ACC AGT AGG CTA TGT   1500
481   Ala Ala Leu Val Gly Ala Val Ile Als Als Als Leu Ile Gly Leu Ser Ser Ars Leu Cys    500
1501  CTT CAG AAG AAG AAC AAG AAG AAC CAA CCC CAG GAC GAA AGG CAG CCA CTG CTC ATG GAC   1540
501   Leu Gln Lys Lys Lys Lys Lys Lys Gln Pro Gln Glu Glu Arg Gla Pro Leu Leu Mat Asp    520
1561  AAA GAC GAC TAC CAC AGC TTG CTG TAT CAG AGC CAT CTC TGA ACA TCC TAG GAA ACA GAG   1620
521   Lys Asp Asp Tyr His Ser Leu Leu Tyr Gln Ser His Leu ---
1621  TGG ACT GAA AGG TTT TAC CTC ACT CGA CCT ATT TGT TGG TCT ITC TAC AAA TTT AAA CTA   1680
1681  GCA TAA AAC ATA GAC CAT AGC TGT TTG CCT TTT TTT CAG ACC CAT GTT TTT TCC TAA GTC   1740
1741  CTA GTT TCT AAG AAA TGa CTG GGA TTT GCT AAA ATA TAT ATA TAT ATA AAT AAT AAC TTA   1800
1801  CTA ATA OCT AAA TAA AAT TTC CTC TTA CAA CTA AAA AAA AAA AAA AA
```

FIG 3.3 Nucleotide sequence of a mouse tyrosinase cDNA and the deduced amino acid sequence. The nucleotide sequence of message strand is numbered in the 5′ to 3′ direction. The noncoding sequence is indicated by negative numbers. Nucleotide residue 1 is the first nucleotide of the ATG initiation codon. Potential glycosylation signals are shown underlined (—); potential transmembrane region is shown double underlined, (=); stop codon is indicated by broken line, (---); From Kwon *et al.,* 1988.

D. Amino Acid Sequence Comparison

As pointed out earlier, *S. glaucescens* tyrosinase is the smallest representative of this group of enzymes known so far. It would seem therefore that in the course of evolution the molecular complexity of the enzyme has gradually increased up to that of mammalian tyrosinases which are close to the hemocyanin copper (II) protein of arthropods and mollusks (for review see Solomon, 1981). Such a phylogenetic relationship is easily apparent from the comparison of the amino acid sequences between all tyrosinases, shown in Fig. 3.4.

As estimated by Muller *et al.* (1988), the homology between mouse tyrosinase and related enzymes in region I is 58% to *S. glaucescens,* 63% to *S. antibioticus,* 47% to *N. crassa,* and 89% to human tyrosinase. A high degree of homology is also seen in region II which contains three potential copper-binding histidines, one of which at position 306 is known to be part of the catalytic center of *N. crassa* tyrosinase (Lerch, 1983; Kupper *et al.,* 1989). In addition, amino acid sequence comparison of the region containing the copper-binding sites in hemocyanin from the spider *Eurypelma californicum* and tyrosinases shows that the three histidine residues are located exactly in the same positions. A more in-depth discussion of the structural relationship between tyrosinases and hemocyanins is given by Lerch (1988).

III. TYROSINASE AS A COPPER MONOOXYGENASE

A. The Active-Site Copper

The involvement of copper as the essential cofactor for the catalytic activity of tyrosinase was established independently by Kubowitz (1938) and Keilin and Mann (1938), working with enzyme preparations from vegetable sources. These studies provided evidence that the catalytic properties of tyrosinase are inhibited by agents which chemically combine with copper, such as cyanide, diethyldithiocarbamate, salicylaldoxime, and carbon monoxide, and are reversibly restored on addition of excess copper. By comparison with hemocyanin, it was suggested that the enzyme contains two copper ions at the active site, shuttling between valence states. After that, it was soon apparent that tyrosinase differs from other copper proteins in several respects, including the lack of visible absorption and the silent response of the metal to electron paramagnetic resonance and magnetic susceptibility. At some time, these spectral features were interpreted in terms of an active-site model involving the presence of one copper ion in the cuprous (I) state, possibly acting as a two-electron donor via a valence change to the Cu(III) state (Robb,

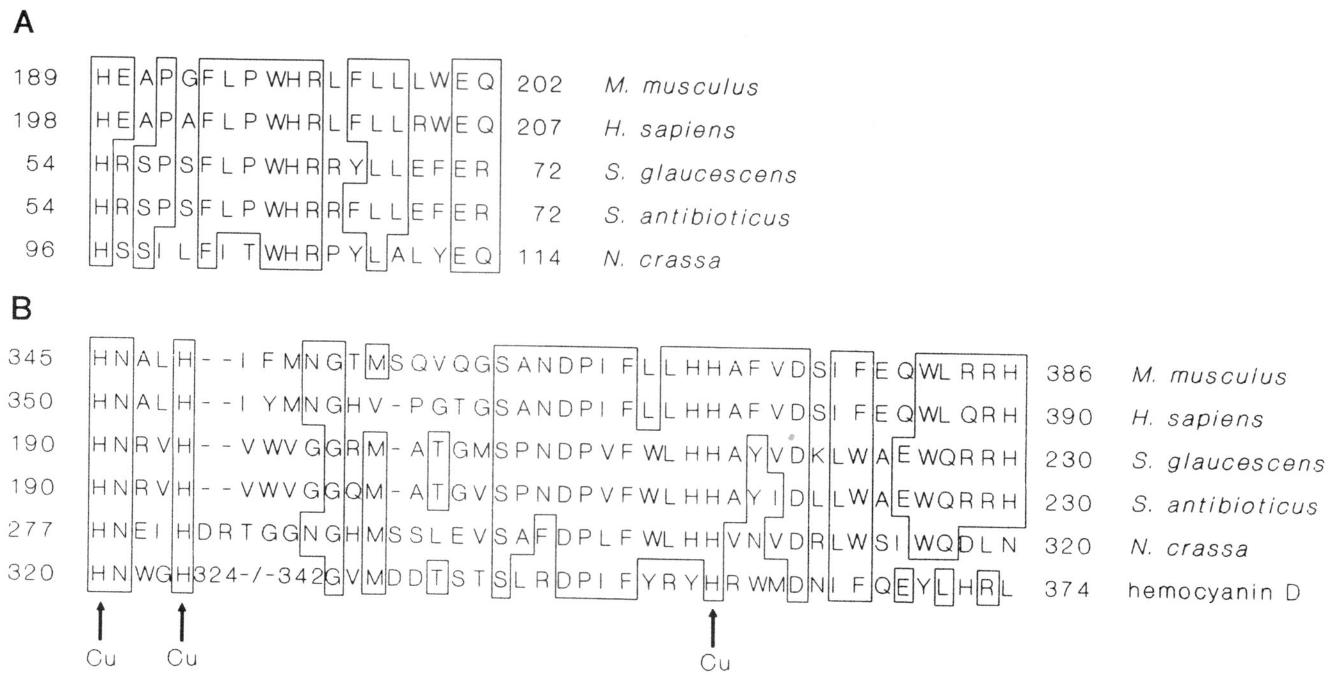

FIG 3.4 Homologous regions I (A) and II (B) of tyrosinases of different species and of hemocyanin D of the spider *E. californicum*. The numbers indicate the position of the amino acids in the respective protein sequence. Conserved amino acids which were found in ≥50% of the cases are boxed. The amino acid sequence is given in single-letter code. Adapted from Müller *et al.*, (1988).

1984). However, recent evidence indicates that in most cases the enzyme contains at the active site two paired copper atoms in the cupric divalent state which are close enough to permit interaction and antiferromagnetic coupling of the single electrons on each atom. The existence of such a functional binuclear copper cluster in tyrosinase has been supported by physical and chemical studies, done mainly on the enzyme from *N. crassa* by Lerch and his associates (Lerch, 1981, 1988).

B. The Ligand Environment

Tyrosinase has proved remarkably resilient to a variety of reagents designed to selectively react with amino acid residues at the active site, including *p*-hydroxymercurybenzoate, iodoacetate, diethylpyrocarbonate, and dimethylsuberimidate. However, insight into the environment surrounding the binuclear copper was gained through investigation of the effects on the enzyme of dye-sensitized photooxidation (Pfiffner and Lerch, 1981). This study showed that three highly photooxidizable histidine residues are ligands to one of the two copper ions at the active site and that ligation is essential for maintaining the activity. The photolabile histidines were identified as His-188, His-193, and His-289.

Besides photochemical inactivation, tyrosinase suffers a gradual irreversible inactivation as the oxidation of the substrate proceeds. While the details and biochemical implications of this reaction will be described in a later paragraph, attention is called here to the fact that this process is accompanied by the selective destruction of one histidine residue, His-306, and this destruction is associated with the loss of 1 g-atom of copper per mole enzyme after complete inactivation (Dietler and Lerch, 1982). Overall, these data have been incorporated into a model for the ligand environment of the active-site copper depicted in Fig. 3.5. In this, ligation of the two copper atoms depends on four histidine residues, three of which act on one atom and one on the other.

C. Copper Electronic Structure: Resting (Met-), Oxy-, and Deoxytyrosinase

Recognition of the binuclear character of the active-site copper in tyrosinase has revealed a close parallelism between the enzyme and the oxygen-transporting hemocyanin. This similarity is apparent from the occurrence of both proteins in three chemically and spectroscopically distinct forms, resting (met-), oxy- and deoxy- forms, which differ in several respects, including redox state of the copper, presence of bound oxygen, etc. These forms are chemically interchangeable through a series of equilibria shown in Fig. 3.6. The first

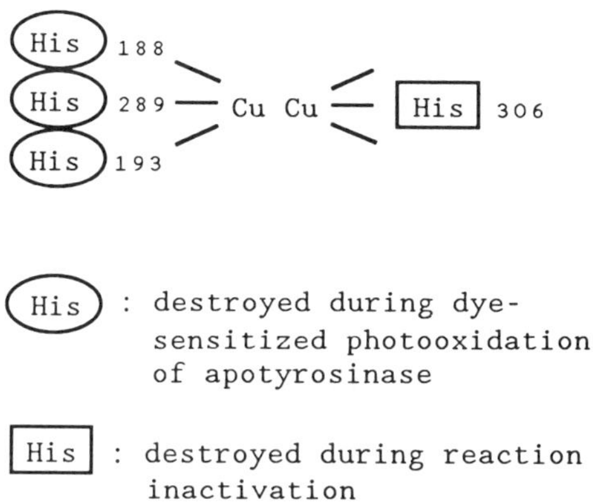

FIG 3.5 Proposed model for the ligand environment of the active-site copper of tyrosinase (Lerch, 1981).

of these forms, termed by Jolley *et al.* (1972) as resting enzyme, refers to the enzyme as isolated from the natural source. However, depending on the source and on the extraction and isolation procedure, small percentages (2–30%) of other forms, particularly the oxy- form, are usually found associated with the resting enzyme. Accordingly, the term mettyrosinase was later introduced in analogy with methemocyanin to specifically indicate the bicupric form of the enzyme completely devoid of the oxy- and other forms (Mason, 1976). On the basis of the lack of an EPR signal and on the stoichiometric reaction with hydrogen peroxide, it was suggested that the copper in mettyrosinase is present in an oxidized, antiferromagnetically coupled bicupric state. A model for mettyrosinase analogous to methemocyanin has accordingly been proposed, involving two tetragonal Cu(II) ions at a distance of about 3.4 Å bridged by an endogenous protein ligand providing an effective pathway for superexchange between the Cu(II) ions (Himmelwright *et al.*, 1980; Woolery *et al.*, 1984).

Two-electron reduction of mettyrosinase leads to the form commonly referred to as deoxytyrosinase. This form is relatively less known than the other forms with respect to the electronic and geometric properties of the active-site copper, owing to the absence of spectral features in the visible region and the lack of EPR signals. By analogy with deoxyhemocyanin, where the copper has conclusively been shown to be present in the reduced cuprous state, deoxytyrosinase has been assigned a bicuprous structure (Lerch, 1988).

Exposure of deoxytyrosinase to molecular oxygen, or reaction of mettyrosinase with hydrogen peroxide, leads to the formation of the oxygen-binding form of the enzyme known as oxytyrosinase. The absorption and circular

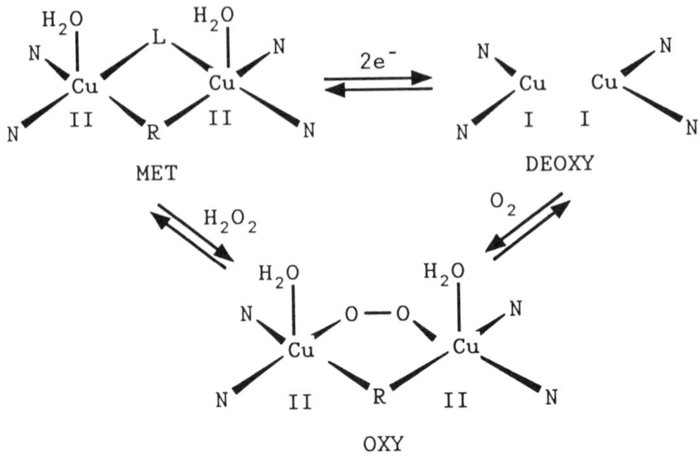

FIG 3.6 Diagram showing the relationships between the three main forms of tyrosinase and the equilibria involved in their interconversion.

dichroism spectra of this form exhibit a very prominent band at 345 nm and several bands in the visible region (Himmelwright *et al.*, 1980). These and other spectral features pointed to a structural model for *Neurospora* oxy-tyrosinase in which two antiferromagnetically coupled tetragonal Cu(II) ions are held together at a distance of 3.63 Å by the endogenous protein ligand R and a peroxide bridge (Woolery *et al.*, 1984). Besides the three major, biologically relevant forms of tyrosinase, other derivatives of the *Neurospora* enzyme have been described exhibiting easily detectable EPR features (Himmelwright *et al.*, 1980). These forms, usually referred to as half-met- forms, have been obtained by treating the resting enzyme, or its met- form, with an excess of sodium nitrite and ascorbic acid, followed by ligand substitution of the half-met-NO_2^- form with different anions (Solomon, 1981). Spectroscopic investigation of these derivatives has thrown additional light on the active-site copper of tyrosinase and the structural arrangement surrounding it.

IV. CATALYTIC PROPERTIES AND MODE OF ACTION

A. SUBSTRATES

Most of what is presently known about the catalytic properties of tyrosinase has been gained through investigation of mushroom tyrosinase, which apparently acts on a relatively broad range of molecules. Some comparative studies with tyrosinase from various animal sources including amphibians, insects, avians, and mammals are also available (Lerner and Fitzpatrick, 1950; Robb, 1984). However, no systematic work has appeared as yet aimed at directly matching the substrate specificity of highly purified enzymes through standardized analytical assays. With these premises, a survey of the catalytic properties of tyrosinase is given in the following, with critical ends prevailing over criteria of comprehensiveness.

1. *Phenols and Catechols*

Typical substrates of tyrosinase are mono- and *o*-diphenols (catechols) devoid of bulky groups or crowded substituent patterns adjacent to the hydroxyl group(s) on the aromatic ring. Thus, antioxidants like 2-*tert*-butyl-4-methoxyphenol (BHA) and 2,6-di-*tert*-butyl-*p*-cresol (BHT) are not oxidized by tyrosinase (Passi *et al.,* 1987). Ethers, like guaiacol, and *m*-diphenols are likewise not acted upon.

As has already been pointed out, tyrosinase exhibits higher affinity for *o*-diphenols than for monophenols. Studies by Duckworth and Coleman (1970) and Passi and Nazzaro-Porro (1981) showed that, within series of 4-substituted monophenols and *o*-diphenols, the rate of oxidation is dependent on the electronic properties of the substituent, decreasing with decreasing electron-donating ability as in the sequence: $-OCH_3 > -OC_6H_5 > -CH_3 > H$. Thus, 4-hydroxyanisole is a better substrate than 4-phenoxyphenol, *p*-cresol, and phenol, in that order. On the other hand, phenols bearing electron-withdrawing groups, such as $-NO_2$, $-CHO$, and $-CO_2H$, are not oxidized at all and act rather as competitive inhibitors of tyrosinase.

In addition to electronic effects, some other factors, including the enzyme primary and tertiary structure, appear to account for tangible differences in the rate of oxidation of seemingly related substrates like dopa, dopamine, and 4-methylcatechol. Thus, L-dopa is reported to be oxidized at a significantly lower rate than catechol with purified mushroom tyrosinase (Duckworth and Coleman, 1970) but seems to be the optimal substrate for human tyrosinase (Karg *et al.,* 1991a). Both mushroom and mammalian tyrosinases exhibit different affinity for the enantiomers of chiral amino acid substrates: Usually the L form is oxidized much faster than the D form, with the notable exception of the enzyme from *Sepia,* mentioned earlier (Section II,C).

The melanin precursor 5,6-dihydroxyindole is generally regarded as a substrate with high affinity for tyrosinase (Mason, 1967; Korner and Pawelek, 1982; Miranda *et al.,* 1985). We will turn back to this point later in Chapter 5. Here, it may be sufficient to say that under the usual assay conditions, the compound is highly susceptible to autooxidation, especially in the presence of metal ions, thus making it difficult to determine the kinetic parameters of the enzymic reaction.

Carboxyethylcatechols, such as dihydrocaffeic acid, and their esters are good substrates of tyrosinase and undergo ready conversion to their *o*-quinones, whose subsequent fate depends on whether the carboxyl group is free or not (Sugumaran *et al.*, 1989a). In the former case, an intramolecular cyclization of the carboxyl onto the electrophilic quinone ring occurs, leading to dihydroxydihydrocoumarins, whereas in the latter case a substantial side-chain desaturation via a quinone methide prevails, affording caffeic acid esters (Fig. 3.7).

Likewise, carboxymethyl-*o*-benzoquinone, readily generated from 3,4-dihydroxyphenylacetic acid by the action of tyrosinase, undergoes intramolecular cyclization of the carboxyl to give trihydroxybenzofuran, along with competing tautomerization to the quinone methide, which rapidly reacts with water to give 3,4-dihydroxymandelic acid (Sugumaran *et al.*, 1989b). Upon oxidation by tyrosinase, 3,4-dihydroxymandelic acid is converted to the corresponding *o*-quinone, which suffers a decarboxylation reaction leading to a quinone methide. This, in turn, is converted to 3,4-dihydroxybenzaldehyde by aromatization-driven tautomerization (Sugumaran, 1986; Bouheroum *et al.*, 1989) (Fig. 3.8).

Another molecule that is oxidized by tyrosinase is 2,4-dihydroxyphenylalanine, which is converted to the highly cytotoxic 2,4,5-trihydroxyphenylalanine

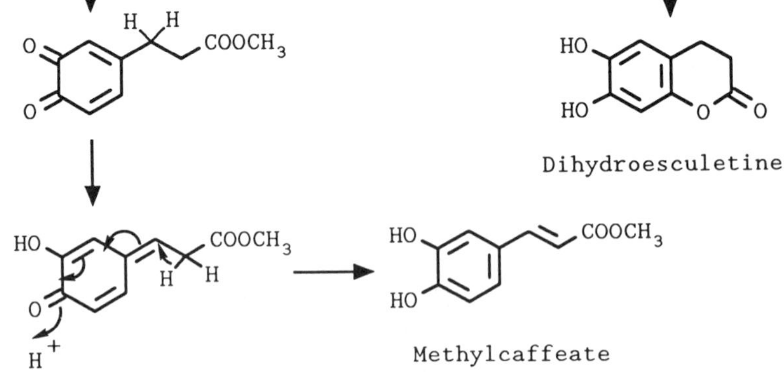

FIG 3.7 Oxidative transformation of dihydrocaffeic acid (path a) and methyl dihydrocaffeate (path b) catalyzed by tyrosinase.

(6-hydroxydopa) (Hansson *et al.,* 1985). In the presence of reducing agents, for example, ascorbic acid, tyrosinase can also induce the further hydroxylation of diphenols with the formation of trihydroxyderivatives. Thus, dopa is converted in substantial amounts to 3,4,5-trihydroxyphenylalanine (5-hydroxydopa) (Hansson *et al.,* 1980).

2. *Tyrosine-Containing Peptides and Proteins*

Tyrosine residues belonging to proteins can be oxidized by tyrosinase. This was demonstrated in an early study on the enzymic oxidation of the model peptide tyrosylglycine, leading to the formation of a red intermediate species that eventually turned brown without deposition of pigment (Bu'Lock and

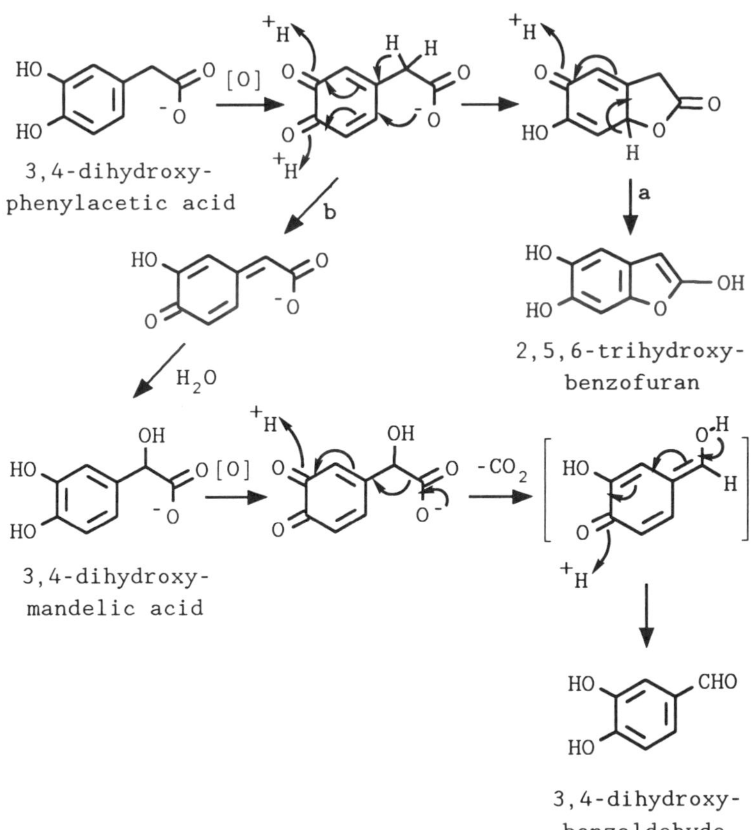

FIG 3.8 Tyrosinase-catalyzed oxidation of 3,4-dihydroxyphenylacetic acid to trihydroxybenzofuran (path a) or 3,4-dihydroxybenzaldehyde (path b).

Harley-Mason, 1951a). These spectral changes were interpreted in terms of a reaction scheme involving the generation of a dopachrome-peptide that rearranges to the corresponding 5,6-dihydroxyindole derivative, which is relatively stable to tyrosinase (Fig. 3.9). Similar results were obtained by Yasunobu *et al.* (1959) on a number of tyrosine-containing peptides. Depending on whether tyrosyl residues are at the N terminus, at the C terminus, or in the middle, a dopachrome absorption curve (in the former case) or a dopaquinone absorption pattern (in the latter cases) was observed, consistent with the proposed oxidation scheme.

The ability of tyrosinase to oxidize tyrosine residues in proteins is of interest in relation to the reported occurrence of dopa and 5-cystein-*S*-yldopa (5-cysdopa) in the hydrolysates of mouse and human melanomas (Takahashi and Fitzpatrick, 1966; Agrup *et al.,* 1978a). High concentrations of protein-bound dopa and 5-cysdopa were also detected in black and yellow hair of mice. Moreover, in a series of related studies, Cory and Frieden (1967a,b) showed that oxidized B chain of insulin, yeast alcohol dehydrogenase, and bovine serum albumin can serve as substrate for tyrosinase. The structural modifications accompanying the oxidation of these proteins by mushroom tyrosinase were investigated by Ito *et al.* (1984b), who found that in bovine serum albumin and in alcohol dehydrogenase, which contain cysteine residues, 5-cysdopa is formed, presumably as indicated in Fig. 3.10, whereas in the case of insulin, which lacks cysteine residues, dopa is the major reaction product. High-molecular-weight (17–130K) dopa proteins have been implicated as the precursors of sclerotized structures in a number of invertebrate animal phyla, such as the byssus of the *Mollusca* (Waite, 1983), the eggcases of *Platyhelminthes,* and the cements of certain *Anellida,* where oxidation products of peptidyldopa may also induce larval settlement and metamorphosis. The exact roles and the molecular mechanisms of dopa proteins and

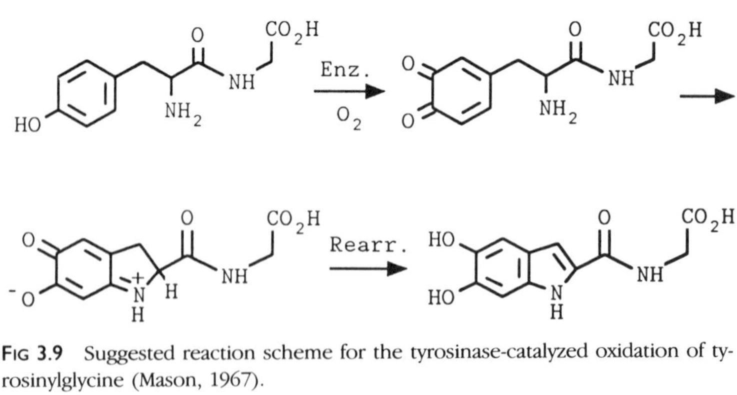

FIG 3.9 Suggested reaction scheme for the tyrosinase-catalyzed oxidation of tyrosinylglycine (Mason, 1967).

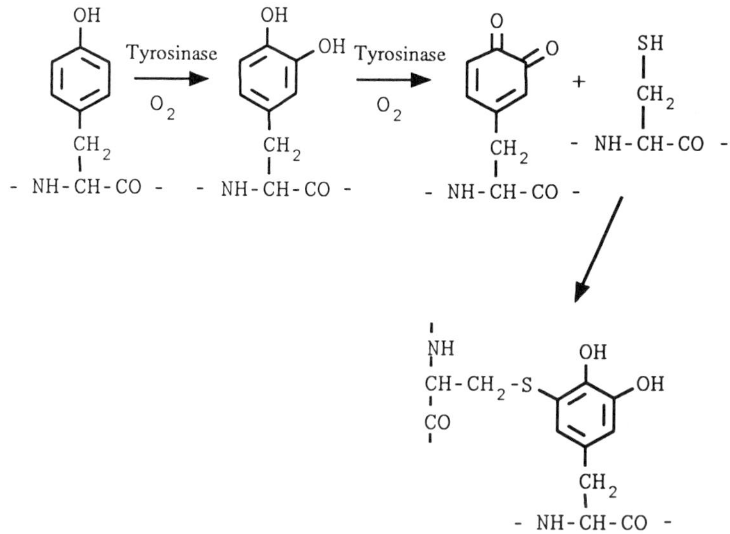

FIG 3.10 Formation of protein-bound cysteinyldopa by tyrosinase-induced modification of tyrosine residue in cysteine-containing proteins according to Ito *et al.* (1984b).

oligopeptides are still in question but presumably rely on the oxidation of the catecholic dopa residues to the corresponding quinones. Some insight into the chemistry of these processes was obtained by Waite and co-workers (Rzepecki and Waite, 1991; Rzepecki *et al.*, 1991) using *N*-acetyldopa ethyl ester as a model. Oxidation with tyrosinase or periodate resulted in the base-catalyzed formation of *N*-acetyl-α,β-dehydrodopa ethyl ester, presumably arising as depicted in Fig. 3.11. A similar reaction was observed upon oxidation of dihydrocaffeic acid, but not of *N*-acetyldopamine.

3. *Other Substrates*

In addition to simple monophenols and catechols, a number of other molecules with phenolic hydroxyl groups are oxidized by tyrosinase (Kalyanaraman *et al.*, 1985). Among these, hydroquinone merits special consideration in view of its marked skin depigmenting properties (Jimbow *et al.*, 1974). Originally regarded as a tyrosinase inhibitor (Chen and Chavin, 1972), this compound was later shown to be a substrate of the enzyme (Passi and Nazzaro-Porro, 1981; Hashimoto *et al.*, 1984, 1988; Havens and Tramposch, 1986). In a more recent study on the effect of hydroquinone on melanogenesis *in vitro* (Palumbo *et al.*, 1991a), evidence was obtained that hydroquinone is per se a poorer substrate of tyrosinase than tyrosine; yet, in the presence of tyrosine, it is preferably oxidized, due to the generation of catalytic amounts of dopa

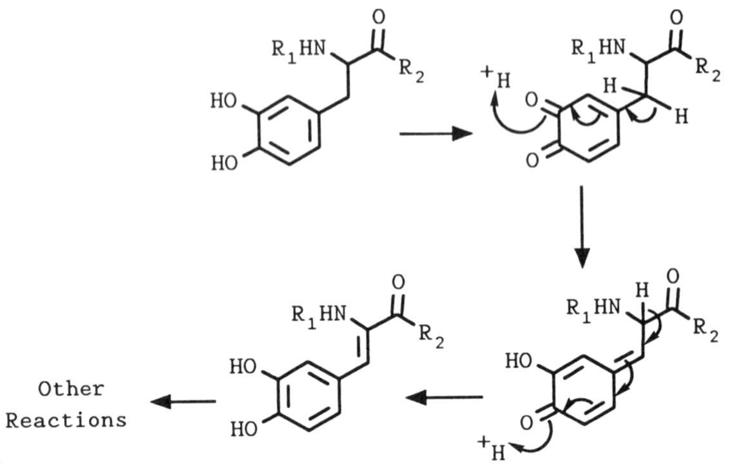

FIG 3.11 Oxidative conversion of *N*-acetyldopa ethyl ester to *N*-acetyl-α,β-dehydrodopa ethyl ester (Rzepecki and Waite, 1991).

acting as cofactor. Under these conditions, the reaction leads to the formation in the early stages of hydroxybenzoquinone as the major product, together with lower amounts of benzoquinone, resulting probably from redox-cycling processes. On continued oxidation, a complex mixture of reddish-brown products is formed, two of which could be isolated after reduction and subsequent acetylation. These were identified as the biphenyl 3.**1**, arising by coupling of hydroxybenzoquinone with benzoquinone, and the dibenzofuran 3.**2**.

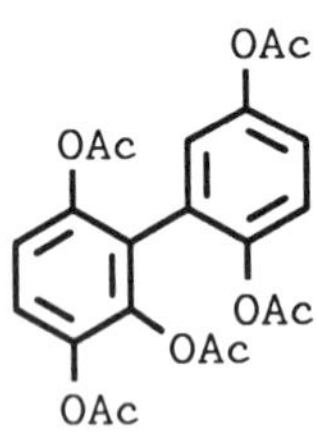

3.1

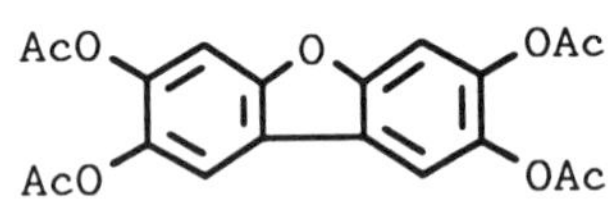

3.2

Electron-rich aromatic amines and *o*-aminophenols are also oxidized by tyrosinase (Toussaint and Lerch, 1987). Arylamines and the corresponding monophenols exhibit similar Michaelis constants, (K_m = 0.11–0.49 m*M*) but the former are oxidized at a markedly slower rate (K_{cat} = 1–3 min^{-1} versus

1320–6960 min^{-1}). In the case of *o*-aminophenols, the K_m values are similar to those of the corresponding *o*-diphenols, with oxidation rates comparable to 2,3-dihydroxybenzoic acid (K_{cat} = 0.2–0.8 s^{-1}).

B. The Lag Period

A distinguishing feature of the kinetics of oxidation of tyrosine and other monophenols by tyrosinase is the occurrence of a lag period, often referred to as the induction time, preceding the actual onset of the reaction (Lerner and Fitzpatrick, 1950). A typical progress curve for the oxidation of tyrosine catalyzed by tyrosinase, exhibiting the induction time, is shown in Fig. 3.12.

This lag time is subjected to the influence of several factors, including substrate and enzyme concentration, pH of the medium, and presence of hydrogen donors, particularly L-dopa (Pomerantz, 1966; Pomerantz and Warner, 1967). Other catechol compounds such as dopamine, adrenalin, and noradrenalin can shorten the lag but cannot eliminate it even when added at high concentrations.

Tetrahydropteridin, a well-known specific cofactor of other aromatic hydroxylases, is also effective in stimulating tyrosinase activity, though to a lesser extent than dopa. The dependence of the induction time on pH has been

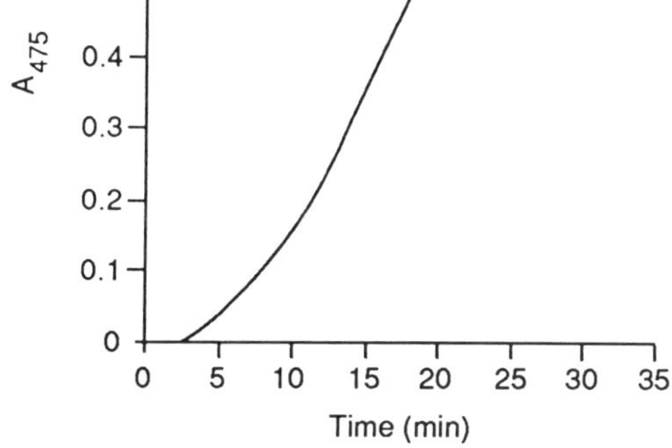

FIG 3.12 Lag period in the tyrosinase-catalyzed oxidation of tyrosine, as evidenced spectrophotometrically by monitoring dopachrome formation at 475 nm.

pointed out occasionally in the past and has been reexamined by Ramaiah and co-workers (Devi *et al.*, 1987, 1989). They found that, when assayed at acidic pH (e.g., 5.0), tyrosinase from murine melanoma and human skin as well as from mushrooms loses its requirement for dopa. Since the pH of the melanosome is thought to be acidic (Moellman *et al.*, 1988), this would provide a mechanism of activation of tyrosinase in melanocytes.

Several hypotheses have been proposed to explain the occurrence of the lag and its dependence on the above factors, especially the marked activating effect of dopa. The most recurrent one envisages the possibility that tyrosinase contains at least two combining sites, one for the substrates tyrosine and dopa, and one specific for dopa, with the latter presiding at the activation of the enzyme toward tyrosine hydroxylation. As soon as a small amount of dopa is formed during the lag time in the oxidation of tyrosine, it would bind to the specific site, thereby eliciting a conformational change in the enzyme which leads to increased affinity for the substrate tyrosine (Hearing *et al.*, 1980). The lag period has also been explained in terms of a hysteretic process, corresponding to a slow aggregating step in response to exposure to the substrate (Garcia-Carmona *et al.*, 1980).

A common assumption in all these studies is that dopa is the natural cofactor for the activation of resting tyrosinase *in vivo*. However, since dopa is not an usual amino acid, the question arises as to how tyrosinase is initially activated to produce its own cofactor. A new entry to this problem follows from the finding that catalytic amounts of ferrous ions are capable of activating tyrosinase in the absence of dopa (Palumbo *et al.*, 1985, 1990b). The effect is highly specific for the reduced form of iron and is not suppressed by scavengers of oxygen radical species, consistent with a mechanism involving reduction of the cupric ions at the active site of the resting enzyme to the cuprous state. Such a redox activation mechanism is well in line with the results of recent studies by Boyer *et al.* (1986) showing that tyrosinase and other copper enzymes catalyze the oxidation of Fe(II) to Fe(III).

At a 1 m*M* concentration Ni^{2+} and Co^{2+} have also been shown to stimulate hydroxylation activity of tyrosinase, whereas Zn^{2+} exhibits an opposite inhibitory effect (Jara *et al.*, 1990).

C. Mechanism of Action

A widely accepted mechanistic scheme accounting for the hydroxylase–oxidase activity of tyrosinase is shown in Fig. 3.13. According to this scheme, the hydroxylation of the phenol is envisaged as being brought about specifically by the active form of tyrosinase, oxytyrosinase, which coordinates the substrate into a ternary complex made up of the bicupric cluster, the phenol, and molecular oxygen. Such a complex would be in a suitable geometric arrangement to permit the *o*-hydroxylation of the correctly positioned phenol substrate, presumably by a mechanism that involves the cupric ion-promoted electrophilic attack on the polarized peroxide bond to the aromatic ring. The resulting *o*-diphenol is then oxidized from the equatorial position, liberating the *o*-quinone and deoxytyrosinase. A similar mechanism has been recently

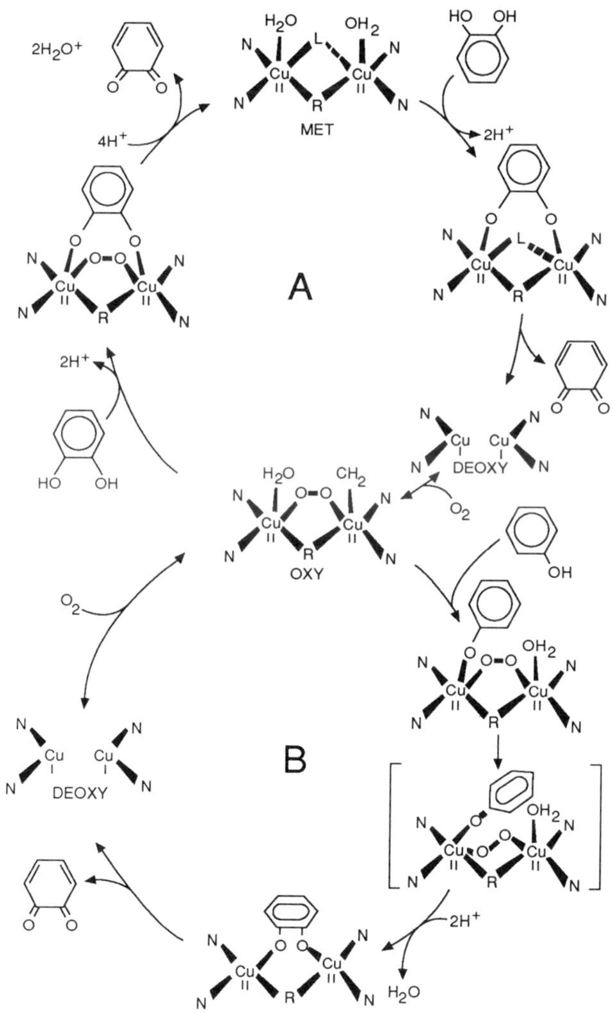

FIG 3.13 Mechanistic scheme of the hydroxylase–oxidase activity of tyrosinase. After Winkler *et al.* (1981).

suggested by Karg *et al.* (1991a) for the hydroxylase–oxidase activity of human tyrosinase. From consideration of the proposed sequence of reactions, it follows that the rate of hydroxylation of tyrosine to dopa and then to dopaquinone is proportional to the amount of enzyme in the reduced form. On these grounds, the occurrence of a lag period would be well justified by the redox potential of copper, ensuring that under resting conditions most of the enzyme is in the bicupric state and is therefore inactive unless a hydrogen donor is present. This function may be served by dopa (or other catechols), which binds to the mettyrosinase to generate the reduced form by a two-electron transfer process. Thus, the presence of dopa in a steady-state concentration is a necessary requirement to sustain a constant monohenolase activity of the enzyme. Indeed, after the induction period is over, the rate of oxidation of tyrosine becomes identical to that of dopa, consistent with the fact that dopa is simultaneously a product of tyrosine oxidation, a substrate, and an activator of the enzyme.

Using a purified tyrosinase preparation from a human melanoma, Wood and Schallreuter (1991) found that the rate of enzymic oxidation of tyrosine was enhanced 40-fold when the incubation mixture was saturated with superoxide anion (at concentrations higher than 5×10^{-3} *M*). In control experiments they found that in the absence of tyrosinase no reaction took place. Accordingly, an alternative mechanism to that described above was proposed, in which superoxide can act both as substrate and as one-electron donor for the rapid formation of the tyrosinase–oxy- complex, which is the rate-limiting step when oxygen is used (Wood and Schallreuter, 1991). However, considering the high local concentrations of superoxide required to initiate the reaction, it is not clear whether the proposed mechanism may take place *in vivo*.

V. INHIBITION

It can confidently be said that few other enzymes have been studied so extensively with regard to inhibition as tyrosinase. This interest has by far exceeded enzymological aspects per se, increasingly steadily in response to motivations of great biological, medical, cosmetic, and practical concern. One such incentive is provided by the exigency of the food industry to prevent the enzymic browning affecting several kinds of fruits and vegetables subject to prolonged storage. However, the major stimulus has derived from the broad spectrum of alterations that affect the human pigmentary system, ranging from unaesthetic cutaneous brown spots to the extensive hyperpigmentation associated with certain physiological or pathological conditions like chloasma or melasma and cutaneous melanoma. The search for inhibitors of tyrosinase has been very active, extending the original small set of noncompetitive inhibitors to a broad range of phenolic compounds, sulfur-containing compounds, metal chelating agents, reductants, carboxylic acids, and so on. In more recent years, the mode of action of these substances has in some cases been reconsidered, and a number of compounds initially regarded as inhibitors have in fact been shown to interfere with melanin formation *in vitro* and *in vivo* by mechanisms other than enzyme inhibition. A typical example is provided by sulfhydryl compounds, such as cysteine and glutathione (see Chapter 7 and 8), which act by scavenging the dopaquinone formed by enzymatic oxidation of tyrosine.

In the following survey we will focus on the most representative examples of tyrosinase inhibitors, which have been divided into two broad groups based on their structural features and/or supposed mode of action. It is clear, however, that such a division is largely arbitrary and responds mainly to criteria of convenience.

A. Copper Chelating Agents

Within this broad category fall a variety of tyrosinase inhibitors displaying as the common structural unit two spacially close nucleophilic groups capable in principle of binding the copper ions at the active site of the enzyme (Fig. 3.14). Examples are mimosine, a naturally occurring pyridone analog of dopa (Hashiguchi and Takahashi, 1977) and the structurally related kojic acid, an antibiotic produced by a *Streptomyces* strain and recently shown to exhibit depigmenting activity *in vivo* (Mishima *et al.,* 1988). Tropolone is one of the most potent inhibitors of tyrosinase ($IC_{50} = 4 \times 10^{-7}$ *M*) and apparently acts both by chelating copper at the active site and by occupying the substrate binding portion of the enzyme (Kahn and Andrawis, 1985). Benzhydroxamic acid, another effective inhibitor, is competitive for catechol substrates and noncompetitive for oxygen (Rich *et al.,* 1978).

Reduced human thioredoxin and 2,3-dithiopropanol have been reported to act as allosteric inhibitors of tyrosinase yielding bis-cysteinate complexes with one of the copper atoms in the active site. Bis-cysteinate tyrosinase activity is down-regulated to 30% of native enzyme activity in the L-dopa assay (Wood and Schallreuter, 1988). A possible mechanism for the observed inhibition would involve the reduction of a specific disulfide bridge at the active site of tyrosinase to give a dithiol involved in the bis-cysteinate reaction (Fig. 3.15) (Wood and Schallreuter, 1991).

Phenylthiourea (PTU) and diethyldithiocarbamate (DDC), which contain a thioamide moiety, have recently been shown to inhibit mouse melanoma tyrosinase in a noncompetitive and a competitive fashion, respectively, sug-

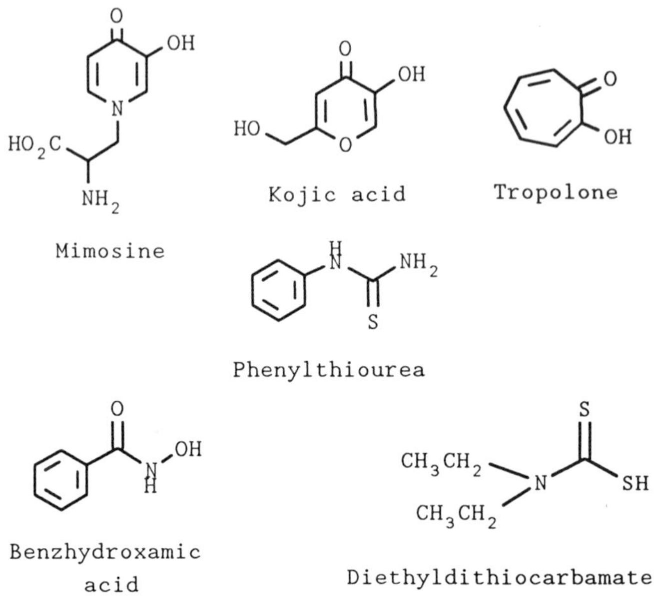

FIG 3.14 Structure of representative copper-chelating inhibitors of tyrosinase.

gesting that they act by different mechanisms in spite of chemical analogy (Laskin and Piccinini, 1986). It is perhaps not superfluous to stress at this point that the mechanism of the inhibitory action of all these compounds has not been completely elucidated, and their classification as copper binding agents is dictated more by consideration of their chemical properties than by direct experimental evidence.

B. Carboxylic Acids

A comprehensive systematic account of early work on the inhibitory effects of carboxylic acids on tyrosinase activity was given by Krüger (1955). A most outstanding feature emerging from those studies was the marked dependence of the inhibitory effects on the pH of the medium, with the inhibition increasing as the pH decreases. This was not due to a change in the dissociation of the inhibitor, since for benzoic acid there was a decrease in the inhibition over a pH range where the acid is fully dissociated, but most likely to a change in the dissociation of a critical protonated group of the enzyme. Moreover, since the midpoint of the range of pH effects is close to the pK_a of the histidine

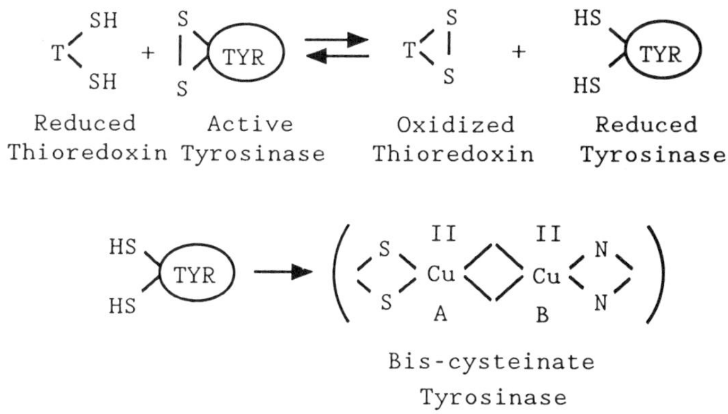

FIG 3.15 Proposed mechanism for the formation of an intrinsic bis-cysteinate complex with human tyrosinase by reaction of reduced thioredoxin with a specific disulfide bridge at the surface of the enzyme (Wood and Schallreuter, 1991).

imidazole groups in proteins, it was concluded that inhibition by carboxylic acids results from an acid–base interaction with a histidine ligand of the active-site copper.

More recent studies on inhibition of *Neurospora* and mushroom enzymes by benzoic acid point to a competitive inhibition with respect to dopa or catechol as the substrate, but a noncompetitive or mixed-type inhibition with respect to oxygen (Duckworth and Coleman, 1970; Gutteridge and Robb, 1975). Chemical and spectroscopic evidence has also been presented for the binding of a broad range of carboxylate competitive inhibitors to the various forms of tyrosinase (Wilcox *et al.,* 1985). These inhibitors have been divided into two groups: (1) poor inhibitors, which show an equilibrium constant for binding to the enzyme similar to that for binding to aqueous Cu(II) complexes, and (2) good inhibitors, which are substrate analogs in that the carboxylate is conjugated into an aromatic ring, producing a planar structure, and which bind with an equilibrium constant higher by an order of magnitude relative to aqueous copper.

A number of straight-chain dicarboxylic acids from C_8 to C_{13}, arising from the metabolism of lipids, have also been reported to inhibit tyrosinase to a varying extent, acting as competitive inhibitors of the enzyme (Nazzaro-Porro and Passi, 1978; Nazzaro-Porro *et al.,* 1979). Kinetic studies showed that enzyme inhibition is strictly dependent on the presence of two free carboxyl groups and on the chain length but is not affected by the presence in the chain of an unsaturated bond. These results furnished the experimental background for the successful exploitation of azelaic acid in the therapy of pathological pigmentary disorders like chloasma and lentigo maligna (Nazzaro-Porro *et al.*, 1980). As a result of more recent work (Breathnach, 1989; Passi *et al.*, 1989), it has become clear that the biological activity of azelaic acid does not depend exclusively on its inhibitory effect on tyrosinase. This molecule acts against basic metabolic pathways in the cell by inhibiting respiratory chain enzymes and DNA synthesis and may therefore have a potential as a general antitumoral agent.

VI. ENZYME INACTIVATION

A most distinctive feature of tyrosinase, that has been recognized almost concomitantly with the discovery of the enzyme, is the tendency to undergo rapid inactivation during the catalytic activity, particularly with catechol substrates. This is an early example of what is commonly referred to as suicide inactivation, but which was originally termed reaction inactivation (Robb, 1984). During the aerobic oxidation of catechols, the rate of oxygen consumption falls off rapidly from the moment the reaction is initiated, and the oxidation stops before it goes to completion unless relatively large amounts of the enzyme are used (Nelson and Dawson, 1944). This effect was initially attributed to the destructive action of products formed during the oxidation of catechols, in particular hydrogen peroxide and *o*-benzoquinones. However, both species were soon ruled out as the cause of tyrosinase inactivation, and the problem remained dormant for some time.

The *o*-quinone hypothesis of inactivation was reconsidered years later by Ingraham (1959) and Brooks and Dawson (1966), who suggested that the process results from an active site-directed attack of a quinone product of tyrosinase to a nucleophilic group essential for the catalytic activity. This view was supported by an apparent incorporation of 1-^{14}C-labeled phenol to yield a covalent radioactive addition product to the enzyme (Wood and Ingraham, 1965). The inactivation of mammalian tyrosinase by dopa was examined by Tomita *et al.* (1980), who concluded that neither oxidation products of dopa nor oxygen radicals are primarily responsible for the process.

Further studies by Dietler and Lerch (1982) indicated that a time-dependent irreversible loss of activity occurs during oxidation of labeled phenol with *Neurospora* tyrosinase but, in contrast with the previous work, there was only negligible incorporation of the substance into the enzyme. Amino acid analysis of the completely inactivated enzyme showed that one out of nine histidine residues, that is, His-306, was lost. It was also demonstrated that 1 g-atom of copper per mole of enzyme was lost after complete inactivation. From these data, it was concluded that the catechol substrate is not the species actually responsible for inactivation and that modification of His-306 is the result of an attack of a highly reactive species originating from molecular oxygen, most likely the hydroxyl radical.

VII. ASSAY

A. Hydroxylase Activity

Selective measurement of the hydroxylase activity of tyrosinase is complicated by two factors: First, a hydrogen donor, often an *o*-diphenol, is sometimes required to initiate hydroxylase activity; and second, the product of the hydroxylation reaction, an *o*-diphenol, is itself a substrate of the enzyme for oxidase activity, implying that as the hydroxylation reaction progresses the oxidase activity, which is one order of magnitude higher, tends to make a significant contribution to the overall enzyme activity. These and other problems in assaying tyrosinase activity have been examined in some detail by Jara *et al.* (1988a).

Basically, three types of techniques are currently used for tyrosinase assay. The first is radiometric and involves two different conceptual approaches: (1) the determination of radioactive melanin formed by oxidation of [U-^{14}C]tyrosine (Chen and Chavin, 1965) and (2) the detection of tritiated water formed by displacement of tritium on hydroxylation of [L-3,5-$^{3}H_2$]tyrosine (Fig. 3.16) (Pomerantz, 1966).

The second technique is spectrophotometric and is based on the detection of dopachrome formation at 475 nm by oxidation of tyrosine (Mason, 1948). Though widely used, this method suffers from some drawbacks, which are mainly related to the detection of a product which is formed nonenzymatically distal to the tyrosine–tyrosinase reaction. This implies that any factor preventing the oxidative cyclization of dopaquinone severely interferes with the assay. Another disadvantage is that the spectrophotometric detection of dopachrome at 475 nm is not very sensitive, owing to the low molar extinction coefficient of the aminochrome (3600 $M^{-1}cm^{-1}$).

A variant of the spectrophotometric assay, adopted especially by commercial houses, takes advantage of the increase in absorbance at 280 nm during the oxidation of tyrosine as an indicator of hydroxylase activity (Duckworth

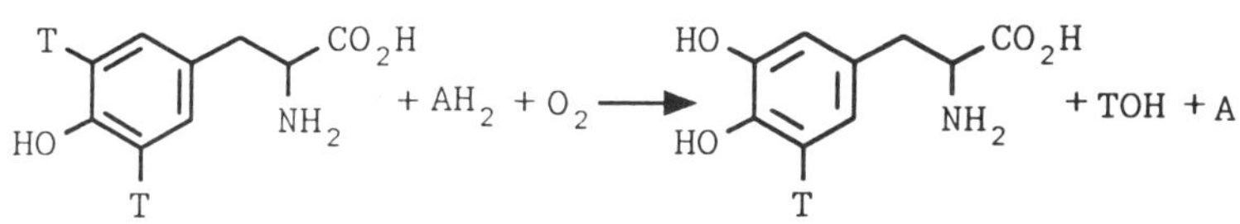

FIG 3.16 Pomerantz's assay of tyrosinase activity using tritiated tyrosine.

and Coleman, 1970), but its validity has been questioned (Robb, 1984). The third technique is based on the monitoring of oxygen consumption, for example, polarographically with Clark-type electrodes, and is often used in conjunction with other methods (Kean, 1964; Palumbo *et al.*, 1990b).

B. Oxidase Activity

Monitoring of oxidase activity of tyrosinase is relatively simpler and can be performed with all three techniques used for determination of hydroxylase activity. Dopa is often used as the substrate of choice, and its oxidation is monitored either spectrophotometrically at 475 nm with the dopachrome assay or radiometrically, measuring the tritium liberated from [6-^{3}H]dopa (Pomerantz, 1976). An electrometric method for determination of tyrosinase activity has also been reported, involving the quantitative measurement of the protons released in the enzymic conversion of dopa to dopachrome (Solano-Munoz *et al.*, 1985). Other methods of monitoring dopa oxidation are based on the trapping of the enzymically generated dopaquinone by suitable agents and quantitative analysis of the stable products thus formed. Among these are the spectrophotometric detection of the pink product formed by reaction with 3-methyl-2-benzothiazolinone hydrazone (Mazzocco and Pifferi, 1976) and the HPLC analysis of the cysdopas which are formed from L-dopa in presence of excess cysteine (Jerbil *et al.*, 1983).

CHAPTER 4

NATURAL AND SYNTHETIC MELANINS

I. OCCURRENCE AND DISTRIBUTION

A. EUMELANINS

As mentioned earlier, eumelanins are of widespread occurrence in mammalian tissues, such as skin, hair, eye, inner ear, and mucous membranes. In addition, they form the pigment, and in some cases, the major component of melanomas frequently encountered in mammals as well as in other animals. Among vertebrates, eumelanins are also commonly found in avians and occasionally in reptiles, amphibians, and fishes (Fox and Vevers, 1960; Needham, 1974; Fox, 1976).

Little is known about the distribution of eumelanins in invertebrates, where, save the ink of cephalopods, much of the older work does little more than record the existence of dark granular pigments in certain organisms. These include the jellyfish *Pelagia noctiluca* (Fox and Millot, 1954), certain coelenterates, such as the dark variant of the plumose anemone *Metridium senile* (Fox and Pantin, 1944), the amoebocytes of the sea urchin *Diadema antillarum* (Jacobson and Millot, 1953), the brittlestar *Ophiocomina nigra* (Fontaine, 1962), various species of worms (Kennedy, 1969), the black slug *Arion ater,* the snail *Limnaca* sp. and other gasteropods (Goodwin, 1972), and the tadpole larvae of ascidians (Whittaker, 1973). However, none of these presumptive melanin pigments have been isolated and analyzed to prove that they are authentic melanins.

A few additional remarks are in order about the dark pigmentation which is often associated with the stiffening and tanning of most insect cuticles (reviews by Kayser, 1985; Sugumaran, 1987; Peter, 1989). During this process, structural proteins are covalently modified and cross-linked by *o*-quinone and quinone methides arising by enzymatic oxidation of *N*-acetyldopamines (4.**1**). Another principal sclerotization agent is *N*-β-alanyldopamine (4.**2**), which contains a free amino group. The exact mechanism of polymerization of these

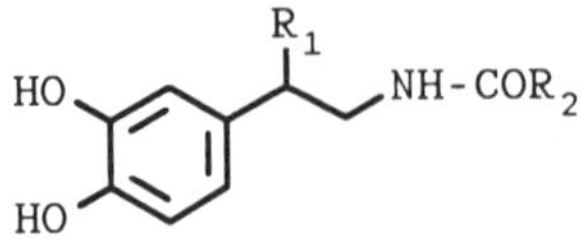

4.1

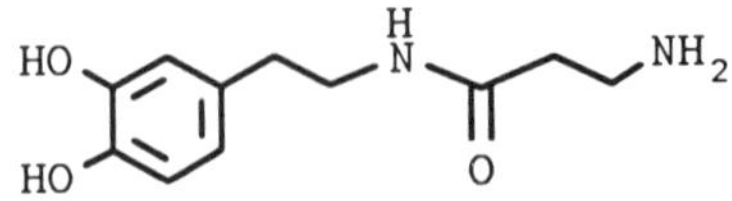

4.2

4.1a : R_1=H, R_2=CH_3
4.1b : R_1=OH, R_2=CH_3

substrates is still in question, but it seems clear that the associated pigments are chemically distinct from vertebrate eumelanins, though they also arise biogenetically from tyrosine. In the tobacco hornworm (*Manduca sexta*) larva, the dark cuticular pigment is formed primarily from dopamine and therefore also requires the enzyme dopadecarboxylase for its formation (Riddiford and Hiruma, 1988). Moreover, it appears that the nonproteinogenic amino acid β-alanine plays an important role in the acylation of dopamine and, hence, in the regulation of the balance between melanization and sclerotization, which otherwise remain two distinct processes in both chemical and functional aspects.

A curious sideline story of melanogenesis in invertebrates followed the presumptive identification by Mazza and Stolfi (1931) of hallachrome, the red pigment of the sea worm *Halla parthenopea,* with dopachrome, which is one of the first intermediates in the tyrosinase-catalyzed conversion of tyrosine to melanin. Mazza and Stolfi's conclusion was later substantiated by other workers, including notably Evans and Raper (1937), who reported that reduction of each of these pigments with palladium and hydrogen led to the formation of leukodopachrome. Thus, for many years the name hallachrome was used in place of dopachrome and vice versa, until evidence began to appear that the two pigments were in fact not identical (Mason, 1959). However, it was only in the 1970s that Prota and his co-workers succeeded in isolating hallachrome in crystalline form. Surprisingly enough, they found that the pigment did not contain nitrogen at all, being a rare example of 1,2-anthraquinone with a substitution pattern as in 4.**3** (Prota *et al.,* 1972).

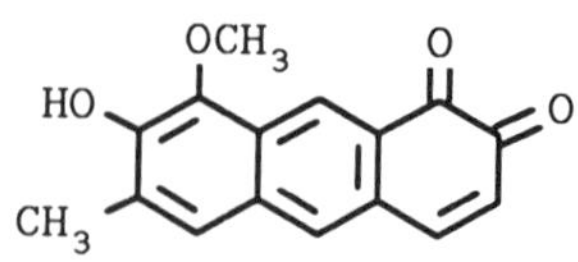

4.3

B. Pheomelanins

Though the literature is replete with papers dealing with yellow, tawny, or reddish-brown melanins in hair and feathers, only a few of these pigments have been isolated and shown to be genuine pheomelanins by chemical analysis. They are listed in Table 4.1.

In addition to these well documented examples, there is strong presumptive evidence for the occurrence of pheomelanins in the red or buff feathers of other breeds of domestic fowl, for example, the jungle fowl *G. gallus,* the Buff Orpington chicken, and the Bourbon red and Jersey Buff varieties of turkeys (Somes and Smyth, 1965), as well as in the hairs of guinea pigs, mice, horses, sheep, and llamas of different breeds and color genotypes (Ito *et al.*, 1984a; Sponenberg *et al.*, 1988a,b; Jimbow *et al.,* 1991).

Pheomelanin has also been detected in the skin of patients with dysplastic nevi (Yamada *et al.,* 1989) and, more recently, in normal epidermis of white Caucasians of skin types I, II, and III (Thody *et al.,* 1991). However, in view of certain pitfalls intrinsic in the microanalytical method used (see Section III), a word of caution is in order about the actual presence of pheomelanins in epidermal tissues. From the standpoint of comparative biochemistry, it is noteworthy that, unlike eumelanins, which are very widely distributed in the animal kingdom, pheomelanins have so far been found only in mammals and birds. Rust-colored instead of the usual black or brown melanosomes have

TABLE 4.1
SOME TYPES OF HAIR AND FEATHERS CONTAINING PHEOMELANINS

Source	Identification criteria[a]	Reference[b]
Mammalian hair:		
Human red hair	DS	a, b
Red haired Rabbit	DS	a
Goat (Sicilian sp.)	DS	a
Goat (Pakistan sp.)	DS	a
Red-haired cattle	EA	c
Avian feathers:		
New Hampshire chicken	EA, DS	a, d, e
Rhode Island chicken	DS	a
Pheasants (Phasianus colchicus)	DS	a
Partridges (Alectoris rufa)	DS	a

[a]DS, degradative studies; EA, elemental analysis.
[b](a) Fattorusso *et al.,* 1970b; (b) Chedekel *et al.,* 1978; (c) Dryja *et al.,* 1979; (d) Minale *et al.,* 1967; (e) Prota and Nicolaus, 1967b.

been occasionally observed in the melanophores of some species of leaf frogs *(Phyllomedusinae),* but on analysis these were found to be due to pteroidin, a red-colored pteridine dimer (Misuraca *et al.,* 1977).

A more relevant example of pheomelanic pigment in lower organisms is adenochrome, which confers the characteristic purple color to the branchial hearth of certain cephalopods, for example, *Octopus vulgaris* (Prota, 1980b). Known since 1906 for its ability to chelate Fe(III), the pigment was eventually isolated in a crude form by Fox and Uppdegraff (1943), who suggested a tentative formula $C_{41}H_{72}N_{12}O_{25}S_2$. In view of a possible relation with vertebrate pheomelanins, the pigment was later reexamined at Naples (Prota *et al.,* 1977a). Using various analytical techniques, including chemical degradation, it was found that adenochrome is in fact a complex mixture of unusual peptides derived from the amino acid glycine and three novel amino acids, adenochromine A, B, and C, which account for the chelate formation with Fe(III) (Fig. 4.1).

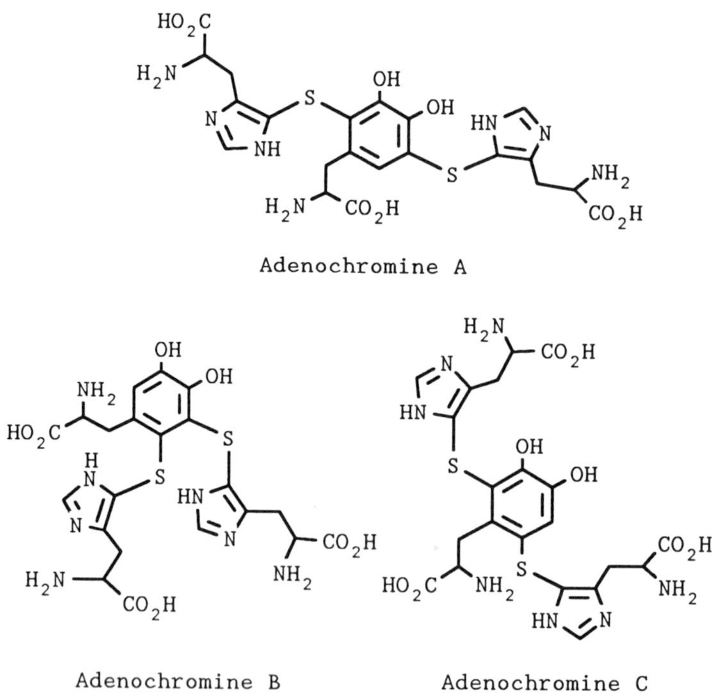

FIG 4.1 Structure of the adenochromines, the iron binding constituents of adenochrome.

There are grounds to believe that adenochromines arise biogenetically by addition of a new thiolhistidine amino acid to dopaquinone, formed by tyrosinase oxidation of tyrosine (Prota, 1980b). The analogy of such a pathway with pheomelanin formation (Chapter 7) is remarkable, especially when compared with the early stages leading to the formation of the intermediary cysdopas. A major difference between the addition of the thiolhistidine and the addition of cysteine to dopaquinone is that in the former case the reaction takes place to a significant extent (about 25%) at C-6, that is by 1,4-addition; this accounts for the presence of substantial amounts of the 5,6-disubstituted isomer among the adenochromines.

As a concluding remark, it is interesting that among invertebrates, cephalopods are unique in that they are capable of forming a typical eumelanin in the ink gland and a pheomelanin-related pigment, adenochrome, in the branchial heart.

C. Trichochromes

Once regarded as typical of bright red human hair, trichochromes have been found in the hair of several other mammals, including red-haired cattle (Misuraca *et al.*, 1974), the golden hamster, New Zealand red rabbit, and red mongrel dog (Flesch *et al.*, 1967; Flesch, 1968, 1970). They also occur in red and buff feathers of domestic fowl (Smyth *et al.*, 1951), especially the New Hampshire and Rhode Island chickens, which provide the most convenient source for their extraction and purification (Prota and Nicolaus, 1967b; Agrup *et al.*, 1978a). As in the case of pheomelanins, there are several types of red hair which do not contain trichochromes, for example, Irish setter dog, Ayrshire calf, or all-red guinea pig (Flesch *et al.*, 1967). Indeed, it seems that the distribution of trichochromes in hair and feathers parallels that of pheomelanins, consistent with the close chemical and biogenetic relationships between the two classes of pigments.

Whether trichochromes are present in skin and in extracutaneous tissues is still uncertain. It is noteworthy, however, that they have been found in the urine of some patients with melanoma metastasis (Prota *et al.*, 1976a; Agrup *et al.*, 1978b). The highest values of trichochromuria have been found in patients with metastasizing melanoma and diffuse melanosis showing highly increased excretion of cysdopas. In any case, no relationship was found between the pigmentation type of the melanoma patient and trichochrome excretion. Light- and electron-microscopic observations indicate that trichochromuria in melanosis is connected with the previously described electron-dense granules in lysosomes of dermal histiocytes (Rorsman *et al.*, 1986). However, it is not clear if trichochromes are produced in melanocytes or if they appear as the result of metabolic events in other cells, like keratinocytes or macrophages, which may incorporate the precursors of trichochromes, that is, cysdopas.

D. Nonsulfur-Containing Pheomelanins

Both in man and in some animals, yellow or reddish pheomelanic pigments exist which are different from those derived from cysteine and dopa. Paradoxically, they occur in the reddest species. Fattorusso *et al.* (1970b), for example, found that the alkali-soluble pigment from red hairs of the Argentinean goat gives on oxidation pyrrole-2,3,5-tricarboxylic acid (PTCA), which is a typical degradation product of eumelanins (see Chapter 5). Similar pigments were reported to occur in the hairs of the Addis Abeba sheep and wild boar.

On several occasions, the author also extracted alkali-soluble pigments from reddish hair, which on analysis did not contain much sulfur, nor did they give degradation products typical of pheomelanins. At present, little is definitely known about the chemistry of these nonsulfur-containing pheomelanic pigments except that they behave like bleached eumelanins (G. Prota, unpublished). Studies are underway to verify the possibility that these "pheomelanin-looking" pigments are structural variants of eumelanins, arising from partial peroxidative cleavage of 5,6-dihydroxyindole units. Such a view is not unlikely, considering the high susceptibility of the eumelanin polymer to hydrogen peroxide which is involved in the later stages of melanogenesis (see Chapter 5). Indeed, *in vivo* modulation of eumelanin color by hydrogen peroxide would provide an alternate mechanism to understand the polychromy of epidermal melanin pigmentation, which is otherwise difficult to explain on the basis of only two basic types of pigments, that is, eumelanins and the sulfur-containing pheomelanins.

In light of the preceding discussion, it is clear that the term *pheomelanin* is liable to become misleading in the absence of a descriptive definition. Therefore, to avoid confusion, the author's suggestion is to retain the term pheomelanin for the cysteine-dopa-derived pigments and use the new term oxymelanin to describe those alkali-soluble red or yellow pigments which are devoid of sulfur. Oxymelanin is a somewhat loose term, but so are pheomelanin and eumelanin. The hope is that in the long run all these terms will be replaced by a more rational terminology.

E. Other Melanins

Certain melanins found in epidermal tissues have physical and chemical properties which do not classify them clearly as part of any of the above-mentioned groups. A typical example was described in my collaborative study with the Lund group, aimed at characterizing the pigments from human melanomas (Prota *et al.*, 1976b). One of the pigments isolated from a patient with auburn hair resembled eumelanins with respect to color and insolubility in alkali, but on analysis it showed a sulfur content of 5.8%, which could not be explained as deriving from protein. A subsequent study of some other melanomas with the same methods led to the isolation of two more examples of such pigments (Rorsman *et al.*, 1979). In any case, the small amounts of material isolated precluded further structural investigation of the pigment.

The problem was then tackled by model experiments, assuming that the sulfur-containing "eumelanins" found in melanomas could arise by a co-polymerization process involving both eumelanin and pheomelanin precursors. In a relevant study (Ito *et al.*, 1980), it was found that tyrosinase-catalyzed oxidation of dopa in the presence of varying amounts of 5-cysdopa gave pigments whose physical and analytical properties stand in between those of synthetic dopamelanin and cysdopa pheomelanin. Similar types of pigments were later reported to occur in mammalian hair and eyes of different colors (Novellino *et al.*, 1981; Sealy *et al.*, 1982a; Ito *et al.*, 1984a). However, whether they are heteropolymers or mixtures of the two types of polymers is not yet clear, although the former view is widely accepted.

II. EXTRACTION AND ISOLATION

Selection of the isolation procedure depends on the type of melanin and the biological source of material. In the case of eumelanins, a procedure widely used involves homogenization of the tissue in a blender followed by extensive hydrolytic treatment with mineral acids, such as concentrated HCl for several days or boiling 6 *M* HCl to solubilize protein and other extraneous materials (Nicolaus, 1968, 1978). The extent of structural modification that results from these rather harsh conditions should be taken into consideration when such samples are used for further studies. As suggested by Blois *et al.* (1964), eumelanins structurally modified by treatment with acid should be referred to as B-type, to distinguish these materials from the native pigments (A-type).

In some favorable cases, for example, sepiomelanin and eye melanin, these can be isolated with minimal alteration by mechanical separation of the pigment granules, followed by a short treatment with 0.5 *M* HCl at room temperature and extensive sonication in deionized water (Benathan and Wyler, 1980). Likewise, melanosomes from melanoma tissues have been purified by sucrose density gradient ultracentrifugation after digestion with detergents or proteolytic enzymes, for example, pronase and papain, at neutral pH (Prota *et al.,* 1976b; Jimbow *et al.,* 1982). Whether these mild treatments efficiently remove proteins and other extraneous materials incorporated or bound to the pigment granules is not clear, but this is undoubtedly a minor drawback with respect to the alternative of having a structurally modified product as in the drastic acid procedure.

The extractability of pheomelanins, as well as oxymelanins, depends on the compactness of the horny structures. During extraction with dilute alkali, the loose chicken feathers yield relatively large amounts of pigment within a short time, and in the process they lose their red color (Prota and Nicolaus, 1967b). Human and mammalian hair requires a long period of extraction, during which part of the pigment may decompose within the hair shaft. This problem can be overcome by using a reducing agent (e.g., Na_2S, thioglycolic acid, dithiothreitol) to break the disulfide bonds of keratin, thus favoring the extraction of the pigment with dilute alkali at ambient temperature (Fattorusso *et al.,* 1970a; Wolfram *et al.,* 1970). This procedure, however, also results in the solubilization of relatively large quantities of proteins which are difficult to remove from the extracted pigment. Another procedure developed more recently involves enzymatic digestion of hair keratin (Goetz *et al.,* 1981). In any case, chromatographic separation of pheomelanins is difficult, and a new method of purification has been proposed based on a combination of selective dialysis and isoelectric focusing on a granulated bed (Deibel and Chedekel, 1982).

Extraction and isolation of trichochromes from avian feathers is usually carried out according to the original procedure developed by Prota and Nicolaus (1967b). This involves acidification of the alkaline extract to pH 1 to remove the bulk of the acid-insoluble pheomelanins, followed by chromatography of the supernatant on a strong cation exchange resin, for example, Dowex 50W-X2, H^+ form. Under these conditions, the pigments are strongly adsorbed by the resin; after washing with 0.1 *M* HCl and water they are eluted with dilute alkali and then separated into individual components by chromatography on Sephadex G-25, using phosphate buffer at pH 11 as the eluent.

A variant of this procedure is based upon the property of trichochromes to undergo decarboxylation when heated in acid solution. Thus, extraction of hair by boiling with 0.1 *M* HCl yields directly the decarboxylated pigments which can be purified readily by chromatography and can be identified by their characteristic pH-dependent spectra (Prota, 1972). However, if cysdopas are present in the pigmented tissue, hot acid extraction might result in artificial formation of trichochromes E and F. Alkaline extraction of cysdopa-containing hair or feathers might also give some trichochrome C as an artifact. However, oxygenation of the alkaline extract before the purification steps proved to be useful for excluding artificial formation of trichochromes (Agrup *et al.,* 1978a).

III. DETECTION AND IDENTIFICATION OF MELANINS

A histochemical method widely used for studies of skin pigmentation is based upon the property of eumelanin granules to reduce ammoniacal silver nitrate to metallic silver (Masson, 1914). Eumelanin granules also form complexes with Fe(II) which can be detected with potassium ferricyanide (Lillie, 1955). Another staining technique takes advantage of the affinity of certain basic dyes, for example, methylene or toluene blue, for eumelanins (Lillie, 1957). No staining method is available for detection of pheomelanins in tissues. Pigment color and the ultrastructure of melanosomes are commonly used to differentiate eumelanin from pheomelanin. The presence of the latter is supported if the granules are soluble in dilute alkali. These criteria, however, may be misleading, for they do not allow one to differentiate authentic pheomelanins from oxymelanins.

Sarna and Swartz (1979) have delineated a series of tests whereby eumelanins can be detected and identified in tissues and body fluids via an EPR assay of their free radical properties. The method has proved useful in identifying, for example, traces of eumelanin in amelanotic melanomas (Sarna and Swartz, 1979) and in showing that the pigment isolated from human blue and brown eyes is predominantly of eumelanic type (Menon *et al.,* 1987). Of particular interest also is the possibility of using EPR spectroscopy for distinguishing eumelanins from pheomelanins (see Section VI,B).

Other semiquantitative and quantitative methods are based on the optical and spectrofluorimetric properties of melanins. These methods, however, require solubilization of the samples, which in the case of eumelanins is critical. So far, two approaches have been exploited, one of which involves treatment of the pigment sample with alkaline hydrogen peroxide under appropriate conditions, for example, pH 9–10 (Wolfram *et al.,* 1970; Hall and Wolfram, 1975). The dark brown soluble pigment thus obtained, commonly referred to as melanin free acid (MFA), proved to be useful for quantitation of eumelanins in biological samples (Crippa *et al.,* 1989).

The other solubilization procedure is based on treatment of the pigment sample with a 0.1 *M* solution of Soluene 100 (dimethyl-*n*-dodecyl-*n*-undecylammonium hydroxide) in toluene at 37°C (Oikawa and Nakayasu, 1975). Crude and wet melanin preparations, as well as pellets of cultured pigmented melanoma cells, became solubilized within 1 hr, whereas vacuum-dried melanins purified by treatment with boiling 6 *M* HCl are solubilized only after several hours. The mechanism of the soluene solubilization is unknown, and a structural modification of the pigment due probably to autooxidation cannot be excluded. This would be supported by the observation that neutralization of the melanin solution in Soluene with acetic acid did not precipitate the pigment (L. J. Wolfram, unpublished).

A microanalytical method has been proposed which purportedly permits direct analysis of melanins in biological samples without the difficult task of the isolation of the pigment from the tissues (Ito and Fujita, 1985). The rationale of the method, outlined in Fig. 4.2, follows from the demonstration that permanganate oxidation of eumelanin yields PTCA as the relevant degradation product, whereas hydroiodic acid hydrolysis of pheomelanin gives mainly aminohydroxyphenylalanine (AHP) along with other phenolic amino acids (see Chapters 5 and 7).

HPLC analysis of PTCA and AHP may therefore provide an indirect means for a rapid and most sensitive determination of the eumelanin and pheomelanin content in tissues. Typically, permanganate oxidation of sepiomelanin gives PTCA in about 2% yield, whereas cysdopa pheomelanin gives ca. 20% yield of APH. Based on these reference values, the tissue contents of eumelanin and pheomelanin may be roughly estimated by multiplying those of PTCA and AHP by factors of 50 and 5, respectively. The method is simple to perform and highly sensitive, and it has been extensively applied for qualitative and quantitative analysis of melanins in mammalian tissues (Jimbow *et al.*, 1991).

However, the results are not unambiguous because PTCA may also arise from degradation of melanin, for example, as DHICA present in pigmented tissues as well as in melanoma cells (Hatta *et al.*, 1988). Moreover, when both

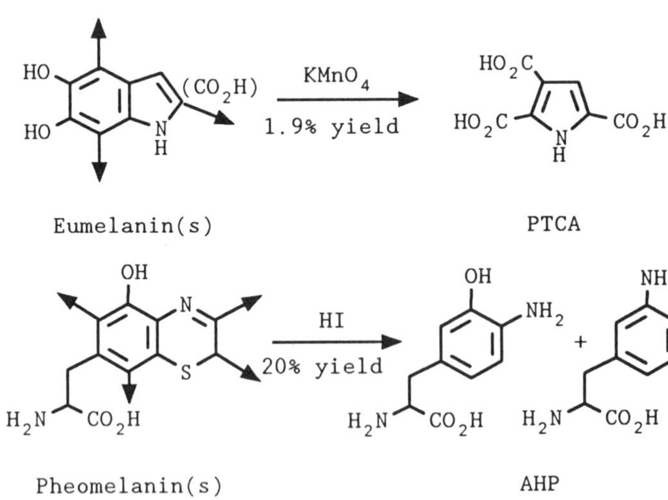

FIG 4.2 Formation and yields of pyrroletricarboxylic acid (PTCA) and aminohydroxyphenylalanines (AHP) by degradation of eumelanins and pheomelanins, respectively. Arrows indicate all possible positions of attachment of the units in the pigment polymer.

PTCA and AHP are obtained, it is difficult to say whether they arise from degradation of a mixed-type melanin or of a mixture of eumelanin and pheomelanins. In view of these limitations, the correlation between thc PTCA/AHP ratio and the type of melanogenesis should be supported by other analytical methods. It may also be necessary to find an indicator other than PTCA for eumelanin analysis, inasmuch as the yield of this degradation product is exceedingly low with respect to that of AHP.

IV. SYNTHETIC MELANINS

A laboratory procedure which has been widely used to obtain eumelaninlike pigments involves oxidation of L-tyrosine or L-dopa at neutral pH in the presence of tyrosinase, usually the commercially available mushroom enzyme. The pigment thus obtained is precipitated by lowering the pH of the solution, collected by centrifugation, and eventually carefully washed to remove the enzyme and side-reaction products. The exact conditions under which the oxidation is performed are extremely important, and variations, even minor ones, can have a marked effect on the reaction sequence and hence on the structure and properties of the final product. The most critical factors which so far have been overlooked are the reaction time and the mode of oxygenation, that is, pure oxygen or air (see Chapter 5).

Variations in the amount and purity of the enzyme used are also bound to affect the results, and it is desirable for the specific activity of each preparation to be estimated before it is used. This particular difficulty is avoided when the pigment is obtained by autooxidation of dopa at a pH above neutrality, for example, 8 (Sealy *et al.*, 1980; Crippa *et al.*, 1989). Usually, a catalytic amount of copper ions is added to the solution to accelerate pigment formation. The convenience of this synthetic procedure is obvious, but there is evidence that the pigment thus obtained differs significantly from enzymically prepared dopamelanin (Benathan and Wyler, 1980; Ito, 1986; Chedekel *et al.*, 1986).

A considerable number of studies have been done on melanins obtained by autooxidation of catecholamines, especially dopamine (Swan, 1974). Inasmuch as natural eumelanins arise biogenetically from tyrosine via dopachrome, the use of these pigment preparations for studies on the structure and biosynthesis of eumelanins may be misleading and should be discouraged. Rather, they can provide useful information on the oxidation chemistry underlying the formation of neuromelanin and related metabolites in dopaminergic cells (Chapter 6). Other substrates which have been used to obtain eumelaninlike pigments include DHI and DHICA (Nicolaus, 1968). Various methods of synthesis have been used, including (1) autooxidation in the presence or in the absence of metal ions; (2) enzymic oxidation with tyrosinase or peroxidase; and (3) anodic oxidation in both acidic and neutral media (see Chapter 5).

As far as pheomelanin preparations are concerned, the original synthesis, developed by Prota and Nicolaus (1967b), involved tyrosinase-catalyzed oxidation of L-dopa (or tyrosine) in the presence of excess L-cysteine at pH 6.8, followed by chromatography of the acid-insoluble fraction on a Sephadex column. This procedure was subsequently improved by Fattorusso *et al.* (1969a), and more recently bettered by Ito (1989), who has evaluated the effects of changes in the dopa/cysteine ratio and other reaction conditions, including the mode of oxygenation and the amount of tyrosinase. Another procedure for obtaining more homogeneous material involves tyrosinase-catalyzed oxidation of intermediary cysteine-dopa adducts, particularly 5-cys-dopa, in the presence of a catalytic amount of dopa (Fattorusso *et al.*, 1969a; Ito *et al.*, 1980; Deibel and Chedekel, 1982).

V. PHYSICOCHEMICAL PROPERTIES

A. Hydration and Aggregation

All types of melanins, both natural and synthetic, contain enormous amount of bound water, which appears to be essential to preserving the structure of the pigment in a solvent-swollen state. Typically, 1 ml of native pigments granules contains no more than 10 mg of dried material (Sealy *et al.*, 1980); however, if resuspended in water or left in air, the dried material does not swell easily and eventually regains only part of the water originally bound to the native pigment granules. This is yet another unusual property of melanins which must be borne in mind when dealing with these pigments. In general, drying of melanins should be avoided because it affects the properties of the pigment polymers, which become much more aggregated and virtually inert in terms of physicochemical interactions (Crippa *et al.*, 1989).

X-ray analysis of wet and dry samples of eumelanins (Chio, 1977) suggests that the bulk of bound water that comes off easily by heating at temperatures below 100°C is roughly similar to liquid water, while the remainder, that is, the tightly bound water, that can be removed at 150°C or higher, has a structure similar to that of amorphous ice. At physiological pH melanins are not truly soluble and at best form colloidal systems which are negatively charged. In the case of eumelanins, the average radius of the suspended pigment particles is on the order of 10–20 Å, as determined by X-ray small-angle scattering measurements (Miyake *et al.*, 1986). The rate of particle sedimentation is generally accelerated by lowering the pH of the solution, which decreases solvation of the ionizable groups and formation of large agglomerates resulting from hydrophobic interactions of pigment subunits.

Aggregation of melanin also occurs in the presence of metal ions (Okazaki *et al.*, 1985) which can act as counterions to acidic residues. Static and dynamic light-scattering measurements (Huang *et al.*, 1990) carried out on synthetic dopamelanin indicate the existence of two regimes of aggregation kinetics; one between pH 3.4 and 7.0, where the aggregation is slow, and the other below pH 3.4, where the aggregation is very fast. Moreover, the aggregation of the melanin particles can be reversed by increasing the pH of the solution, which rapidly deaggregates the precipitate into smaller-sized clusters.

B. Light Absorption and Scattering

In surveying the optical properties of melanins it may be useful to distinguish those which depend on their physical state of insoluble, discrete particles from those other features which are related to their chemical constitution and reactivity. The broad variety of colors and hues that melanin pigments can produce *in vivo* is one familiar example of a phenomenon of physical origin.

Exact Mie calculations of the scattering and absorption cross-sections for individual pigment granules have shown that the colors of melanin dispersions are strongly dependent on both the pigment granule size and the Mott–Davis optical energy gap parameter E_o, which controls the dispersion of the optical constant k in an amorphous semiconductor (Kurtz *et al.*, 1987, 1989). This implies that the size of the melanin granules is at least as important in determining the color as the molecular structure of the pigment polymer. As a rule, the lighter colors are associated with smaller particle sizes and the darker colors with larger ones.

Besides the light-scattering properties, natural and synthetic melanins exhibit also intrinsic absorption throughout the ultraviolet and visible regions (Kollias *et al.*, 1991). Published spectra of eumelanins (Blois *et al.*, 1964; Das *et al.*, 1976) are characterized by a monotonic increase in the absorbance with decreasing wavelength, with a barely detectable shoulder between 290 and 320 nm. The absorption spectra of pheomelanins are similar, showing no definite bands in either the visible or the UV region. Based upon diffuse reflection spectra in the visible region, Kollias and Bager (1985) found that native melanin in human skin displays quite different absorption characteristics in the ranges 550–750 and 400–550 nm, with the strength of the absorption in one range being apparently independent of the other. In another study, the absorption coefficient (μ_a) of melanosomes in the skin was estimated to be on the order of 10^3 cm^{-1} at 400 nm, dropping to 10^2 cm^{-1} by 700 nm (Jacques and McAuliffe, 1991). This can only be an order-of-magnitude calculation because of the variation in size and composition of melanosomes.

The origin of the "chromophoric" features of melanins, especially eumelanins, has been the subject of a longstanding debate. According to Blois (1978), organic substances that are black in color can arise where the material has low-lying conduction bands or where charge–transfer interactions predominate, as in quinhydrones. A somewhat different view of the interaction of melanin with light has been offered by McGinness and Proctor (1973) in terms of the properties of an amorphous semiconductor. In short, melanin would be black because the absorbed light is not reirradiated but instead captured and converted to rotational and vibrational energy. This view, however, does not take into account the ability of melanin to act as an efficient light scatterer. Indeed, both the light optical density of eumelanins and the overall shape of the absorption spectra can be accounted for in terms of Rayleigh scattering (by the molecules) and Mie scattering (by the pigment granules), as suggested by Wolbarsht *et al.* (1981).

C. MELANINS AS AMORPHOUS SEMICONDUCTORS

The possibility that melanins could act as intrinsic semiconductors was first suggested by Longuet-Higgins (1960) on theoretical grounds, receiving additional support through the work of Pullman and Pullman (1961) on the electronic structure for an idealized ordered 5,6-indolequinone polymer of type I (4.**4**). This enabled the prediction of an exceptional electron-accepting ability arising from extension of the lowest empty band in the bonding energy region. Consequently, excitation of an electron from the highest filled to the lowest empty band should be very easy.

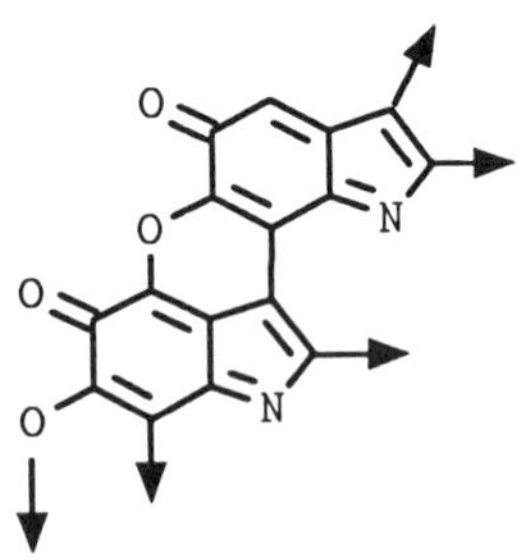

4.**4**

Pullman's model was later rejected by Blois and co-workers (1964) on the basis of EPR data, which seemed to rule out the presence of π-electron delocalized bands. Recently, however, Galvao and Caldas (1988, 1990a,b) have presented new modeling studies that apparently reconcile the EPR properties of eumelanins with the electronic structure of a semiconductor polymer of 5,6-indolequinone with structural defects of limited order acting as electron-capture centers. In keeping with these theoretical studies, some natural and synthetically prepared eumelanins were shown to act as amorphous semiconductors at biological electrical field strengths (McGinness *et al.,* 1974). Of a number of other bioorganic molecules tested, only cytochrome *c* behaved similarly, but at the high potential gradients required for inorganic thin films. Both threshold and pseudomemory switching were also observed in eumelanins (Filatovs *et al.,* 1976). The memory state was found to return slowly to the off state when the field is removed; however, Culp *et al.* (1975) were unable to restore the on state electrically. Data have been presented showing that eumelanins also exhibit a large linear term in the specific heat with a phase transition near 1.9K, which is of magnetic origin rather than due solely to the amorphous structure (Mizutani *et al.,* 1976). The observed effects occur in both natural and synthetic melanins, but they are most pronounced in the melanosomes isolated from a human melanoma (Mizutani *et al.,* 1976, Strzelecka, 1982a,b). These experimental observations are of interest in relation to a possible functional role of melanins in areas where energy transduction occurs, for example, the inner ear, retina, or leptomeninges. An attractive hypothesis, developed by McGinness *et al.* (1974), is that melanins may de-excite certain biological molecules by converting electronic energy into heat. Strictly related to this view is a study by Kono *et al.* (1979) showing that melanins and melanosomes are exceptionally "black" materials with respect to absorption and dispersion of ultrasound waves.

D. Metal Binding

Natural melanins, especially eumelanins, behave like weak cation exchange polymers with a capacity for binding substantial amounts of metal ions (Sarna *et al.,* 1981). EPR studies carried out mainly with bovine eye melanin and dopa melanin indicate that different binding mechanisms of ions are operative, depending upon the pH and the metal concentration (Froncisz *et al.,* 1980; Bogacz *et al.,* 1989). At neutral or slightly acidic pH, the binding involves mainly carboxylate groups available at the surface of the melanin polymer, while at pH above 7 chelation of the metal takes place at the catecholic moiety of the 5,6-dihydroxyindole units.

Several papers have appeared suggesting a correlation between the binding properties of melanins and the pathogenesis of chronic lesions in the skin and eye secondary to metal intoxications (Potts and Au, 1976). However, in the absence of a proper history of intake, the demonstration of a specific binding to melanin is difficult unless highly sophisticated techniques are used, such as neutron activation analysis or administration of radionuclides followed by whole-body autoradiography (Larsson and Tjälve, 1978). These have shown that a metal affinity *in vitro* does not necessarily imply that a binding will also takes place *in vivo*. Thus, metals such as barium, radium, and zinc are strongly bound to eumelanins *in vitro*, whereas after *in vivo* administration a strong labeling of pigmented structures is observed with ^{140}Ba, ^{226}Ra, and ^{201}Tl, but not with ^{63}Ni or ^{63}Zn. It seems therefore that the ability of cations to pass the melanocyte membrane is a most important controlling factor for melanin binding to occur *in vivo*. Using ^{201}Tl as a probe, Tjälve *et al.* (1982) have obtained evidence by whole-body autoradiography of pigmented mice that the metal has a biological fate similar to that of potassium and is taken up into the intracellular compartments probably using potassium transport mechanisms.

E. Drug Binding

In addition to metal ions, many drugs and xenobiotics are taken up and retained for a long period of time in melanin-containing tissues, such as, the eye, inner ear, and skin (Lindquist, 1986). Among these, chlorpromazine has been given much space in the ophthalmologic literature because of findings indicating that in certain instances the use of this drug is responsible for retinopathy. Potts (1962) found that the uveal tract of pigmented eye concentrated *N*-substituted phenothiazines as much as 50 times the mean distribution value. In more refined studies by whole-body autoradiography it was shown that labeled [^{35}S]chlorpromazine as well as [^{14}C]chloroquine were selectively concentrated in the pigmented layer of the eye, whereas the incorporation in the cornea and lens was negligible (Lindquist, 1973). Moreover, no incorporation of these drugs was found in whole-body autoradiographies of albino mice.

Similar selective uptakes in pigmented tissues (eye, skin, inner ear, melanomas) have been observed with chloroquine, haloperidol, cocaine, and a miscellanea of chemicals, including polycyclic aromatic hydrocarbons, aflatoxins, and the potent herbicide paraquat (Larsson and Tjälve, 1979; Lyden *et al.*, 1982; Larsson *et al.*, 1988). *In vitro* studies of the binding parameters by the Scatchard method have shown that the dominant interaction between positively charged drugs and the melanin polymer is ionic in nature and reversible. However, a nonelectrostatic interaction seems also involved in the melanin binding of chlorpromazine and chloroquine (Larsson and Tjälve, 1979; Stepien and Wilczok, 1982). This would be consistent with the marked electron density of these molecules, whereby they would act as electron donors with the melanin as the acceptor. More recently, conformational analysis and molecular graphics have been exploited to model a representative melanin structure to estimate the *in vitro* affinity of different compounds (Raghavan *et al.*, 1990). The lower binding energies were typically obtained for compounds with high melanin affinity such as methylene blue and chloroquine, whereas the higher binding energies were found for phenol and glucose, which do not bind to melanins.

A number of drugs with high melanin affinity have been tested as carriers of radioisotopes in the diagnosis and possibly in the treatment of malignant melanomas (Packer, 1984). In most cases, however, a high uptake of the radiolabeled material in normal pigmented tissues, especially in the eye uvea, and the associated side effects at the doses needed for targeted chemotherapy exclude the possibility of clinical application. This drawback is apparently overcome by using methylene blue, a relatively harmless melanin-binding drug, which has a strong therapeutic action on melanotic melanoma cells in mice when labeled with the short-range α-particle emitter ^{211}At (Link *et al.*, 1989; Link and Carpenter, 1990).

A related strategy for melanoma is based upon the boron neutron capture therapy (BNCT), which involves selective loading of tumor with an enriched ^{10}B compound and subsequent irradiation with low energy (0.025 eV) or thermal neutrons to yield stripped-down He and ^{7}Li nuclei (Barth *et al.*, 1990). The therapeutic potential of this approach was first recognized by Mishima (1973) using as a capture agent a [^{10}B]chlorpromazine derivative. Since then a variety of different ^{10}B compounds have been synthesized and evaluated for their *in vivo* ability to accumulate in melanoma tissues. Among these, *p*-boronophenylalanine (4.**5**) seems to be a most promising candidate for BNCT treatment of melanoma, as indicated by the results of a pilot clinical trial in Japan (Mishima *et al.*, 1989a,b).

Among melanoma seekers, 2-thiouracil (TU) (4.**6**) and some related thioureylenes, for example, methymazole (4.**7**), are attracting increasing interest because of their remarkable ability to be selectively incorporated into melanin-synthesizing cells. This was shown by Whittaker (1971), who incubated slices of chick-embryo retinal pigment epithelium in medium containing [^{14}C]TU and found that incorporation was dependent on functioning tyrosinase activ-

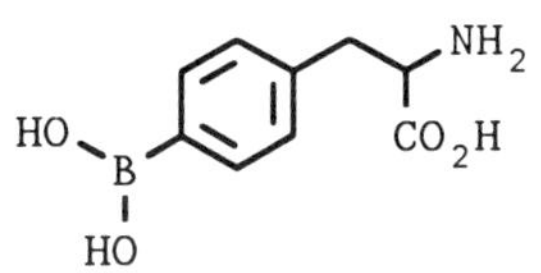

4.**5**

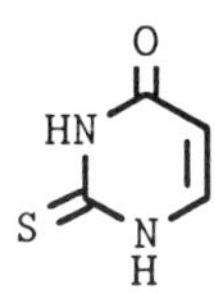

4.**6**

ity. Whole-body autoradiographic studies (Dencker *et al.*, 1979) of the distribution of [^{14}C]TU in pregnant pigmented mice showed that the radioactivity uptake in the fetal eyes, where the rate of melanin synthesis is abnormally high, was about 60 times greater than that in maternal eyes (Fig. 4.3). Remarkably high incorporation of TU was observed in melanotic Harding-Passey melanomas transplanted to mice, S91 murine melanomas, lung metastases of B16 murine melanomas, and human melanoma cells *in vitro,* pointing to the potential of TU as a selective marker for tumor growth (Yamada *et al.*, 1988).

The conclusion of all these studies was that TU serves as a false melanin precursor capable of covalently binding to a quinonoid intermediate of the melanogenetic pathway by a mechanism reminiscent of that leading to the sulfur-containing pheomelanins. Such a view has been recently substantiated by a chemical study (Palumbo *et al.*, 1990a) showing that, under the usual conditions of melanogenesis *in vitro,* TU combines covalently with dopaquinone to give a colorless product that has been isolated and identified as the 6-*S*-thiouracil-dopa adduct (4.**8**).

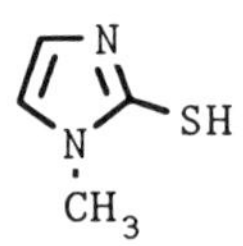

4.**7**

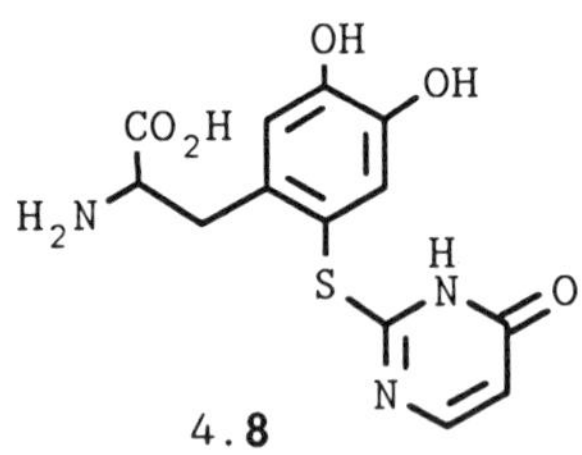

4.**8**

An extensive account of the current exploitation of TU and other thioureylene compounds as carriers for radioisotopes in the diagnosis and treatment of melanoma is given by Larsson (1991).

VI. SPECTROSCOPIC CHARACTERIZATION

A. X-Ray Diffraction

The first attempt to characterize melanins by X-ray analysis was made by Spiegel and Henny in 1939. They studied and compared the diffuse rings on film-recorded diffraction patterns of *Sepia* and human melanoma melanins on one hand and tyrosine and other synthetic melanins on the other. After this study no research of any significance appeared until the intensive studies by Thathachari (1976), which provided more insight into the spatial structure of the pigment polymer. In a large number of melanins studied, both natural and synthetic, he observed only a short-range order in the arrangement of the

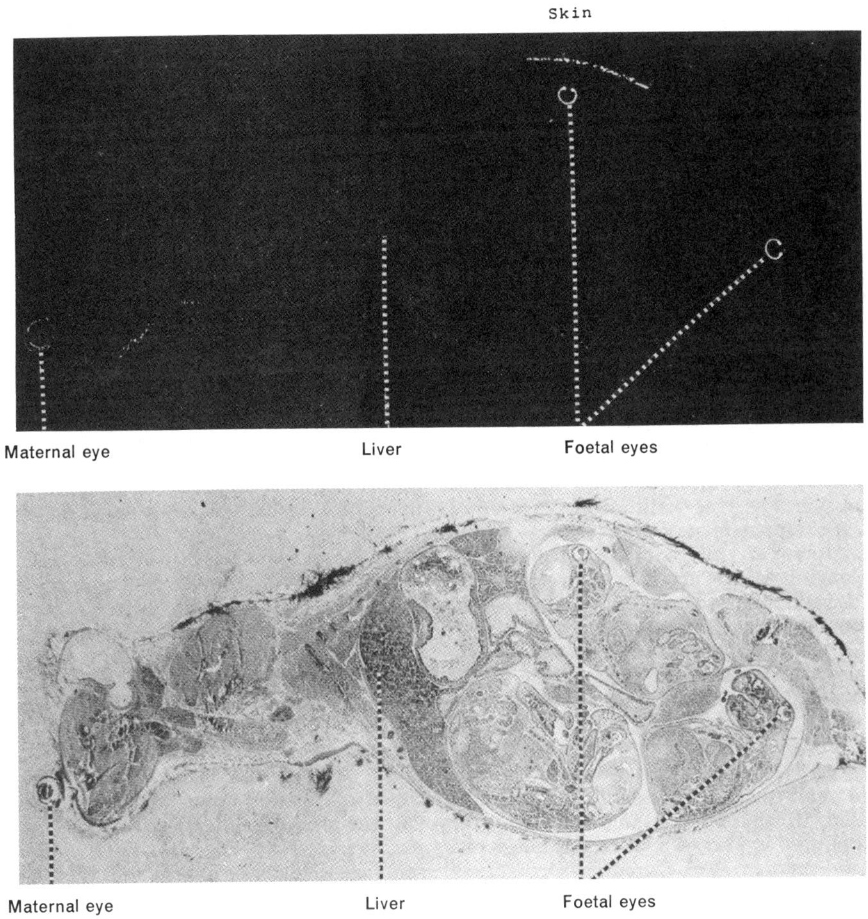

FIG 4.3 Autoradiograms (top) and the corresponding sections (bottom) showing the distribution of radioactivity in pregnant mouse at 18 days gestation, 24 hr after intravenous injection of [^{14}C]Tu. Activity is high in fetal eyes and low in maternal eyes. Courtesy of Professor B. Larsson.

monomeric units, aligned in a parallel fashion with an average interlayer spacing of about 3.4 Å. Surprisingly enough, a sample of pheomelanin from chicken feathers gave a similar diffraction pattern with a maximum at about 3.3 Å (Thathachari, 1976).

The close similarity of the Bragg spacing of the structurally different pigments examined was interpreted as arising from nearly parallel stacking of some of the monomeric groups to form a local domain that is randomly ordered. This has been substantiated by more recent investigations conducted with various highly sophisticated techniques, including Rayleigh scattering of Mossbauer radiation (Crippa *et al.,* 1989).

A series of experiments has been performed on lyophilized melanosomes from various sources in order to analyze the pigment in the natural state as polymerized into these organelles and to verify on such a biological amorphous material the possibility of obtaining intensity-scattering curves from which Bragg distances could be calculated (L'vov *et al.,* 1983; Bridelli *et al.,* 1991). These investigations suggest that the pigment in melanosomes maintains many structural features of the purified form and that the biochemical composition of the organelles can be responsible for the observed differences in the diffractograms.

B. EPR SPECTROSCOPY

A most unusual property of melanins is their characteristic electron paramagnetic resonance (EPR) signal. Since it was first reported (Commoner *et al.,* 1954), the origin of this paramagnetism has engendered one of the most fertile fields of research that has maintained a high share of interest as well as much controversy (Blois *et al.,* 1964; Sealy *et al.,* 1980; Sarna, 1991). Efforts have been also made to correlate the free radical character with some of the alleged functions of melanins, such as in the physiology of vision and in photoprotection, but the evidence is not entirely unambiguous (Crippa *et al.,* 1989; Sarna, 1991).

1. Eumelanins

Most EPR studies have been carried out on hydrated suspensions of eumelanins, natural and synthetic, the spectra of which are all remarkably similar. They consist of a single rather featureless signal with a line width of about 4–6 G and a g-value close to 2.004. No hyperfine coupling is detected and the spin concentration is very small, usually within the range $4–10 \times 10^{17}$ spins/g. Another remarkable feature of the EPR signal is its stability under a variety of conditions, including reduction with ascorbic acid, treatment with hot mineral acid, and solubilization with alkaline hydrogen peroxide. Semiconductor models (Longuet-Higgins, 1960; Pullman and Pullman, 1961; Galvao and Caldas, 1990a,b) and charge–transfer complexes (Schultz, 1986) through the stacked monomer units of the eumelanin polymer have been proposed to account for such an unusual type of stable organic radicals, but most workers favor Blois' interpretation that the unpaired electron is due to a small fraction of semiquinonoid units which are deeply embedded within the polymer, thus being chemically inaccessible.

At one time it was thought that EPR was useful for elucidating the structure of eumelanin as well as its mode of formation, that is, radical versus ionic mechanism. However, it should be emphasized that eumelanins usually contain about one free radical per 800–1000 units of monomer (Mason *et al.*, 1960) or, at most, one free radical per 200 units (Blois *et al.*, 1964). Therefore, the free radical represents a very small part of the eumelanin polymer, implying that study of the EPR properties does not probe the structure of the molecule as a whole. Notwithstanding this limitation, EPR proved to be a most useful tool in investigating the reactivity of melanins under a variety of conditions of chemical and biological significance (Sarna and Swartz, 1992).

Sarna *et al.* (1980) have shown that the free radical population in hydrated melanin suspensions is markedly affected by the pH of the medium, owing to the operation of an equilibrium that establishes a stable fraction of semiquinonoid subunits within the polymer.

$$Q + QH_2 = 2Q^{-\bullet} + 2H^+$$

where Q and QH_2 represent 5,6-indolequinone and 5,6-dihydroxyindole units, respectively, of the melanin polymer. As expected, the main spectral changes are observed between pH 8 and 12, where ionization of the phenoxyl groups should occur. Diamagnetic ions, such as cadmium, zinc, aluminum, scandium, and lanthanum, also enhance the EPR signal of eumelanins by a factor varying between 1.6 and 9, presumably by a mechanism involving complex formation (Felix *et al.*, 1978b).

Another possibility is that attachment of metal ions to nonradical sites on the polymer lowers the effective pK_a of the free radicals. This might account for the observation that the EPR spectrum in the presence of diamagnetic metal ions resembles that observed at higher pH values, where the semiquinone anion is likely to dominate (Sealy *et al.*, 1980). By contrast, addition of paramagnetic metal ions, for example, Cu^{2+}, results in a decrease of the apparent free-radical intensity almost to zero, without any noticeable change in the signal line shape. This effect, first described by Blois and co-workers (1964), was attributed to a chemical redox reaction occurring between semiquinone radicals and the metal ion. However, a different explanation follows from the work of Sarna *et al.* (1976), showing that the decrease in signal intensity is magnetic in character and is presumably due to complex formation. This places the ion in close proximity to the free radical and the consequent strong magnetic interaction quenches the signal drastically.

2. *Pheomelanins*

In an early study, Mason *et al.* (1960) found that the free-radical content of intact human red hair was lower than that in black hair, but no difference was observed in the EPR spectra of the two samples examined. Similar results were later obtained by Chedekel *et al.* (1977), who reported that the EPR signal in intact red hair and extracted pheomelanin was similar to that of eumelanins. In another study by Hansson *et al.* (1979a), the solid-state EPR spectrum of a pheomelanin isolated from a human melanoma showed a very broad signal with a hyperfine structure that apparently was due to the presence of manganese ions. Trichochromes C and F were also analyzed and gave a broader signal than dopamelanin and, in addition, a hyperfine structure that was also attributed to manganese ions.

More recently, the EPR properties of natural and synthetic pheomelanins have been examined anew under optimal analytical conditions (Sealy *et al.*, 1982a,b). Accurate measurements on hydrated suspensions of 5-cysdopa pheomelanin at varying pH in D_2O and in the presence of complexing agents gave a spectrum which was different from the single featureless line obtained from dopamelanin under similar conditions. In fact, the spectrum has three features characteristic of an immobilized radical with hyperfine splitting due to nitrogen ($I = 1$). Given the presence of hydroxy-1,4-benzothiazine units in the pigment polymer, this new kind of free radicals has been tentatively assigned to semiquinoneimines such as 4.**9** and the protonated form of 4.**10** that predominates at low pH.

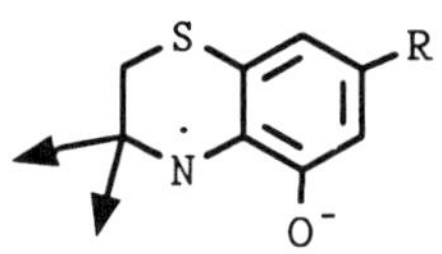

4.**9**

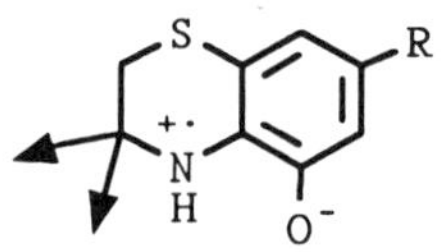

4.**10**

As with flavin systems, the free-radical concentration is affected by complexation and increases with decreasing pH. A sample of pheomelanin isolated from chicken feathers gave an EPR spectrum almost identical to that of synthetic 5-cysdopa pheomelanin (Sealy *et al.*, 1982b). The alkali-soluble pigment from red human hair, however, had a major spectral component similar to that found in eumelanins, with a smaller component corresponding to that found in natural and synthetic pheomelanins. Two other spectra, obtained from eye melanoma and brown eye melanin, also showed signals characteristic of both eumelanin and pheomelanin-type free radicals (Fig. 4.4).

These results would be in accord with the notion that these pigments, as well as other melanins, are heteropolymers derived by a cooxidation process involving 5,6-dihydroxyindole and 1,4-benzothiazine intermediates. And, indeed, the EPR spectra of model copolymers obtained by enzymic cooxidation of dopa and 5-cysdopa in various proportions were found to exhibit similar features, with changes in the intensity of the two types of free radicals

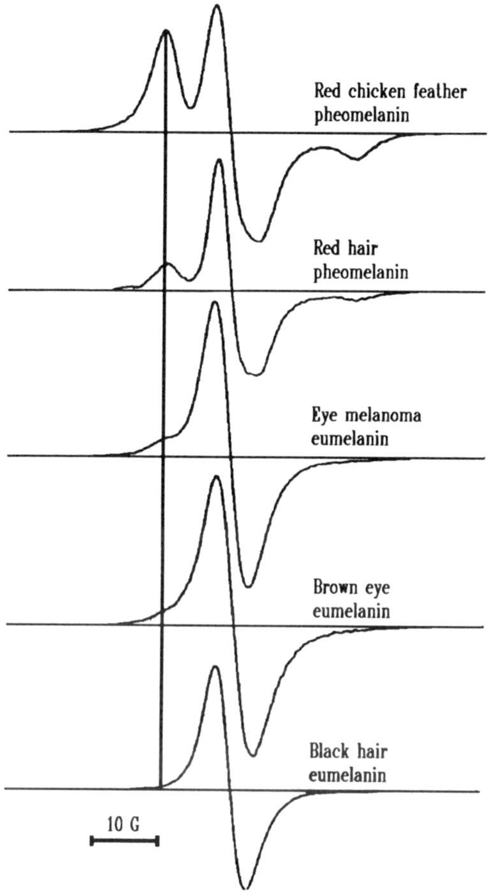

FIG 4.4 EPR spectra of suspensions of natural melanins (1 mg/ml) containing 3 m*M* Zn^{2+}, pH 4.5, at −196°C.

reflecting the molar ratio of the substrates in the original oxidation mixtures (Sealy *et al.,* 1982a). It should be noted, however, that the EPR spectra of simple mixtures of cysdopa and dopamelanins closely resemble those of the copolymers, except for some minor details such as, for example, separation of the extrema in the signals associated with the semiquinoneimine free radicals. Thus, in the absence of additional data, EPR spectroscopy may not allow a clearcut differentiation between copolymers and mixtures of eumelanin and pheomelanin. Another difficulty in the interpretation of the EPR data is due to the possibility, not considered at the time the previously mentioned studies were conducted, that the pigment preparations could contain oxymelanins rather than, or in addition to, pheomelanins.

C. Other Analytical Methods

The IR spectra of melanins show absorption bands expected for an aromatic or heteroaromatic system with OH, NH, and conjugated C=O groups, but line broadening and overlapping preclude a detailed analysis. Reported spectra even of chemically different melanins, for example, sepiomelanin and catechol melanin, are remarkably similar (Bonner and Duncan, 1962; Blois *et al.,* 1964). Likewise, the IR spectrum of reduced and/or methylated dopa/melanin was found to show no distinct differences when compared with chemically unmodified samples (Wilczok *et al.,* 1984).

Some differences between natural and synthetic melanins were apparent in the fingerprint region using very dilute KBr pellets (Baradani *et al.,* 1982). The sensitivity of the method proved to be useful for studying protonation and deprotonation of ionizable groups at different pH and for studying as well the binding of iron and the increase of carboxyl group content in oxidized dopamelanin (Bridelli *et al.,* 1986). Electron spectroscopy for chemical analysis (ESCA), or X-ray photoelectron spectroscopy (XPS), has also been shown to possess a considerable potential for structural investigation of eumelanins in the solid state (Williams-Smith and Dunne, 1976; Clark *et al.,* 1990). Using this technique, significant differences in the elemental composition and overall functional group distribution have been detected, for example, between natural and synthetic eumelanins. In addition, analysis of the spectra seems to support the notion that eumelanins are heterogenous polymers containing both intact and partially oxidized 5,6-dihydroxyindole units.

New perspectives in the structural analysis of melanins have been offered by the availability of solid-phase ^{13}C and ^{15}N NMR spectrometers operating in the cross-polarization/magic-angle spinning mode (CP-MAS). Analysis of the naturally abundant ^{13}C and ^{15}N spectra has allowed the assignment of several functionalities and the comparison of melanins of various origin (Aime and Crippa, 1988; Duff *et al.,* 1988). However, the line broadening and inordinate number of resonance signals make the interpretation challenging. More recently, Aime *et al.* (1990) have used CP-MAS ^{13}C spectroscopy to monitor the structural changes of eumelanins following solubilization with hydrogen peroxide. Comparison of the ^{13}C spectra revealed that sepiomelanin is structurally more similar to oxidized dopamelanin than the parent dopamelanin obtained by tyrosinase oxidation of dopa in the presence of catalase. Using the same technique, Peter and Förster (1989) have obtained supporting evidence that dopamine melanin, unlike dopamelanin, has a high content of aliphatic carbon atoms arising from incorporation of uncyclized dopamine units in the pigment polymer (see Chapter 6).

^{13}C NMR spectroscopy has also been used to study the autooxidative and tyrosinase-catalyzed conversion of specifically labeled L-dopa and DHI to melanin (Chedekel *et al*, 1988). In the enzymically produced pigment, analysis of the ^{13}C NMR spectrum showed the presence of the benzylic carbon of dopa in the indolic unit of the pigment backbone. By contrast, in the melanin obtained by autoxidation, the C-3 carbon appeared to be in the form of both a pyrrole residue and a carbonyl functionality, implying a substantial difference between the two pigment preparations.

Interest in mass spectrometry as an analytical tool in melanin chemistry has increased steadily in recent years, following the development of new ionization techniques, especially fast atom bombardment (FAB), which is particularly suitable for high-molecular-weight molecules. However, contrary to expectations, all types of melanins so far examined, including the low-molecular-weight trichochromes (G. Prota, unpublished), did not give rise to significant FAB mass spectra under a variety of experimental conditions. Consequently, two main approaches involving pyrolysis and subsequent identification of specific degradation products have been attempted to characterize melanins. The first approach is based upon analysis of the pyrolysis products by gas chromatography coupled with mass spectrometry (Dworzanski *et al.*, 1984; Dworzanski and Debowski, 1985). In the second one, the sample is introduced directly into the ion source of the mass spectrometer and heated with a linear temperature gradient up to 450°C (Allegri *et al.*, 1987, 1988, 1990). In both cases, analysis of degradation products of melanin preparations proved to be difficult from the structural viewpoint. However, the qualitative and quantitative patterns of products provide a useful fingerprint for the analysis and identification of natural and synthetic melanins.

CHAPTER 5

EUMELANINS

I. INTRODUCTION

The preceding chapter provided the reader with a general overview of the main types of natural and synthetic melanins, along with an account of their physicochemical properties. In this chapter and the following two, we shall discuss in more detail the chemistry of each group of pigments. This is a difficult task since in the past few years many of the classical concepts which had guided structural work have been questioned, and the new picture of the chemistry of melanins emerging from these studies is still being assessed. In such a situation, the best one can do is to critically present the state of the art in the field, with a view toward separating facts from speculations and thus putting the subject into perspective for future work. Our coverage therefore will be illustrative rather than comprehensive.

In this chapter we will describe the chemical composition and reactivity of eumelanins first. We will then go on to discuss the critical issue of the structure of the pigment polymer, as derived from the three major conceptual approaches based on (1) chemical degradation, (2) incorporation of labeled precursors in melanogenesis both *in vivo* and *in vitro,* and (3) structure and mechanism of formation of oligomeric intermediates in the oxidative conversion of 5,6-dihydroxyindole(s) to melanin(s).

II. GENERAL PROPERTIES

A. EUMELANINS AS MELANOPROTEINS

As obtained, native eumelanins contain a certain amount of proteins, as estimated by the amino acids recovered after acid hydrolysis. The protein can be removed, with difficulty, but the nature of the pigment makes it impossible to establish whether and how they are linked together. Yet, most investigators favor the view that natural eumelanins are melanoproteins. This concept was first introduced by Görtner (1911) to describe the pigment isolated from black wool by extraction with NaOH followed by precipitation with acid. Other workers have since used some variation of this procedure to obtain eumelanin from hair, feathers, and melanoma tissues (e.g., Greenstein, 1948; Hack and Helmy, 1964; Bolt, 1967; Jimbow *et al.,* 1982). All have found protein associated with the purified pigment and have assumed this protein to be covalently linked to the pigment polymer. Greenstein (1948) suggested that when melanogenesis occurs in the presence of protein a sulfur-linked melanoprotein will result by addition of cysteine residues to indolequinone units of polymer. This would explain the difficulty in obtaining sulfur-free eumelanins and the formation of cysteic acid and taurine by peracetic oxidation of sepiomelanin which had been treated extensively with 6 *M* HCl (Piattelli *et al.,* 1963).

However, the same data would also be in keeping with the addition of cysteine or glutathione to dopaquinone or other melanogenic precursors, such as dopachrome or 5,6-indolequinone (Prota *et al.,* 1988). There is also evidence suggesting that the sulfur found in *Sepia,* B16, and Harding-Passey melanoma melanins may be derived artificially from the reaction of the pigment with cysteine or cystine in the course of the HCl hydrolysis (Ito *et al.,* 1988).

A causal relationship between synthesis of melanosomal proteins and melanogenesis is suggested by the ultrastructural observations, discussed in Chapter 2, showing that melanins are synthesized on the filamentous matrix of premelanosomes, so that one of the many proteins of these organelles may eventually become bound to the pigment. If so, formation of melanoprotein should then originate from an interaction between two independently synthesized entities in the melanosome, the pigment moiety and an apoprotein moiety. This hypothesis was elegantly tested by Whittaker (1979) by incorporation experiments during early differentiation of chick retinal pigment cells, which have a very high rate of melanin synthesis. Using [DL-U-^{14}C]tyrosine he found that short-term formation of labeled melanin was not affected by inhibitors of protein synthesis, such as puromycin and cycloheximide. Moreover, melanin isolated from tissues incubated previously in [L-U-^{14}C]valine had no associated radioactivity when compared with the radioactivity of melanin prepared from tissues in the same series incubated with labeled tyrosine.

These results strongly suggest that coordinated synthesis of protein is unnecessary for melanogenesis to occur and that no significant amount of new protein(s) is bound covalently to the pigment. Thus, melanoproteins, if they occur, are either minor components of the melanosome or must involve weak bonding associations between the pigment and protein. In line with this view, Benathan and Wyler (1980) have reported that native sepiomelanin, as

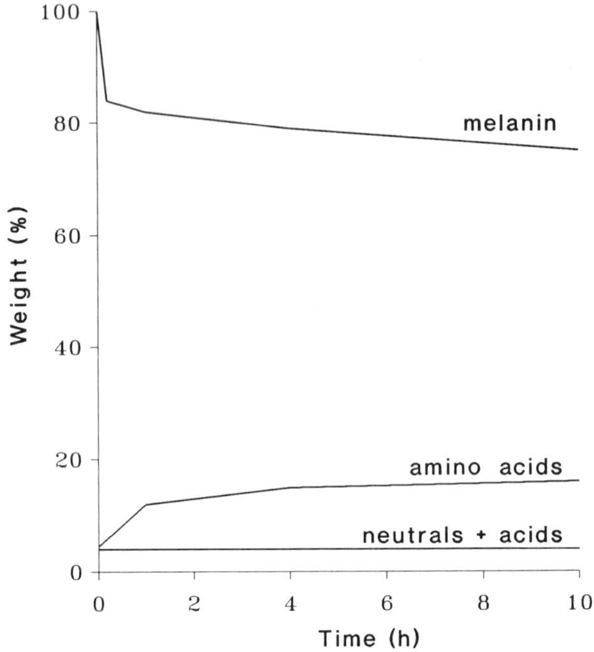

FIG 5.1 Gravimetric control of the hydrolysis of sepiomelanosomes with 6*M* HCl at 100°C (Benathan and Wyler, 1980).

well as bovine eye melanin, lose in hot 6 *M* HCl most of their "proteic" components during the first 15 minutes, and the remainder are removed in about an hour (Fig. 5.1). Analysis of the solubilized fraction revealed a mixture of amino acids (11%) along with small amounts of glucosamine (0.04%). It seems therefore that a considerable portion of the amino acids and peptides is very loosely bound to the melanin granules, which raises some doubts about the commonly held view of natural eumelanins as melanoproteins.

B. Molecular Weight

The thought is rather widespread that eumelanins are unusually large-sized polymers or mixtures of polymers with a tridimensional structure consisting of several hundreds of monomeric units. The exceedingly low solubility of the pigments and the black "chromophore," reminiscent of graphite, support this assumption. It should be noted, however, that a number of insoluble dyes are known which in the solid state appear very dark in color and yet are of relatively low molecular weight.

Unfortunately, the insolubility of eumelanins makes determination of their molecular weight an almost impossible task. To overcome this difficulty, some approaches have been attempted using solubilized samples of both natural and synthetic melanins (Miyake *et al.,* 1986). The molecular weights of such pigment preparations, determined by viscosity, vapor pressure osmometry, and X-ray scattering, were surprisingly lower than expected, ranging between 1100 and 6000, irrespective of the sample origin (*Sepia* melanin, enzymic or autooxidative dopa, and DHI melanin). In another study (Wolfram *et al.,* 1970), the solubilized pigment (MFA) obtained from dark human hair (see Section II,G) was found to have a molecular weight of about 10,000 (vapor pressure osmometry), the value being unaffected by the duration of the oxidation time in the range from 10 to 1440 min. However, the lack of a precise characterization of the solubilized melanin samples used makes it difficult to draw any conclusion regarding the average molecular weight of native eumelanins.

C. Analysis and Carboxyl Content

There is a wide range of variation in the elemental analytical results for eumelanins in the literature (Nicolaus, 1968; Swan, 1974). Even repeated analyses of the same melanin, for example, sepiomelanin, from different laboratories are significantly divergent (Table 5.1). One of the reasons for the lack of reproducibility is associated with the hydration state of the sample, and a suitable protocol to obtain consistent results has been provided by Benathan and Wyler (1980) and Zeise (1991). Other contributing factors include the presence of protein, salts, and other impurities, as well as structural modifications of the pigment associated with the methods of isolation or preparation used (see Section II,F).

In view of these limitations, elemental analyses of melanins can at best serve to estimate an approximate agreement of the sample composition with the requirements of a model DHI polymer. To this end, the C/N ratio of the pigment under investigation is preferably used since it is independent of the degree of hydration and is therefore much more reproducible (Ito, 1986). Low C/N ratios are indicative of pigment degradation and loss of carbon dioxide, while high values (greater than 8) suggest lower degradation or the participation of C_9 units like DHICA.

A further characterization of eumelanins is provided by the determination of the carboxyl groups in the pigment polymer. An early method, developed by Piattelli *et al.* (1962), is based upon the property of both natural and synthetic eumelanins to undergo decarboxylation when heated at 190–200°C in Vaseline. In the case of sepiomelanin, the amount of carbon dioxide

TABLE 5.1
ELEMENTAL COMPOSITION (%) OF SOME NATURAL EUMELANINS

Source	C	H	N	S	C/N	Reference
Sepia ink[a]	64.7	2.5	9.0	0.3	8.4	Piattelli *et al.* (1962)
Sepia ink[b]	60.9	3.4	8.4	0.3	8.4	Piattelli *et al.* (1962)
Sepia ink[a]	57.7	2.6	6.2		10.8	Benathan and Wyler (1980)
Sepia ink[a]	52.2	3.4	7.9	0.2	7.7	Ito (1986)
Squid ink[b]	60.9	3.1	6.0		11.8	Nicolaus *et al.* (1964)
Octopus ink[b]	59.0	3.3	8.0		8.6	Nicolaus *et al.* (1964)
Bovine eye[a]	53.6	2.7	7.5		8.3	Benathan and Wyler (1980)
Bovine eye[b]	54.9	4.2	9.3	0.8	6.9	Dryja *et al.* (1979)
Bovine eye[b]	60.4	4.6	8.7	0.9	8.1	Nicolaus *et al.* (1964)
B-16 Melanoma[b]	64.5	5.5	7.1		10.6	Nicolaus *et al.* (1964)
B-16 Melanoma[b]	52.3	4.5	7.3	0.6	8.4	Ito (1986)
Mouse hair[b]	56.5	4.6	7.6	0.8	8.7	Ito (1986)
Horse hair[b]	64.6	5.1	6.0	1.3	12.6	Nicolaus *et al.* (1964)
Dog hair[b]	61.2	4.0	5.1		14.0	Nicolaus *et al.* (1964)
Human hair[b]	61.1	4.6	8.4	3.7	8.5	Nicolaus *et al.* (1964)
Bovine hair[b]	52.8	4.0	8.3	0.9	7.4	Dryja *et al.* (1979)
Bovine hair[b]	65.5	5.5	5.4	1.6	14.0	Nicolaus *et al.* (1964)
Blackbird feathers[b]	55.8	3.6	8.2	1.0	7.9	Fattorusso *et al.* (1965)

[a]Purified by refluxing with 6*M* HCl.
[b]Treated with concentrated HCl at room temperature.

evolved, measured as barium carbonate, was 9.1%, and this was shown to correlate the content of carboxyl groups determined separately by methoxyl group analysis of the melanin methyl ester. Other methods which proved to be useful include acidimetric titration (Benathan and Wyler, 1980) and measurement of the amount of carbon dioxide liberated by heating in 6 *M* HCl (Ito, 1986; Palumbo *et al.*, 1988). The results of a number of such determinations are reported in Table 5.2. The consistent finding is that all natural eumelanins contain a substantial amount of carboxyl groups, pointing to the presence of structural units other than DHI in the pigment polymer.

D. REDOX STATE

There are several lines of evidence suggesting that eumelanins may exist in different states of oxidation. Figge (1939), for example, found that sodium hydrosulfite turns the color of dopamelanin from black to light brown and that the reduced pigment could be reoxidized by the addition of potassium ferricyanide. Moreover, studies carried out with manometric methods provided evidence that melanins consume oxygen at a rate and to an extent which are

TABLE 5.2
CO_2 EVOLUTION FROM SOME NATURAL EUMELANINS

Sample	% CO_2 (w/w)	References
Sepiomelanin[a]	9.1[b]	Piattelli *et al.* (1962)
Sepiomelanin[c]	9.5[d]	G. Prota *et al.* (unpublished)
Sepiomelanin[c]	10.6[b]	Ito (1986)
B-16 melanosomes[e]	9.6[d]	Ito (1986)
B-16 melanosomes[a]	4.1[b]	Nicolaus *et al.* (1964)
Black human hair melanin[a]	4.1[b]	Nicolaus *et al.* (1964)

[a]Acid treated.
[b]Thermal decarboxylation in mineral oil.
[c]Native sepiomelanosomes.
[d]Refluxing in 6*N* HCl, 24 hr.
[e]Native melanosomes.

dependent on the pH of the medium. The phenomenon is enhanced by the addition of NADH or other reducing agents, which evidently convert melanin to a more oxidizable form (Van Woert, 1967; Gan *et al.*, 1974; Menon *et al.*, 1977). EPR oximetry experiments showed that on passing from pH 5.5 to 11 the rate of oxygen consumption increases several thousand times, with an abrupt rise in the pH range 9–10.5 (Sarna *et al.*, 1980). Using bipyridinium quaternary salts as a redox probe, the one-electron redox potential of two synthetic melanins was studied by the pulse radiolysis method (Sarna and Swartz, 1991). Even though accurate determination of the potential of the fully oxidized/semireduced melanin moieties was not possible, approximate estimates suggested that the one-electron reduction potential of the major reactive site of synthetic eumelanin was between −450 and −550 mV.

Other redox systems studied include ascorbate, sodium borohydride, benzoquinone, permanganate, chlorophenolindophenol, and nitro blue tetrazolium (Crippa *et al.*, 1989). In these reactions, eumelanins can act either as electron acceptor or as electron donors, according to a biphasic mechanism which is reminiscent of the electron transfer processes in redox-conducting films deposited at solid electrodes (Manimala and Horak, 1986). This has been tested by the preparation of a melanin film electrode by anodic deposition of DHI in both acidic and neutral media (V. Horak and G. Weeks, unpublished).

Despite these investigations, the exact proportion of reduced (H_2Q) and oxidized (Q) indole units within the eumelanin polymer is not known with certainty. Optical measurements (Gan *et al.*, 1976) of the reaction of ferricyanide with dopamelanin indicate a reducing capacity of 5.3 mEq of oxidant per gram of melanin. This value, however, is susceptible to variation depending on the state of oxidation of the dopamelanin preparation used. One would also anticipate that natural eumelanins with their associated protein would be less reactive, and there is evidence that this is so (Gan *et al.,* 1974; Menon *et al.,* 1976).

In kinetic studies of the redox reactions of dopamelanin with Ti(III) and with Fe(III), Horak and Gillette (1971) found that a fast first electron-exchange step is followed by a slow second step. While the quinone–quinol interconversion reaction appeared to involve an exchange of two electrons, only 0.5 electrons were found to account for the fast reaction step between dopa and Ti(III); similarly, 0.02 electrons per indole unit were exchanged with Fe(III). From these and other data, they concluded that dopamelanin in air atmosphere exists predominantly in the quinonoid form. Such a view is in contrast with the work of Nicolaus and Piattelli (1962), which showed that eumelanins, both natural and synthetic, contain a high percentage of phenolic groups. These were estimated by methylation of the pigments with an excess of diazomethane and subsequent analysis of the resultant products by the Zeisel method. In the case of sepiomelanin, the difference between the methoxy value of specimens treated with methanolic hydrogen chloride (6.0%) and with diazomethane (18.8%) gave a content of phenolic hydroxy groups of 13.1%. Comparable results were obtained by Piattelli *et al.* (1962) for other eumelanins investigated, including enzymic dopa and DHI melanin, which gave similar percentages of methoxy groups (20.6 and 21.6%, respectively).

This would suggest that the pigments are not fully quinonoid but are approximately halfway between the quinonoid and diphenolic stages. However, a reliable estimation of the relative proportion of oxidized and reduced indole units within the polymer may be difficult due to the complex number of functionalities, other than phenolic groups and carboxyl groups, that may react with diazomethane. This is apparent from the formation of methylenedioxy groups and *N*-methylation of the NH of pyrrole rings (Piattelli *et al.,* 1962; Swan and Waggott, 1970). In any case, the state of oxidation of the individual types of units in the polymer remains unknown, since determination of the methoxyl groups gives only an estimation of the relative numbers of quinonoid and phenolic units as a whole.

E. Stability to Acids

The frequently cited resistance of eumelanins to chemical agents has led over the years to an image of these pigments as virtually inert materials. In connection, most authors quote the discovery of melanin, apparently unmodified, in a specimen of a fossil decapod 150 million years old. It has also been stated that "certain melanins may be heated to 600°C without decomposition"

(Nicolaus, 1968). Given the alleged inertness of eumelanins, studies on the chemistry of these pigments have usually been carried out on natural or synthetic samples isolated or purified by extensive hydrolytic treatments with mineral acids, such as concentrated HCl at room temperature for 7 days or boiling 6 *M* HCl. Whether such drastic conditions could cause alteration of the pigment backbone has until very recently been overlooked due to the fact that on exposure to acids as well as other reagents the material remains apparently unchanged with respect to its color and solubility properties. New evidence, however, indicates that such apparent inertness is due to the inadequate methods used for detecting structural modifications following chemical treatments rather than to the lack of intrinsic reactivity of the pigments to acids. This has been documented by Benathan and Wyler (1980), who found that sepiomelanin as well as other eumelanins, upon heating in acids, undergo extensive degradation with concomitant decarboxylation, accounting for a weight loss of as much as 23% of the original material. Synthetic melanins prepared from dopa, DHI, and DHICA behave similarly, with the extent of the decomposition depending upon the experimental conditions (Ito, 1986; Palumbo *et al.*, 1988), hence the discrepancies in the elemental composition and carboxyl content of acid-treated and native eumelanins.

Interestingly enough, it seems that some 90 years ago Landolt (1899) had already recognized the decomposing effect of acids. Thus, to isolate the melanin of the eye coat he removed the granules by rubbing them loose with water, precipitating the pigment with saturated ammonium sulfate, washing with water, drying, and extracting with ether and alcohol in a Soxhlet apparatus. Landolt makes a very significant statement when he adds "when boiled with acid the pigment is modified, the nitrogen content being reduced, and of course, the oxygen content increased." The observed instability of melanin in acid was later confirmed by Görtner (1910), who obtained evidence that melanin was also unstable in alkali. He then concluded that, in order to obtain the pigment "in the form that it exists in living tissues, great care is needed to avoid decomposition." It is unfortunate that these recommendations have escaped the attention of workers in the field. The image of melanins as inert materials has undoubtedly contributed to the slow progress in the study of their chemistry.

F. Autooxidation

A major consequence of the redox interaction of eumelanins with oxygen, previously mentioned in Section II,D, is the generation of hydrogen peroxide, arising conceivably by reversible electron transfer processes from autooxidizable catechol groups of the pigment. Van Woert (1967) postulated a direct two-electron transfer reaction; recent evidence (Korytowski *et al.*, 1985), however, points to a more plausible one-electron reduction of oxygen to superoxide followed by spontaneous dismutation and/or further reduction by melanin, yielding hydrogen peroxide.

Whether the interaction of eumelanins with oxygen has any effect on the structure of the pigment polymer has been investigated recently at Naples (G. Prota, O. Crescenzi, and M. d'Ischia, unpublished). A probe melanin sample, prepared enzymatically from dopa under conditions of minimal autooxidation, was finely suspended in neutral buffer and was allowed to autooxidize for 26 hr under a stream of oxygen. Analysis of the pigment by acid decarboxylation revealed a marked rise in the carboxyl content (as determined from the carbon dioxide evolved), from 4.8 to 8.9% by weight. In a similar experiment, dopa was oxidized enzymatically over a prolonged period, and the pigment formed was collected periodically and analyzed for its carboxyl content. The results, shown in Fig. 5.2, clearly indicate that the acidity of the pigment increases significantly with increasing oxygenation time. As expected, the rate of autooxidation is pH dependent, being faster at higher pH, for example, 8, and it is also affected by catalase. Preliminary experiments indicate that natural eumelanins such as sepiomelanin are susceptible to autooxidation as well.

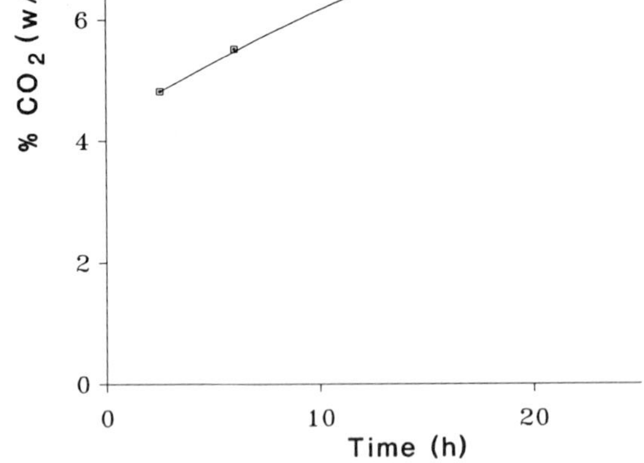

FIG 5.2 Carboxyl content of enzymic dopamelanin as a function of the incubation time.

Chemically, the observed modifications are akin to those that occur during autooxidative conversion of dopa to melanin at pH 8 (Swan, 1974; Nicolaus, 1978). Under these conditions hydrogen peroxide is generated, which attacks the pigment polymer being formed probably at the site of the labile indolequinone units, as shown in Fig. 5.3.

Given the presence of hydrogen peroxide in biological systems and the marked affinity of eumelanins for this oxidant, it is conceivable that autooxidative breakdown processes analogous to those observed *in vitro* also occur *in vivo* within pigmented tissues. This would be in keeping with the presence of carboxylated pyrrolic units in naturally occurring eumelanins (see Section II,A) as well as the occurrence of oxymelanins in certain types of red hair and feathers. Indeed, the role of hydrogen peroxide in melanogenesis is a most critical one both for what concerns the formation of melanins and for their subsequent structural modifications.

Though this has long been known, the practical consequences of the preparation of synthetic melanins have been little considered, as evidenced by the different oxidation conditions used, for example, air or pure oxygen, from 4 hr up to 3 days. One could therefore expect that melanins obtained after long oxygenation times are more acidic than those prepared under conditions of minimal autooxidation. In support of this, G. Prota *et al.* (unpublished) re-

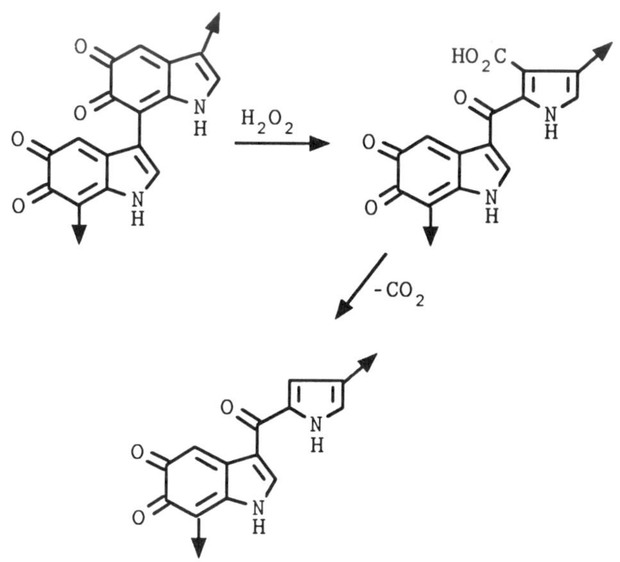

FIG 5.3 Proposed mode of attack of hydrogen peroxide at the indolequinone units in the eumelanin polymer (Swan and Waggott, 1970).

TABLE 5.3
CARBOXYL CONTENT OF SOME ENZYMIC AND AUTOOXIDATIVE DOPAMELANINS PREPARED ACCORDING TO LITERATURE PROCEDURES

Oxidative conditions	Substrate concentration (m*M*)	Reaction time (hr)	Melanin % (w/w)[a]	CO_2 % (w/w)[b]	Reference
Autooxidation					
Aqueous ammonia, pH 8.0, 25°C	25.5	72	25	4.2	Sealy *et al.* (1980)
Phosphate buffer, pH 8.0, 25°C	15.3	72	47	6.4	Binns *et al.* (1970)
Phosphate buffer, pH 8.0, 37°C	12.5	24	53	7.3	Ito (1986)
Phosphate buffer, pH 8.0, 37°C	2.5	24	52	9.5	Ito (1986)
Enzymic oxidation (tyrosinase, O_2)					
Phosphate buffer, pH 6.8, 37°C	12.5	4	66	4.7	Ito (1986)
Phosphate buffer, pH 6.8, 37°C	2.5	4	50	6.2	Ito (1986)
Phosphate buffer, pH 7.0, 25°C	2.5	2.5	59	4.8	G. Prota *et al.* (unpublished)
Phosphate buffer, pH 6.8, 35°C	3.8	6	35	5.6	Piattelli *et al.* (1962)
Phosphate buffer, pH 6.6, 25°C	1.2	4	35	6.9	Mason (1967)

[a]After equilibration with saturated aqueous $CaCl_2$.
[b]By refluxing in 6*N* HCl (24 hr). The CO_2 evolved was collected in standard $Ba(OH)_2$ and determined by back titration with standard HCl under nitrogen.

cently investigated a series of dopamelanins, both autooxidative and enzymatic, prepared according to "classical" procedures (Table 5.3).

Analysis of the carboxyl content revealed that the melanins examined are in fact markedly different materials, reflecting the various conditions under which they were prepared, especially incubation time and mode of oxidation (air or oxygen). Hopefully current awareness of the instability of melanins to oxidation may aid in developing suitable procedures for the preparation of standard samples of melanins for structural and practical purposes.

G. BLEACHING

The degradation and bleaching of melanin by hydrogen peroxide has been long known and widely employed as a cosmetic treatment for toning down the natural color of hair (Zviak and Dawber, 1986). In view of its practical importance, the chemistry of this reaction has received a good deal of attention, yet several aspects are still poorly understood. Apparently, two distinct processes occur in sequence: a fast solubilization of the pigment, due to disruption of the granules, followed by a slower bleaching, resulting eventually in a pale yellow solution. When the oxidation is stopped in the early stages, just after complete solubilization, a brown pigment can be recovered in about 60% yield by acidification of the mixture (Wolfram *et al.*, 1970; Hall and Wolfram, 1975). The pigment thus obtained, MFA, exhibits a high carboxyl content, as evidenced by IR, ESCA, and ^{13}C spectra (Chedekel *et al.*, 1986) as well as by its solubility in alkali. Chemical analysis suggests that modification of the basic eumelanin structure is not excessive in MFA, which would make this pigment a useful soluble derivative of eumelanin for structural investigation (Clark *et al.*, 1990).

Kinetic studies have shown that the rate of bleaching depends linearly on hydrogen peroxide concentration and increases with increasing pH, an optimal value being just below the pK_a of the oxidant (11.75); at neutral or slightly acidic pH hydrogen peroxide is quite unreactive toward melanin. The reaction is markedly susceptible to catalysis by light and transition metal ions. Even at pH 9, dopamelanin having metal ions removed by previous treatment with Chelex reacts only sluggishly with 0.1 *M* hydrogen peroxide, while addition of redox active cations such as Cu^{2+} dramatically enhances the bleaching process at neutral and alkaline pH. The effect of cupric ions on the rate of melanin bleaching is dependent on the oxidant concentration and is likely to be due to promotion of Fenton-type processes involving hydroxyl radicals (Korytowski and Sarna, 1990).

Inorganic oxidants such as ammonium persulfate and potassium ferricyanide, though not very active toward the pigment, have also been found to accelerate remarkably the rate of eumelanin bleaching by alkaline hydrogen peroxide (G. Prota *et al.*, unpublished). Available evidence from all these studies as well as from investigation of the chemiluminescence accompanying melanin bleaching (Slawinska and Slawinski, 1982) suggests that pigment degradation follows either from nucleophilic attack of OOH^- ions to electrophilic quinonoid units or from a radical path initiated by metal ions and oxygen radicals.

III. STRUCTURAL STUDIES

A. Chemical Degradation

Chemical degradation of eumelanins is notoriously difficult. However, some success has been obtained by Nicolaus and his associates by oxidation with either peracetic acid or alkaline potassium permanganate. Under these conditions, sepiomelanin gave, besides much oxalic and oxamic acid, pyrrole-2,3,5-tricarboxylic acid (PTCA, 5.**1**) in about 2% yield, along with trace amounts of pyrrole-2,3-dicarboxylicacid (PDCA, 5.**2**), 2,3,4-tricarboxylic acid (5.**3**), and the 2,3,4,5-tetracarboxylic acid (5.**4**) identified by paper chromatography (Piattelli *et al.*, 1962, 1963). The yields of these latter acids were exceedingly low; in one experiment, oxidation of 10 g of sepiomelanin with permanganate gave 0.0001 mg of PDCA (estimated), 3.2 mg of 5.**3**, and 61 mg of 5.**4** (Nicolaus, 1968).

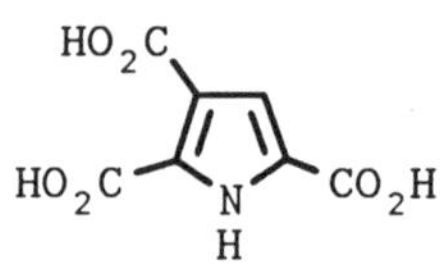

5.**1**

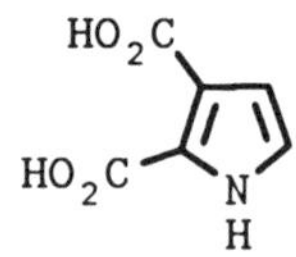

5.**2**

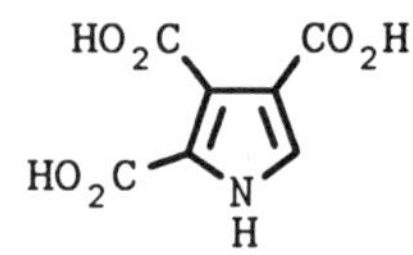

5.**3**

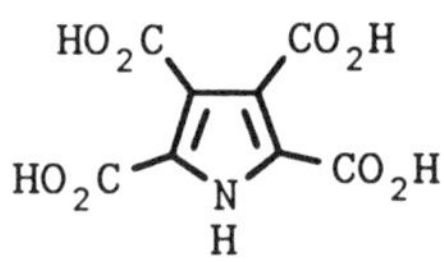

5.**4**

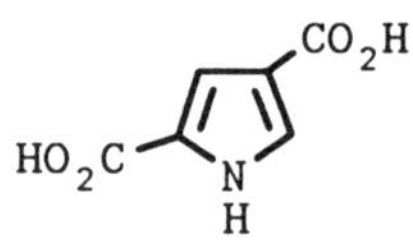

5.**5**

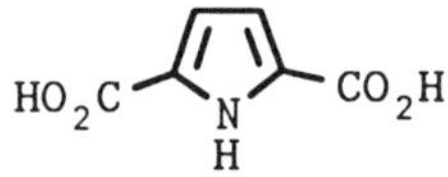

5.**6**

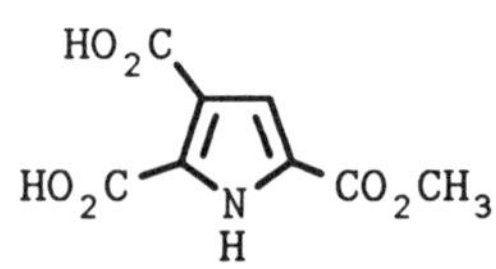

5.**7**

However, in these oxidation experiments some polymeric material (known as "melanic acid") can be recovered, and further oxidation of this usually affords additional quantities of the same pyrrolic acids. When decarboxylated sepiomelanin is oxidized, the yield of PDCA increases while that of PTCA decreases; moreover, pyrrole-2,4-dicarboxylic acid (5.**5**) and pyrrole-2,5-dicarboxylic acid (5.**6**) are also obtained which are absent among the oxidation products of the undecarboxylated pigment. Similar results (Piattelli *et al.*, 1963) were obtained by degradation of other natural eumelanins as well as some synthetic pigment preparations, including DHI melanin (Table 5.4).

TABLE 5.4
PYRROLIC ACIDS OBTAINED BY OXIDATION WITH ALKALINE POTASSIUM PERMANGANATE[a]

	Position of the carboxyl groups[b]		
Source	2,3	2,3,5	2,3,4,5
Natural pigments			
B16 melanoma melanin	+	+++	+
Human hair melanin	+	+++	+
Ox choroid melanin	+	+++	+
Sepiomelanin	+	++++	+
Squid ink melanin	+	++++	+
Synthetic pigments			
Dopamelanin (tyrosinase)	+	+++	+
Dopamelanin (tyrosinase + catalase)	+	+++	+
DHI melanin (tyrosinase)	++	+++	+
DHI melanin (tyrosinase + catalase)	++	+++	+
DHI melanin (autooxidation)	++	+++	+

[a]Piattelli *et al.* (1962, 1963); Fattorusso *et al.* (1965).
[b]The yields of pyrrolic acids were roughly estimated from the intensity and size of the spots developed in paper chromatograms after spraying with diazotized sulfanilic acid.

The origin of the pyrrolic acids was interpreted in terms of oxidative breakdown of the indole units in the pigment polymer, some of which contain a carboxyl group in position 2 of the pyrrole ring (for example, DHICA, dopachrome). Further evidence for the presence of indole-2-carboxylic units in sepiomelanin was obtained by oxidation of a pigment sample methylated with diazomethane, which afforded 5-carbomethoxy-pyrrole-2,3-dicarboxylic acid (5.7). Moreover, alkali fusion of sepiomelanin gave both DHI and DHICA, although in trace amounts.

More insight into the structural units present in sepiomelanin followed from the observation that boiling the pigment with 4% aqueous NaOH also afforded some PTCA (Fig. 5.4). This is probably the result of hydrolysis of some carboxylated pyrrolic units present in the pigment polymer, derived from a peroxidative cleavage of the indole rings similar to that which occurs during melanogenesis *in vitro* (see Fig. 5.3).

Taken together, these studies by Nicolaus' group led to a concept of sepiomelanin as a very complex macromolecule (or mixture of macromolecules) made up of different structural units linked in a random fashion by different types of bonds, as pictorially represented in Fig. 5.5. In this, the indole units in the polymer may be at various oxidation levels, including the leucodopachrome and dopachrome forms.

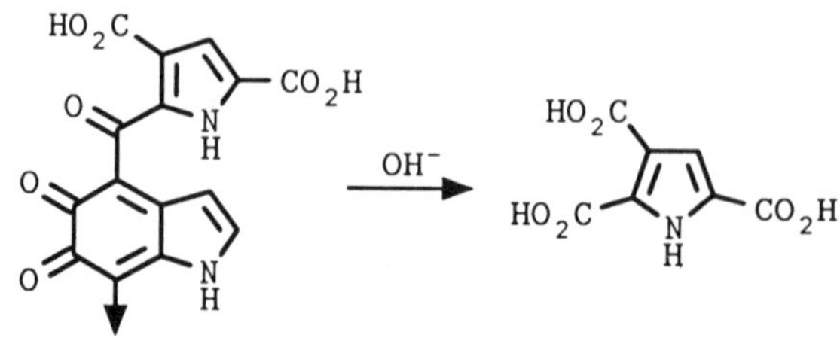

FIG 5.4 Proposed mechanism of formation of PTCA during alkaline hydrolysis of melanin (Piattelli *et al.*, 1962,1963).

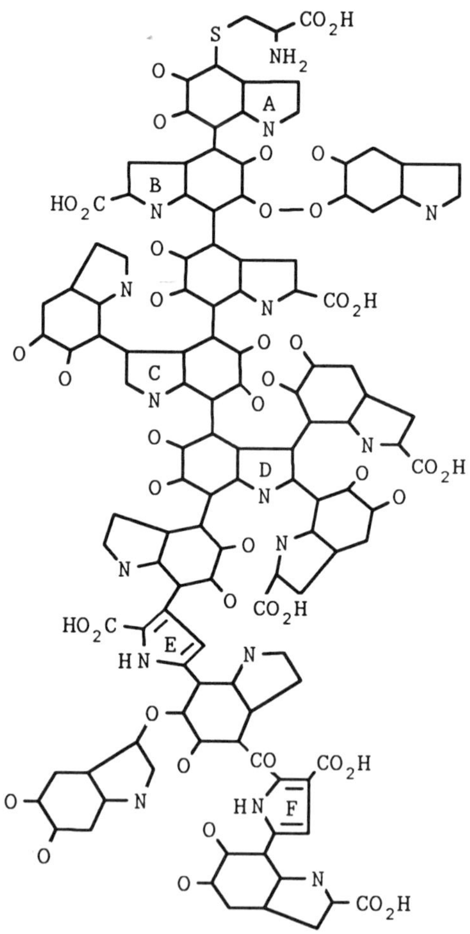

FIG 5.5 Nicolaus' model of melanin structure. The indole units in the pigment polymer may be at various oxidation levels (Nicolaus, 1962).

Nicolaus' model of eumelanin structure, as he freely admitted, is a provocative attempt to describe at the molecular level the apparent highly heterogeneous character of the pigment polymer, as suggested also by the work of Swan and other investigators (see the following sections). It is therefore clear that some of the concepts incorporated into the model are purely speculative and need to be reconsidered. Among these is the view that pigment formation proceeds randomly so that most or all of the early biosynthetic intermediates, including dopa, leucodopachrome, and dopachrome, can be incorporated, though to a different extent, into the final polymeric product.

We will return to this point later. Here it may be noted that the interpretation of the degradative work was undoubtedly thwarted by the exceedingly low yields of the small fragments identified, due partly to their relative instability under the harsh conditions used (Mason, 1967; Ito, 1986). The picture is further complicated by the possibility that some of the degradation products arise at least in part from minor components strongly retained in the pigment granules (d'Ischia *et al.,* 1985). Indeed, it appears that the only significant fragment of the chemical degradation of eumelanins is PTCA, whose origin however is not unequivocal. The general view is that it arises mainly by peroxidative cleavage of DHI units substituted at position 2, as originally suggested by Nicolaus and Piattelli (1962).

Attempts have been made to assess the relative proportions and state of oxidation of these structural units in both natural and synthetic melanins, but the results are far from conclusive. Ito (1986) has compared the yields of PTCA obtained from a number of natural and synthetic melanins prepared both enzymatically and autooxidatively under different conditions. The results, coupled with data of carboxyl content and other considerations, led him to conclude that units derived from DHICA comprise only 10% of synthetic melanins from dopa, but as much as one-half of intact natural eumelanins.

B. Isotopic Studies

1. *Melanoma Melanin*

Hempel (1966) attempted to investigate the structure of melanoma melanin by studying the incorporation of labeled precursors. He injected tritiated dopa into mice bearing Harding-Passey melanoma and by subsequent histoautoradiography of the melanoma tissues found that a considerable part of the radioactivity became incorporated over the area containing melanosomes. In further experiments, Hempel (1967) used *dl*-dopa labeled with tritium at a number of specific ring and side-chain positions, as indicated in Fig. 5.6.

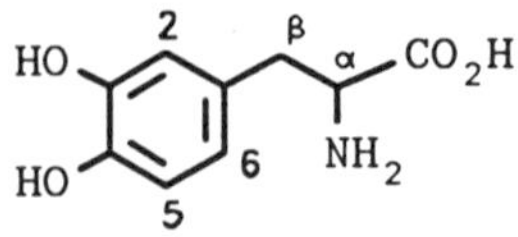

[α-T]dopa	[6-T]dopa
[β-T]dopa	[2,5,6-T_3]dopa
[α,β-T_2]dopa	[carboxyl-^{14}C]dopa
[2-T]dopa	[α-^{14}C]dopa
[5-T]dopa	

FIG 5.6 Dopas labeled with tritium or ^{14}C at specific ring and side-chain positions.

Each of these compounds was injected into melanoma-bearing mice together with [*dl*-α-^{14}C]dopa and the tritium and ^{14}C activities of the melanin pigment were measured by a scintillation counter. Given the ^{14}C activity as a measure of the number of dopa molecules which had been incorporated into the newly formed melanin, from the ratio of tritium to ^{14}C Hempel (1967) calculated the extent of incorporation of the particular hydrogen atom labeled by tritium in the precursor. The conclusion was that none of the specifically labeled hydrogen atoms was completely incorporated, nor were any completely excluded, so it seemed that all positions in the dopa molecule were involved in melanin formation, though to a varying extent.

Analysis of the incorporation values led Hempel to conclude that at least 49% of the melanin units retain the hydrogen at position 6 of dopa and that 71% of the units retain the carboxylic group of the original amino acid. From this it was inferred that the melanoma was a copolymer consisting of dopaquinone, indole-5,6-quinone, and indole-5,6-quinone-2-carboxylic acid in the approximate ratio 3:2:1, as tentatively represented in the partial structure in Fig. 5.7.

Though widely accepted in biomedical circles, Hempel's structure of melanoma melanin is chemically untenable from both the theoretical and the experimental viewpoints. First, it must be noted that the highly reactive 6 position of the dopaquinone units is unlikely to remain unsubstituted in the presence of the free nucleophilic amino group of the alanine chain. Moreover, the sequence, state of oxidation, and positions of attachment of the units in the proposed polymer chain are arbitrary, since they are not deducible from the incorporation studies. Additionally, consider the fact that the content of

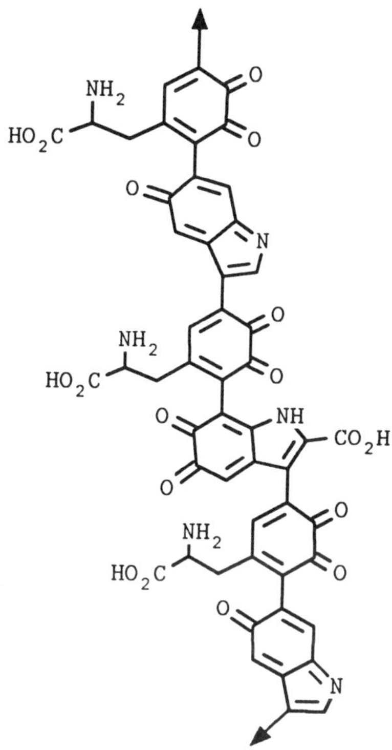

FIG 5.7 Model structure of melanoma melanin according to Hempel (1966).

amino groups of some melanins, including melanoma melanin, was determined by Bloch and Schaaf (1925) using the Van Slyke method, and from the values obtained it was concluded that only about 1% of the total nitrogen is in the form of NH_2. This is in accord with the results of a more recent study by X-ray photoelectron spectroscopy indicating that the proportion of nitrogen present as NH_2 in horse melanoma melanin does not exceed 7% of the total nitrogen content (Williams-Smith and Dunne, 1976). Finally, Hempel's conclusions are based entirely on the assumption that all the measured radioactivity originated from the melanin. However, the results can likewise be interpreted in terms of a nonmelanic pool of unchanged dopa or related metabolites, for example, cysteinyldopas, into the pigment preparation. For these reasons, and others concerning exchangeability of tritium from labeled quinones, Hempel's structure of melanoma melanin should now be discounted.

2. Dopamelanin

A considerable effort toward understanding the structure of the dopamelanin was made by Swan and his associates at Newcastle using ^{14}C- or deuterium-labeled dopa under both enzymic and autooxidative conditions. One of the outcomes of these studies, mentioned earlier, is the demonstration that part of the carboxyl groups in dopamelanin arise by oxidative fission of the indole unit by hydrogen peroxide. In the same series of experiments, Swan and Waggott (1970) found that dopamelanin(s) prepared both autooxidatively and enzymatically from [1-^{14}C]dopa retain(s) some radioactivity, indicating that approximately one in every five units contains a carboxyl group derived from the carboxyl group of dopa. Heating the melanins at 200°C in a vacuum resulted in decarboxylation of all carboxyl groups of the pigments except those pertaining to uncyclized amino acid side chains. From the residual radioactivity retained by decarboxylated melanins, about a 10% proportion of uncyclized dopa units was inferred. The total amount and specific activity of carbon dioxide evolved during decarboxylation indicated an estimated 15% pyrrolecarboxylic acid units. These results, coupled with analytical data of melanins and methylated melanins, led the authors to conclude that autooxidative dopamelanin is made up of indolic, pyrrolic, and uncyclized dopa units in the proportions shown in Fig. 5.8.

In another study by the Newcastle group (Binns *et al.*, 1970a) [*dl*-1-^{14}C]dopa was enzymically or autooxidatively converted into melanin. The resulting pigments were oxidized, to give pyrrolecarboxylic acids, and decarboxylated. In each case, PDCA was not radioactive, while PTCA was radioactive, consistent with the presence in the pigment of DHICA units. Swan and Waggott (1970) also found that the specific activity of PTCA was lower than would be expected if it was all derived by degradation of carboxylated indole units. This indicated the formation of inactive acid, apparently in the ratio 1:3:1 to the active PTCA, derived presumably by oxidation of noncarboxylated indole units linked through their 2 position to other units.

Further experiments (King *et al.*, 1970) were aimed at determining the relative involvement of the various positions of the monomer units in the polymerization and at assessing the proportion of indoline units. Samples of dopa specifically deuterated at each of the α and β positions of the alanyl side chain and at the 2, 4, and 6 positions of the benzene ring, and were converted to melanin both by autooxidation and enzymically, and the pigments thus obtained were analyzed for their deuterium retention.

Melanins prepared from α and β deuterated dopas were boiled with 2 *M* HCl for various periods, after which their deuterium contents were measured. Two kinetically distinct exchange processes apparently occurred, which were interpreted as due to an initial fast exchange at the expenses of deuterium atoms at

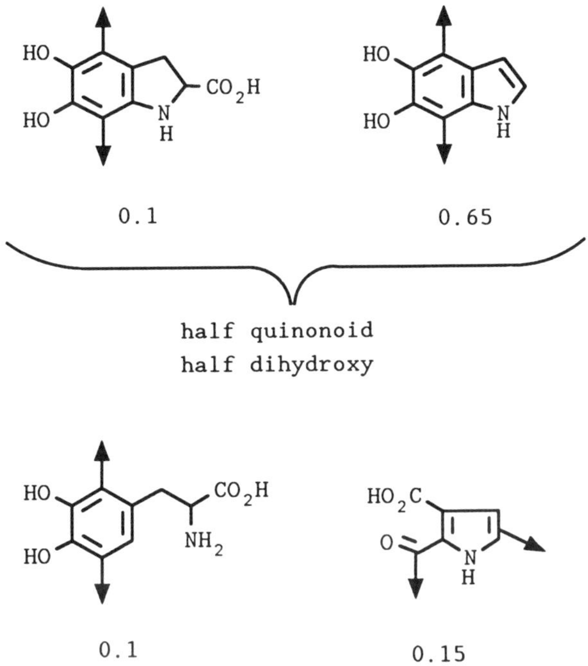

FIG 5.8 Structure and distribution of the main types of units present in dopamelanin (Swan and Waggott, 1970). Arrows indicate the position of attachment of the units in the pigment polymer.

the 2 and 3 positions of indolic and pyrrolic units, followed by a slow exchange involving indoline and dopa units. The deuterium retained after the first exchange process gave a combined amount of the latter units estimated at 37%.

To proceed further in the analysis of the data, Swan introduced the concept of average monomer elemental composition of dopamelanin, corresponding to $C_{7.9}H_{5.55}NO_{3.05}$, which was deduced from the above distribution of units, corrected for a fractional amount of water. On this ground, the amount of deuterium retained in the native melanin was used to assess the relative involvement of the ring positions 2, 3, 4, and 7 in DHI-derived units:

Position	2	3	4	7
Fraction	0.36	0.37	0 .34	0.28

Similar results were obtained by Kirby and Ogunkoya (1965), who investigated the structure of dopamelanin by a similar approach using tritium instead of deuterium. Melanin samples were prepared by enzymic oxidation of a mixture

TABLE 5.5

COMPARISON OF THE RESULTS OBTAINED ON AUTOOXIDATIVE MELANINS BY KIRBY AND OGUNKOYA WITH THOSE OF SWAN'S GROUP EXPRESSED AS FRACTIONAL RETENTION OF ISOTOPE

Position of label in dopa	α	β	2	5	6
Fractional retention of ^{3}H[a]	0.52	0.54	0.37	0.42	0.13
Fractional retention of ^{2}H[b]	0.49	0.41	0.31	0.37	0.10

[a]Kirby and Ogunkoya (1965).
[b]King *et al.* (1970).

of specifically tritiated *dl*-dopa with [*dl*-α-^{14}C]dopa, so that by measuring the tritium and ^{14}C activities separately the fraction of tritium incorporated into the pigment could be determined.

Table 5.5 shows a comparison of the data obtained from autooxidative melanin with those of the Newcastle group, expressed as fractional retention of deuterium. Considering that the conditions used by Kirby and Ogunkoya for the preparation of dopamelanin are different from those used by Swan, the results are surprisingly similar. More recently, Hearing *et al.* (1980) reported that there is no incorporation of noncyclized or carboxylated units (i.e., dopa, dopachrome, leucodopachrome, and DHICA) into dopamelanin obtained from ^{14}C- and ^{3}H-labeled tyrosine with either mushroom or mammalian tyrosinase. Moreover, using labeled dopa approximately 50% of the polymer consisted of carboxylated precursors, presumably as the result of the more rapid formation of melanin. However, a direct comparison of these results with those of Kirby and Ogunkoya is difficult due to the different procedures adopted for melanin preparation (e.g., 4 hr with air versus 24 hr with pure oxygen).

C. OXIDATIVE POLYMERIZATION OF 5,6-DIHYDROXYINDOLE(S)

Since the biosynthetic precursors are known, another way to inquire into the structure of melanins has been to investigate their reactivity and mode of interaction to form the pigment. This approach is intimately related to the biosynthesis of melanins, discussed in Chapter 8. However, because of the extent of the subject and its relevance for the structure of the eumelanin polymer(s), we choose to discuss here the longstanding issue of the mode of polymerization of the postulated ultimate precursors (DHI and DHICA).

1. *Model Studies*

Early studies in this area were largely speculative and were mainly concerned with the mechanism of polymerization of DHI, at that time regarded as the main reacting species. Given the known tendency of the indole system to undergo oxidation at position 3, in considering the structure of the melanin polymer Cohen (1945) indicated the possibility that the polymerizing species could be 5,6-dihydroxyindoxyl (5.**8**), which has a reactive methylene group and may therefore be capable of self-condensation. A slightly different view, of Clemo and Weiss (1945), was that DHI could couple at position 2 and then be oxidized at position 3, and that the resulting indigoid structure 5.**9** would then undergo further polymerization. This was soon disproved by Harley-Mason (1950), who prepared the postulated intermediate and found that it does not yield melanin.

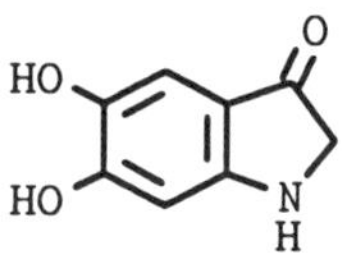

5.**8**

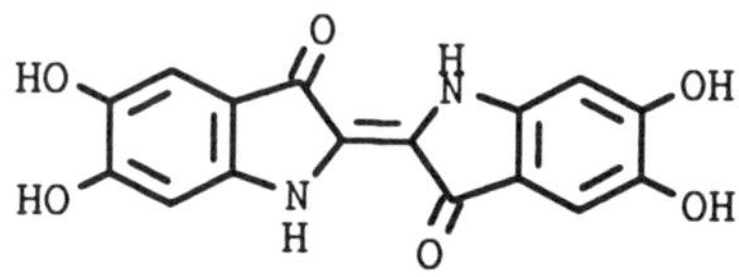

5.**9**

In an extension of this study, Bu'Lock and Harley-Mason (1951a) considered the possibility that polymerization of DHI proceeded via self-condensation of the labile 5,6-indolequinone which could possibly behave as a quinone and as an indole, the anionoid center in the pyrrole ring of one molecule linking up to a cationoid position of another. On this assumption, they performed model experiments to show that *o*-benzoquinone readily reacts with indole to form the intensely blue 3-indolylquinone (5.**10**). If position 3 is blocked, the reaction takes a quite different course, proceeding with difficulty to give colorless products, if any. It was then suggested that melanin formation could proceed by repeated condensation of 5,6-indolequinone between position 3 of one molecule and position 4 or 7 of another. As the product obtained by 3,4-linkage could not be planar and, hence, by analogy with the simple indole–quinone reactions would not form readily, a 3,7-linkage was considered more likely, giving a coplanar structure (5.**11**). The extensive conjugation was thought to account for the general light absorption of the polymer. To explain

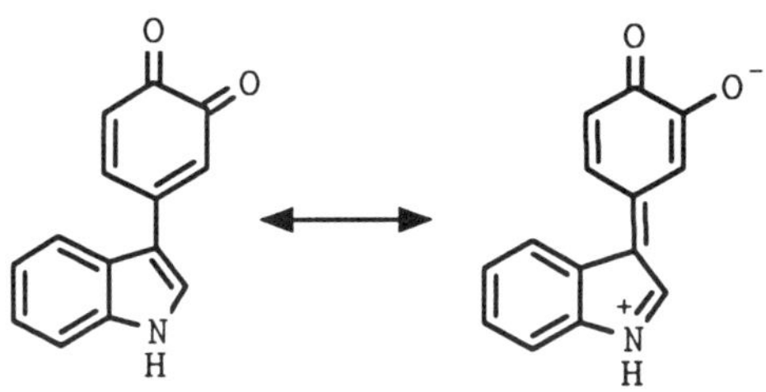

5.**10**

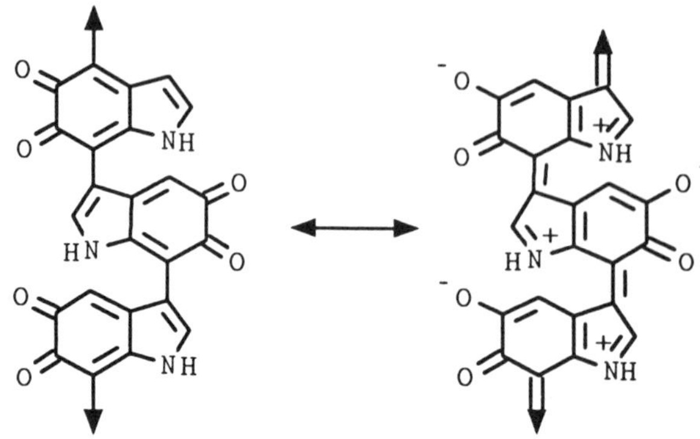

5.11

the extreme insolubility of melanin, it was further suggested that occasional condensation might occur at position 2 with position 4 or 7 of another molecule, the cross-links leading to the formation of a highly irregular, three-dimensional polymer.

Further studies at Cambridge (Cromartie and Harley-Mason, 1953, 1957) and other centers (e.g., Liverpool; see Beer *et al.*, 1954) were aimed at comparing the tendency of a series of 5,6-dihydroxyindoles methylated at various positions to give rise to melanin on autooxidation. The results, summarized in Table 5.6, showed that 1-, 2-, 4- and 7-methyl-5,6-dihydroxyindole, like DHI, give black solids, insoluble in the common organic solvents but slightly soluble in pyridine. On the other hand, 3-methyl- and 4,7-dimethyl-5;6-dihydroxyindole give highly colored solids soluble in ethanol. After considering all the above experiments, it was concluded that the 3 position is

TABLE 5.6

AUTOOXIDATION OF METHYLATED 5,6-DIHYDROXYINDOLES IN AQUEOUS BUFFER, pH 6.85[a]

Positions of methyl substituents	Appearance of precipitate	Solubility of precipitate
None	Black, flocculent	Insoluble in pyridine
1	Black, fine	Dispersed in hot pyridine
2	Black, fine	Insoluble in pyridine
3	Blue, flocculent	Soluble in ethanol and alkali
2,3	None	None
4	Black, fine	Dispersed in pyridine
7	Black, fine	Readily dispersed in pyridine
4,7	Purple, fine	Extracted by ethyl acetate

[a]Cromartie and Harley-Mason (1957).

essential for the formation of a true melanin, together with a 4 or 7 position free, though all positions (2, 3, 4, and 7) are probably concerned to some extent in the building up of a three-dimensional polymer.

These conclusions, however, are questionable as they are based on the assumption that the pigmented material obtained from the variously substituted indoles is similar to that formed from DHI. And moreover, the relative ability of the various substituted indoles to form a "melanin" was mainly studied under conditions which are far from being biomimetic.

2. The Structure of Melanochrome

In a careful spectrophotometric study of the enzymatic oxidation of dopa to melanin, Mason (1948) reported the occurrence of a transient chromophoric phase starting after dopachrome had begun to rearrange and preceding the formation of the final black pigment. This chromophoric phase, characterized by a broad absorption at 540 nm, was apparently due to the generation of a purple intermediate pigment, referred to as "melanochrome." In parallel experiments, the same chromophore was also obtained by oxidation of DHI with tyrosinase. This led Mason to suggest that melanochrome could be the monomeric indole-5,6-quinone, but later Beer *et al.* (1954) pointed out that the light absorption of melanochrome was similar to that of some model indolylbenzoquinones described by Bu'Lock and Harley-Mason (1951a). In a subsequent study, Bu'Lock (1960) reported that in the spectrum of the purple solution formed by enzymic oxidation of DHI at pH 6.8, the initial maximum at 530 nm may represent a dimer, but this soon flattened out to give a maximum at 540 nm, perhaps corresponding to a mixture of oligomers.

In a reexamination of the early studies at Naples, it was found that oxidative polymerization of DHI is markedly susceptible to metal catalysis. An improved procedure was then developed to obtain sufficient amounts of melanochrome for chemical characterization. This involves aerial oxidation of DHI in Hepes buffer at pH 6.8 containing $Zn(OAc)_2$ or $NiSO_4$ (Napolitano *et al.,* 1985b). When melanochrome had attained the maximum concentration, the oxidation was stopped with sodium dithionite, and the ethyl acetate-extractable fraction was acetylated with acetic anhydride/pyridine. The product thus obtained was found to consist of a mixture of oligomers of DHI, the major one of which could be isolated in crystalline form and was identified as the 2,2′-biindolyl (5.**12**) on chemical and spectral grounds. Mild alkaline hydrolysis under anaerobic conditions gave the parent tetrahydroxyderivative 5.**13** (λ_{max} 369 and 350 nm), which rapidly autooxidized to a purple-blue pigment exhibiting the same chromophore of the original oxidation mixture of DHI. On this basis, it seems likely that the purple pigment is a quinone derived from 5.**13.** Unfortunately, the instability and high degree of insolubility of melanochrome have so far prevented any direct analysis aimed at establishing the exact structure of the chromophore.

When the oxidation of DHI was performed with tyrosinase under Mason's conditions of melanogenesis *in vitro,* a work-up of the reaction mixture at the melanochrome stage gave a more complex pattern of products, characterized by the presence of a relatively major component (5–10% yield) identified as the 2,4′-biindolyl (5.**14**). Notably, the symmetrical dimer 5.**12** was not formed. It seems therefore that metal ions affect not only the kinetics of DHI oxidation but also the mode of dimerization, favoring coupling at the 2 and 2′ positions.

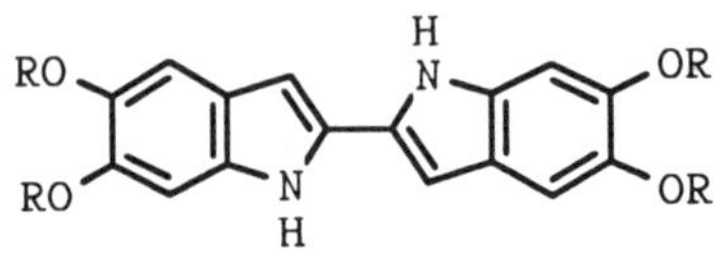

5.**12** (R = Ac)
5.**13** (R = H)

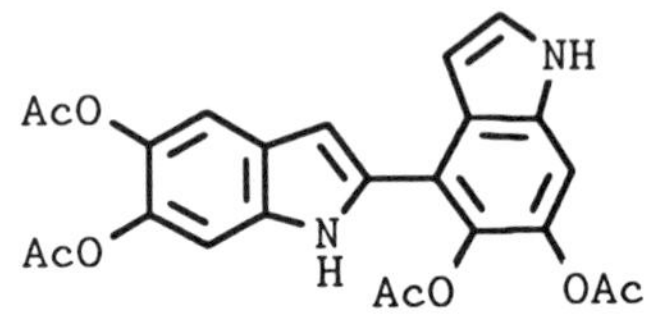

5.**14**

Similar results were obtained using as substrate the 1-methyl derivative, readily available in large quantity by oxidative cyclization of epinephrine. On oxidation and subsequent acetylation, this gave, besides dimers 5.**15** and 5.**16,** a small amount of the related trimer 5.**17** (Corradini *et al.,* 1986).

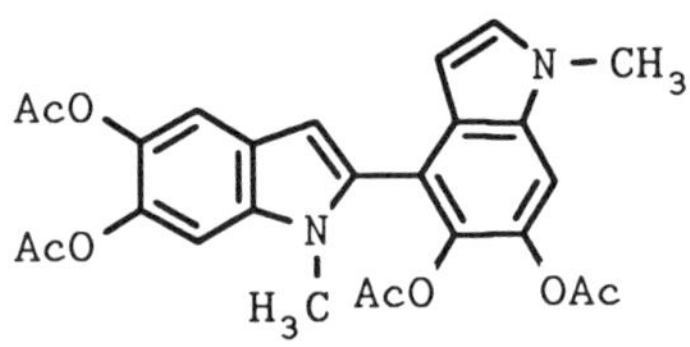

5.**15**

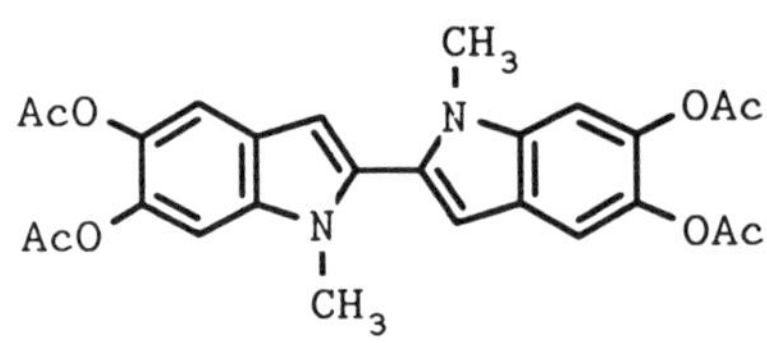

5.**16**

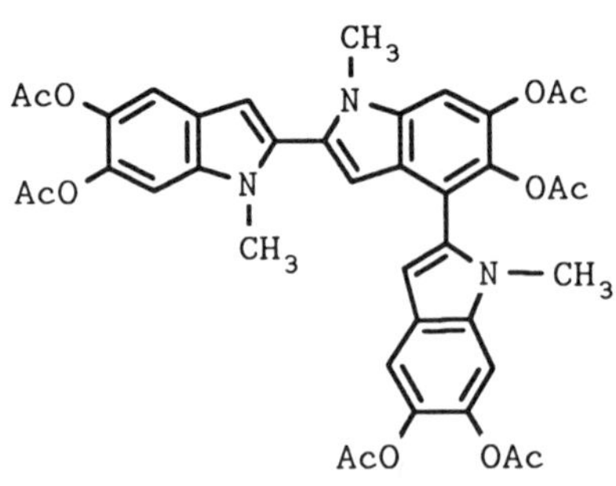

5.**17**

The mechanism of the polymerization of 5,6-dihydroxyindole(s) has not yet been adequately characterized; the reaction is extremely rapid and the intermediates are fugitive, although some have been detected by time-resolved pulse radiolysis (Land, 1988). Available evidence suggests that in the tyrosinase catalyzed oxidation the labile indole-5,6-quinone is formed, as first indicated by Raper (1928). Subsequent addition of the nucleophilic 2 position of DHI to the electron-deficient 4 position of 5,6-indolequinone would then afford the 2,4′-dimer (Fig. 5.9). On the other hand, formation of the symmetrical dimer in the metal-catalyzed oxidation can be accounted for in terms of self-coupling of C-2 centered phenoxy radicals. Radical delocalization at C-2 requires that the C-6 hydroxyl group be specifically involved in the electron transfer process, and this is reasonable considering that the C-6 OH is more acidic than the C-5 OH due to the greater resonance stabilization of the anion.

In more recent studies at Naples, it was found that the peroxidase/hydrogen peroxide system is far more effective than tyrosinase in inducing a fast polymerization of DHI (d'Ischia *et al.,* 1990, 1991). If oxidation is stopped at the very beginning (i.e., 1–3 min), a well-defined pattern of products is obtained, consisting mainly of the 2,4′-dimer and three new oligomers. These were isolated by preparative HPLC and identified as the 2,7′-dimer 5.**18** and the trimers 5.**19** and 5.**20** by careful analysis of their NMR spectra. The estimated yields for the dimers were 20–25% each and 3–5% for the isomeric trimers.

From consideration of the structure of the oligomers identified, including those obtained by tyrosinase or metal-catalyzed oxidation, that is, 5.**12,** 5.**14** and 5.**18** through 5.**20,** it appears that the dominant mode of coupling of DHI is via the 2 and 4 positions of the indole ring, with a minor contribution of position 7. The relevance of these results for the structure of eumelanins is clear, especially for what concerns the alleged involvement of position 3 in the building up of the pigment polymer(s). We will return to this point in the concluding section of this chapter.

As a part of their studies on the chemistry of the melanogenic intermediates, Truscott, Land, and associates at Manchester (Lambert *et al.,* 1989) have examined the oxidation of DHI using pulse radiolysis. The object was to consider the possible ways in which indole-5,6-quinone might undergo poly-

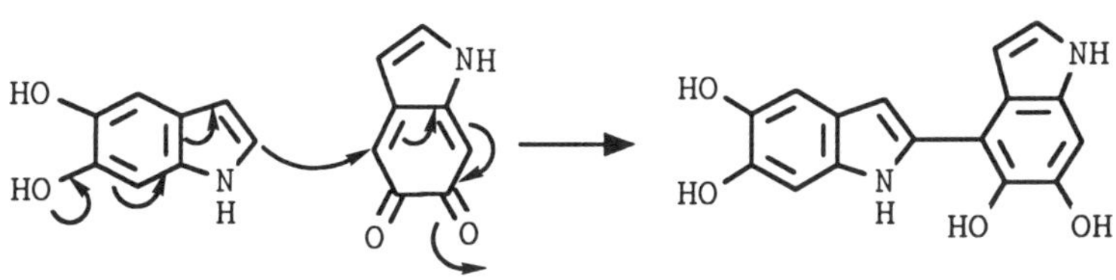

FIG 5.9 Mechanism of formation of the 2,4′-dimer of DHI.

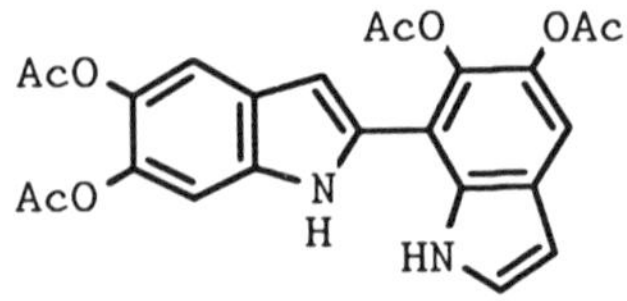

5.**18**

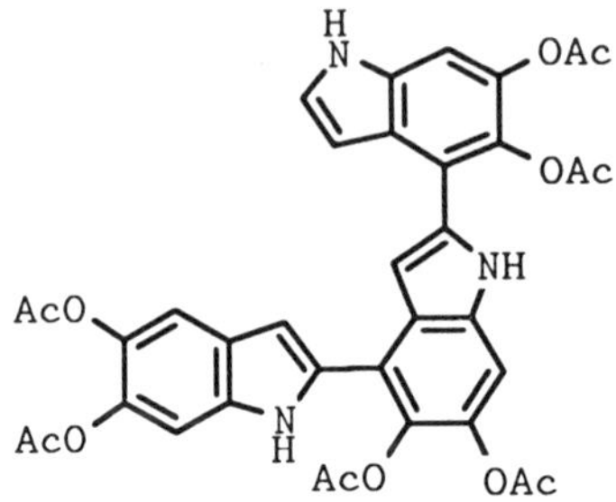

5.**19**

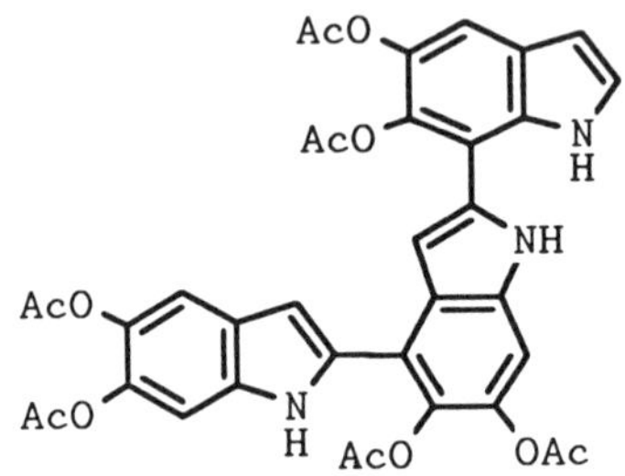

5.**20**

merization. They found that the rate of decay of the transiently generated 5,6-indolequinone, as monitored at 420 nm, is halved when the reaction is carried out in D_2O instead of H_2O. This and other observations were taken to show that the polymerizing species is probably a trihydroxyindole derived by addition of water to 5,6-indolequinone or its tautomers, as illustrated in Fig. 5.10. It will be interesting to see how the proposed pathway fares as research on melanogenesis continues.

3. Oxidative Polymerization of DHICA

The oxidation chemistry of this indole is relatively simpler than that of DHI, because the reactivity of the pyrrole ring is reduced by the presence of the carboxyl group. Accordingly, polymerization of DHICA must inevitably involve both position 4 and position 7. Experimental evidence in support of this view was first obtained by Fattorusso and Cimino (1971) from experiments in which silver oxide was used to generate the quinone of DHICA. Some years later, Ito and Nicol (1974) reported the occurrence of some of the postulated DHICA oligomers, for example, 5.**21,** in the tapetum lucidum of catfish, and showed that a similar pattern of oligomers is formed when DHICA is oxidized with mushroom tyrosinase in the presence of a catalytic amount of dopa. However, none of these oligomers could be adequately characterized due to the adverse chromatographic properties of the materials.

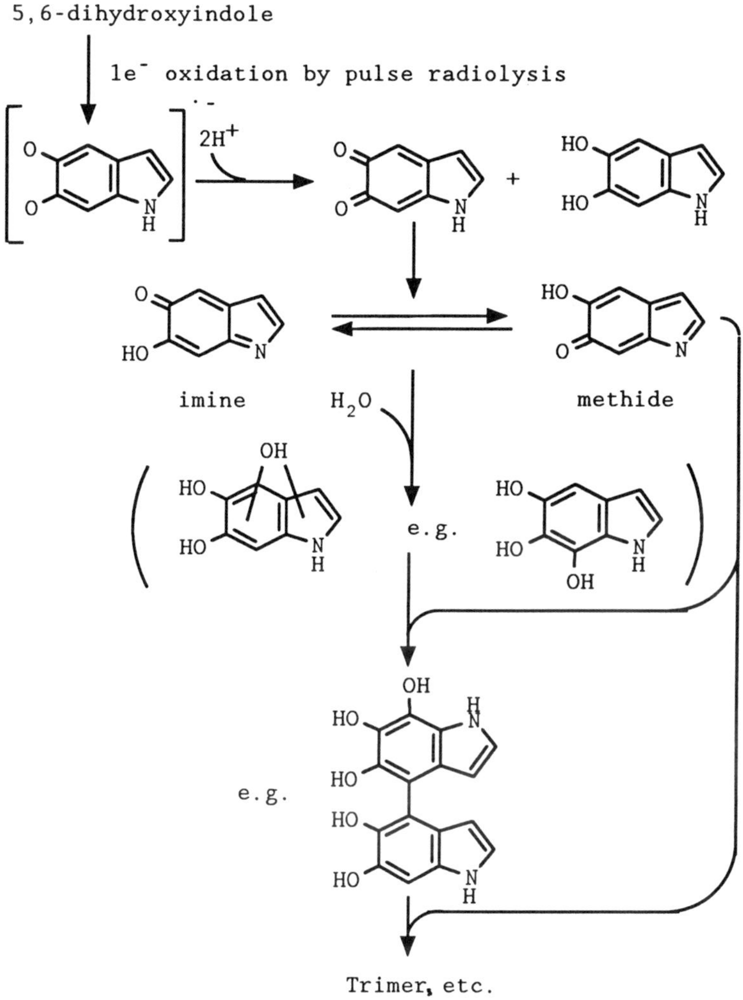

FIG 5.10 Proposed scheme of polymerization of DHI (Lamber *et al.*, 1989).

Recently, the tyrosinase oxidation of DHICA has been reexamined using an approach basically similar to that developed for the investigation of melanochrome (Palumbo *et al.*, 1987b,c). HPLC analysis of the ethyl acetate-extractable mixture after esterification and acetylation gave an ill-defined pattern of products, three of which, present in somewhat greater amount, could be identified as the symmetrical dimer 5.**22,** the 4.7′-dimer 5.**23,** and the trimer 5.**24.**

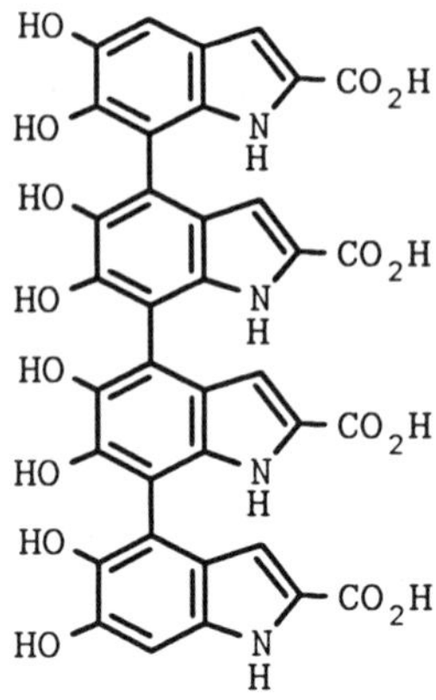

5.**21**

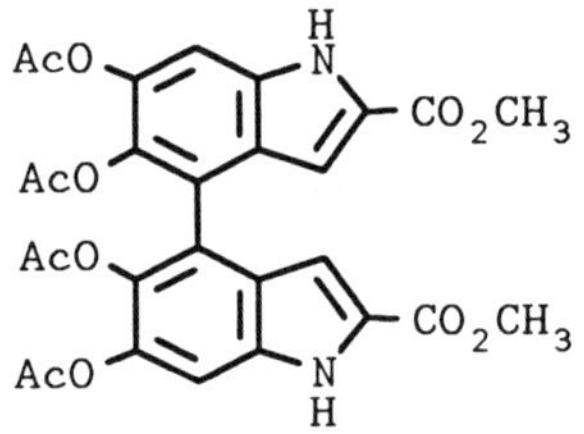

5.**22**

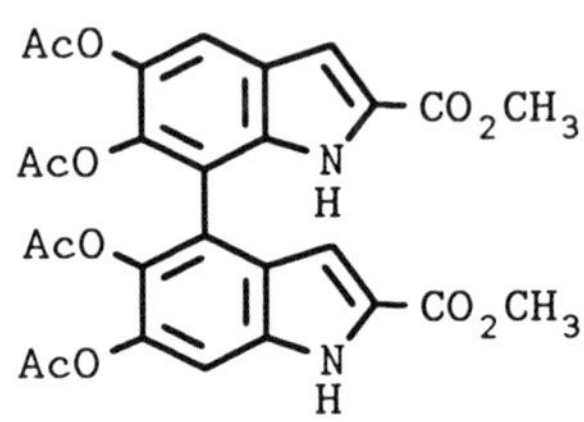

5.**23**

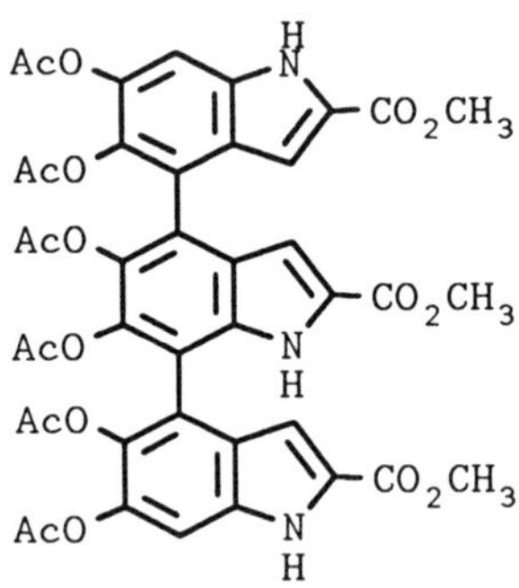

5.**24**

As we shall see later, DHICA is a very poor substrate of tyrosinase, so that polymerization of the indole under the above-mentioned conditions is in fact brought about by dopaquinone formed by enzymic oxidation of the dopa added. In view of these and other limitations, including the exceedingly low yields of oligomers 5.**22** through 5.**24** (less than 3%), oxidation of DHICA was also performed with the peroxidase/hydrogen peroxide system (d'Ischia *et al.*, 1991). Work-up of the reaction mixture as mentioned showed a more distinct pattern of oligomers, consisting mainly of the 4,7′-dimer and the related trimer 5.**24,** in about 10 and 25% yields, respectively.

The elucidation of the structure of the oligomers 5.**22** through 5.**24** reveals a marked tendency of DHICA to polymerize through the 4 and 7 positions, the former being involved to a greater extent. The limited number of reactive sites available on the indole ring prevents a clear cut distinction of possible mechanisms based solely on considerations of positional reactivity. It is likely, however, that the polymerization of DHICA proceeds through indole–indolequinone couplings similar to those envisaged for DHI.

IV. CONCLUSIONS ABOUT THE STRUCTURE OF EUMELANINS

During the 1960s two basically different models of eumelanin structure were proposed and heartily debated: Mason's "homopolymer" view, envisaging a highly regular polymer made up exclusively of DHI units sequentially linked through the 3 and 7 positions, and Nicolaus' "poikilopolymer" concept, supported by Swan and other investigators, which pointed to a highly heterogeneous macromolecule consisting of different units linked fairly randomly. In fact, in retrospect, it is apparent that the question was not properly placed. The homopolymer view developed entirely from the study of the mechanism of formation of a synthetic dopamelanin prepared under experimentally controlled conditions (pH, temperature, oxygen supply, reaction time). On the other hand, the poikilopolymer concept emerged mainly from direct investigations of natural eumelanins or synthetic pigments prepared under acid conditions, which led to significant alteration of the pigment backbone.

At that time, the difference between natural and synthetic melanins, as well as between various synthetic melanins, was not fully appreciated, so that the experimental data were compared regardless of their origin. As we know now (see also the discussion of dopachrome rearrangement in Chapter 8), a dopamelanin prepared enzymatically under conditions of sufficient dilution is indeed made up mainly of DHI, as held by Mason, but, in contrast with the original homopolymer view, the DHI units can be linked through either 2,4 or 2,7 bonds, which accounts for the apparent heterogeneity of the pigment. Much less is known about sepiomelanin and other natural melanins, though an increasing body of evidence would point to a significant participation of DHICA in the build-up of the pigment polymer. In the case of both natural and synthetic pigments, the marked susceptibility to autooxidation further adds to the structural complexity, inasmuch as pyrrole carboxylic acids and other residues can be formed randomly at various sites of the pigment backbone as a result of postsynthetic peroxidative breakdown of indole units. It seems that this has led supporters of the poikilopolymer view to somewhat overemphasize the heterogeneous nature of eumelanins as arising from copolymerization of several melanogenic precursors. While dopachrome and leucodopachrome can be ruled out merely on the basis of their reactivity, there is now fresh evidence (G. Prota, unpublished) that DHI and DHICA, when allowed to co-oxidize under suitable biomimetic conditions, give mainly mixtures of the corresponding homopolymers, in keeping with their markedly different redox potential. Thus, eumelanins would show now more as mixtures of oligomers of DHI and/or DHICA, partially oxidized, rather than as intimate copolymers of various precursors.

In conclusion, though many structural details of eumelanins are uncertain, new evidence is rapidly accumulating that is expected to lead to a fairly realistic picture of such unfathomable materials. Verification of this preview is expected in the very near future.

CHAPTER 6

NEUROMELANIN

I. INTRODUCTION

Neuromelanin is a granular, dark brown pigment which is typical of the nigrostrial neurons in the brain stem of humans. Since the early report by d'Azur in 1786, interest in this pigment has continued unabated, especially in view of its possible role in Parkinson's disease. Though the structure of neuromelanin is not known, there are several lines of evidence which suggest that this pigment is chemically different from the melanins formed in epidermal melanocytes. This fact, coupled with the intrinsic importance of the subject, warrants our devotion of this chapter to neuromelanin.

II. OCCURRENCE AND DISTRIBUTION

A comparative survey of the occurrence of neuromelanin in mammals has been reported by Marsden (1961) to involve 49 species representing 11 orders. Though the validity of this study has been questioned by Barden (1981) on the basis of a supposedly incorrect use of the Masson–Fontana staining technique, there seems to be sufficient evidence of the presence of neuromelanin in, besides man, *Hominoidea, Cercopithecoidea, Ceboidea*, and a member of *Lemuroidea*. In general, the amount of neuromelanin in the brain varies in the different animal species, being greatest in man and progressively lower as the relation to man becomes lower, for example, in carnivores and horses. Laboratory animals such as rats, mice, guinea pigs, and rabbits have little or no neuromelanin. Amphibians have been reported to have melanin-containing nerve cells, which may functionally correspond to those in mammals (Kemali and Gioffré, 1985). Several studies of the distribution of neuromelanin in adult human brain, carried out in various laboratories, have shown that the pigment occurs almost exclusively in catecholaminergic neurons (Fig. 6.1), which are involved in conscious perception, movement, emotions, and memory (Bogerts, 1981; Cowen, 1986).

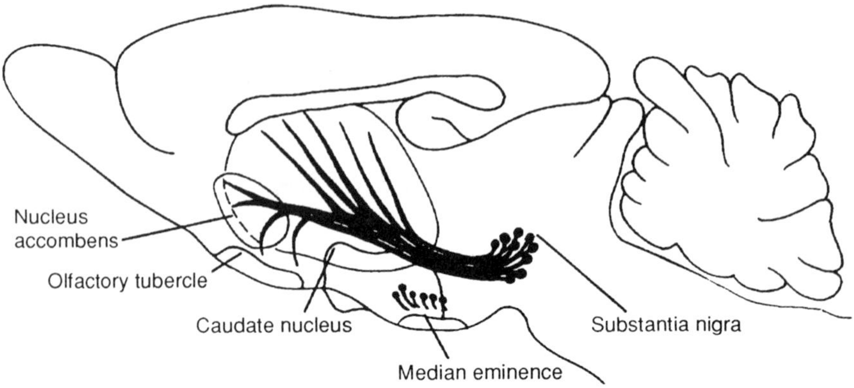

FIG 6.1 Schematic drawing of the brain highlighting dopamine tracts and the neuromelanin-containing *substantia nigra*.

In heavily pigmented neurons, neuromelanin granules are closely packed, dispersed through the cytoplasm, and extend into the axon hillock and initial portions of the axon. Neuromelanin is less abundant in the dendrites and is absent in the neuronal processes remote from the cell bodies, the synaptic nerve endings, and the glia cells. The highest levels of neuromelanin are found in the neurons of the *substantia nigra* and *locus coeruleus*, which are known to contain relatively high concentrations of dopamine and norepinephrine, respectively.

III. ONTOGENESIS AND AGE DECLINE

There is some disagreement as to when neuromelanin makes its appearance in the human brain. The pigment has been described in the *substantia nigra* as early as midterm gestation through 4 years of age, while in the *locus coeruleus* it appears between 5 months gestation and 3 years of age. In the dorsal nucleus of the vagus it appears at 34 weeks gestation (Barden, 1981). During life there is a gradual accumulation of neuromelanin in the dopaminergic neurons with a maximum around 60 years, followed by a decline in the senium (Mann and Yates, 1974). The concentration of neuromelanin around the late eighties is about one-half to two-thirds that of the 60-year level (Fig. 6.2). A similar change in neuromelanin concentration with age has been observed in the brains of rats by Schally and co-workers (Kastin *et al.*, 1979). In general, the heavily pigmented neurons are preferentially lost, suggesting a close link between synthesis or accumulation of neuromelanin and cell degeneration.

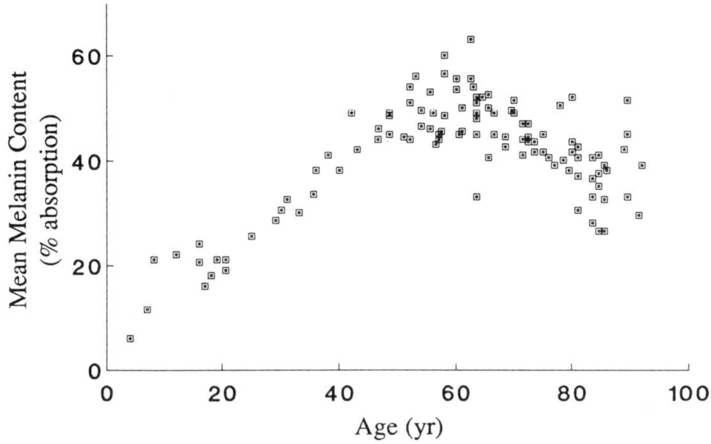

FIG 6.2 The melanin content of nerve cells from the human *substantia nigra* in 110 control subjects between ages 18 months and 91 yr. After Mann and Yates (1974).

The reasons for the decrease in neuromelanin with age are not clear, though it is evidently dependent on a selective loss of dopaminergic neurons, as in Parkinson's disease. Neuromelanin seems to be very stable and is not normally secreted from the cells where it is formed. It may only be released from degenerating nerve cells and phagocytosed by glial cells (Marsden, 1969). Mann and Yates (1977) showed that accumulation of neuromelanin is accompanied first by a reduction in cytoplasmic RNA content and finally by a decrease in nuclear volume in *substantia nigra* neurons. These authors proposed a mechanical disruption of the microanatomy of intracellular membranes by neuromelanin as a mechanism of degeneration of the pigmented dopaminergic cells. Graham (1979) postulated that synthesis of neuromelanin, rather than accumulating, has deleterious consequences for the cells. This view would be in agreement with findings showing that dopamine oxidation occurs more rapidly and the *o*-quinone formed cyclizes more slowly compared to other catecholamines and therefore is highly toxic to the well-being of the dopaminergic cells (Graham, 1978; Cohen, 1983).

IV. MORPHOLOGY AND HISTOCHEMISTRY

Ultrastructural and histochemical studies of neuromelanin granules in the *substantia nigra* and *locus coeruleus* of the human brain indicate similarity to the granules of the brown age pigment lipofuscin with respect to skin and hair melanosomes. Neuromelanin intraneuronal granules appear to be membrane bound with a lobular configuration and a gross tripartite substructure, composed of vacuolar bodies protruding from a granular or linear matrix on which a very dense particulate material is irregularly deposited. Apparently, lipofuscin granules exhibit a dipartite structure composed of a matrix containing vocuolar bodies but lack the dense particulate component typical of neuromelanin (Barden, 1981).

Both lipofuscin and neuromelanin are positive for acid hydrolase activity and are thought to be lysosomal residual bodies containing end products of oxidative damage of lipids (Barden and Brizzee, 1987). Such a common biochemical feature is supported by the observation that neuromelanin bleached with 10% hydrogen peroxide becomes fluorescent under UV light just like lipofuscin. Moreover, melanin-free lipofuscin from the inferior olive can be melanized if impregnated with ferrous sulfide and incubated with L-dopa. These observations led Barden (1969) to conclude that lipofuscin has the properties of bleached neuromelanin and that neuromelanin is melanized lipofuscin. Histochemical experiments, reviewed by Barden (1981), showed that the tinctorial characteristics of neuromelanin reflect fairly well the tripartite structure of the pigment granule as apparent from electron microscopy. Thus, neuromelanin is histochemically divisible into three components: (1) the melanin component, detectable by ferrous and cupric ion uptake reactions, a green staining with Nile blue sulfate, and the bleaching reaction; (2) the lipoproteinlike lipofuscin component, corresponding to the matrix, characterized by several staining reactions all duplicated in inferior olive lipofuscin; and (3) the unbound lipid lipofuscin component, corresponding to the vacuolar component, characterized by staining with naphthol sudans, Sudan III and IV and Oil red 0.

V. BIOSYNTHESIS

A. Precursors

The apparent association of neuromelanin with the catecholamine tracts of the brain is generally regarded as indirect proof of the involvement of dopamine and related metabolites in the biosynthesis of the pigment. Under normal circumstances, this oxidative pathway appears to be only a minor part of the overall metabolism of catecholamines, which involves mainly oxidative deamination by monoamine oxidase (MAO) and *O*-methylation by catechol *O*-methyl transferase (COMT).

A major question that had to be resolved by the first investigators was which of the two main catecholamines in the brain, namely dopamine and norepi-

TABLE 6.1
DOPAMINE AND 5-S-CYSTEINYLDOPAMINE IN HUMAN BRAIN[a]

	Dopamine[b]		5-S-cysteinyldopamine[b]	
	Mean	(Range)	Mean	(Range)
Caudate nucleus	1.2	(0.75–1.6)	0.05	(0.03–0.09)
Putamen	1.3	(0.99–1.6)	0.06	(0.02–0.09)
Globus pallidus	0.74	(0.58–0.95)	0.01	(0.008–0.016)
Substantia nigra	0.31	(0.11–0.46)	0.02	(0.007–0.024)
Cerebellum	0.06	(0.02–0.13)	0	
Occipital cortex	0.13	(0.08–0.20)	0	

[a]From Rosengren *et al.* (1985).
[b]μg/g fresh weight

nephrine, is the actual precursor of neuromelanin. Some useful information was obtained by Das *et al.* (1978), who compared the absorption spectra of various melanin preparations before and after solubilization with sodium borohydride. They found that the spectroscopic characteristics of neuromelanin from human *substantia nigra* matched fairly well with those of synthetic dopamine melanin, but not with those of dopamelanin or other catecholamine melanins. Likewise, Bridelli *et al.* (1982) reported that the IR spectrum of neuromelanin has features similar to those of synthetic dopamine melanin.

More recently, Rosengren *et al.* (1985) have reported the occurrence of a new dopamine metabolite, 5-cystein-S-yldopamine (5-cysdopamine), in several dopaminergic regions of the brain (Table 6.1). The identification was secured by direct comparison with a synthetic sample prepared by addition of cysteine to dopaminequinone, as outlined in Fig. 6.3.

This finding, coupled with the high levels of sulfur in neuromelanin (Barden and Martin, 1972), would support the hypothesis that dopamine is oxidized to dopaminequinone which is in part converted to cysteinyldopamine, acting as an additional pigment precursor (see Section VI).

B. OXIDATIVE SYSTEMS

Very little is currently known regarding the nature of the enzymatic system(s) involved in the oxidative process of neuromelanin formation. The presence of tyrosinase in melanin-producing tissues led Marsden (1965) to invoke a role of this enzyme in neuromelanogenesis in human *substantia nigra*. Support for this view was provided by *in vitro* experiments showing that cells of the

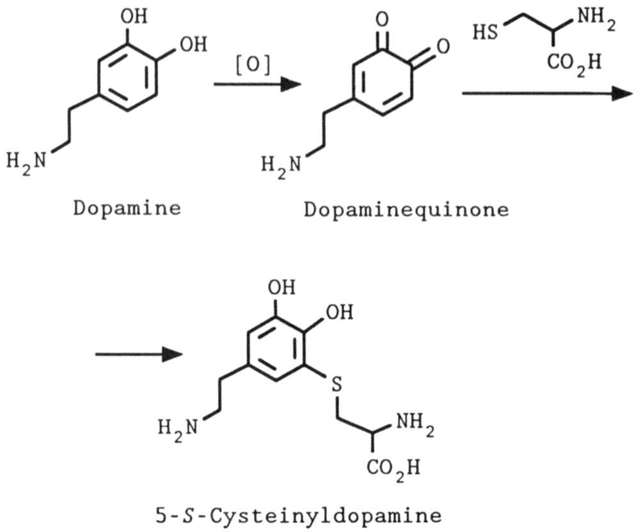

FIG 6.3 Formation of cysteinyldopamine by addition of cysteine to enzymatically generated dopaminequinone.

substantia nigra of adult monkey and cat are blackened on incubation with L-dopa. Histochemical studies by various other workers, however, were unable to confirm tyrosinase in the brain, and all biochemical estimations of this enzyme were unequivocally negative (Barden, 1981). The normal pigmentation of the *substantia nigra* in human albinos (Foley and Baxter, 1958), who lack all peripheral tyrosinase, is also consistent with tyrosinase not being involved in neuromelanin formation.

Another enzyme which has been implicated is peroxidase, which *in vitro* rapidly converts dopamine into a black melanin pigment (Okun *et al.*, 1972). Moreover, in one isolated case of unilateral Parkinsonism a 20-fold decrease in peroxidase activity was observed in affected basal ganglia. Histochemical and biochemical detection of peroxidase in cells of the dopaminergic system have also been reported (Ambani *et al.*, 1975); nonetheless, its role in neuromelanin synthesis has been questioned by Rodgers and Curzon (1975) and, more recently, by Rabey and Hefti (1990), who found that addition of hydrogen peroxide to *substantia nigra* homogenates produces modest increases in melanin formation. Likewise, addition of catalase, which destroys endogenous hydrogen peroxide, failed to significantly inhibit melanin formation below values measured in control homogenates.

The role of MAO appears more relevant to neuromelanogenesis. At a time when tyrosinase was ruled out as the enzyme responsible for pigment production in the brain, Cote (cited in Marsden, 1969) found an apparently similar melanin-forming enzyme activity in the *substantia nigra* of *Rhesus* monkey and identified it as MAO by inhibition experiments. This view received alternate consideration until recently, when Rabey and Hefti (1990) reported new evidence that MAO, coupled with nonenzymatic oxidative processes, is mainly responsible for neuromelanin formation. It is clear, however, that the definitive assessment of these results awaits further experimental support. In this context, more attention should perhaps be placed on autooxidative processes catalyzed by redox metal ions, especially iron, which is known to accumulate in the *substantia nigra* (Youdim *et al.*, 1989).

VI. ISOLATION AND STRUCTURAL INVESTIGATIONS

At present little is definitely known about the structure of neuromelanin: Even the elemental composition is uncertain. This is largely due to the relative inaccessibility of the *substantia nigra*, the exceedingly low amounts of neuromelanin present, and the difficulties connected with liberating the pigment granules from the surrounding lipid matrix in which they are deeply embedded. In early studies (e.g., Van Woert *et al.*, 1967; Das *et al.*, 1978) a complex purification procedure was used involving prolonged proteolytic digestion with trypsin or pronase, treatment with lysing solutions, extensive extractions with organic solvents, and in some cases drastic acid hydrolysis with 6 *M* HCl. Spectroscopic studies of the pigment preparations thus obtained did not provide evidence for any distinguishing structural feature except a vague similarity of the absorption spectrum to that of dopamine melanin (Fig. 6.4).

Recently, an improved procedure for isolating human *substantia nigra* neuromelanin has been developed by Rorsman, Rosengren, and co-workers (Carstam *et al.*, 1991). This involves extraction of lipids and proteins with 2% sodium cholate in 30% ethanol followed by 2% SDS in 10% glycerol. Reductive hydrolysis of the pigment thus obtained with hydroiodic acid and real phosphorous yielded 4-amino-3-hydroxyphenylethylamine in large quantities along with trace amounts of 4-amino-3-hydroxypenylalanine (AHP), indicating that 5-cysdopamine and 5-cysdopa, to a lesser extent, both concur to form the polymer. Dopamine and small quantities of dopa were also obtained by hydrolysis, whereas permanganate oxidation afforded some PTCA. These data would convey an image of neuromelanin as a "mixed-type" melanin derived from dopamine, in which pheomelanic- and eumelanic-type components coexist, probably as a result of an exhaustion of the GSH–cysteine defense system of the *substantia nigra* neurons.

Ongoing work in other laboratories, however, does not seem to provide confirming evidence for this structural view. Wakamatsu *et al.* (1991) failed to

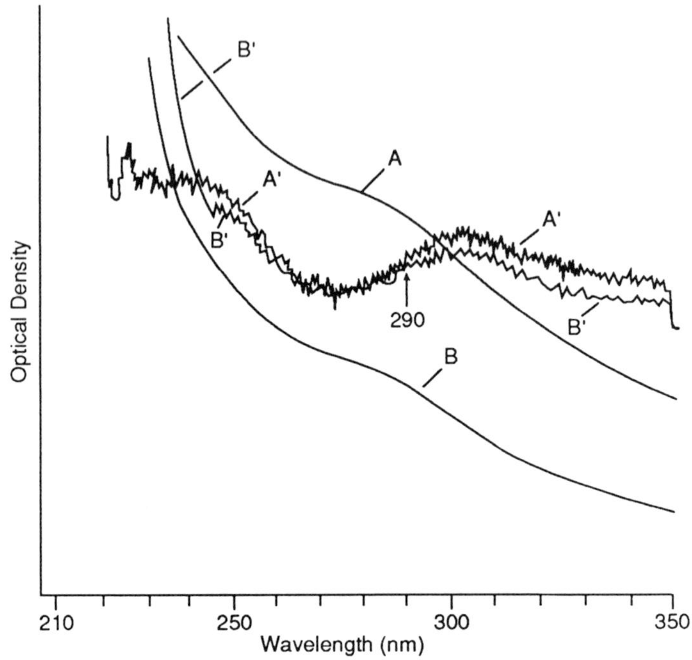

FIG 6.4 Absorption spectra of dopamine melanin (A) and *substantia nigra* melanin (B) solubilized in sodium borohydride. A′ and B′ represent the first derivative spectra of A and B respectively.

detect cysdopamine-derived units in human *substantia nigra* melanin. On the other hand, Zecca *et al.* (1992) reported analytical and spectroscopic data suggesting that 5-cysdopa or 5-cysdopamine linked to a palmityl residue are the most abundant constituents of human neuromelanin. It is therefore clear that the current picture of neuromelanin structure is still very uncertain and susceptible to profound modifications in the near future.

VII. MODEL STUDIES

A. Dopamine Melanin

Most of what is currently known about the structure of dopamine melanin is due to the intensive studies of Swan (1974) on the pigment obtained by autooxidation of dopamine. This was found to be soluble in dilute akali, and when the resulting solution was shaken with benzoyl chloride a brownish product was obtained which was partly soluble in chloroform. By using ^{14}COCl-benzoyl chloride, the product was shown to contain about one benzoyl group per 8-carbon unit in the polymer. Moreover, the presence of two carbonyl stretching bands in the IR spectrum (1746 cm^{-1}, strong, and 1646 cm^{-1}, weak) indicated that benzoylation had occurred mainly at the phenolic hydroxyl groups and, to a lesser extent, at amino groups.

More insight into the structure of dopamine melanin was obtained by oxidation of samples of dopamine, specifically deuterated at the α and β positions of the side chain, and at the 2, 5, and 6 positions of the benzene ring (Binns *et al.*, 1970b). In all cases, the ratio of deuterium enrichment in the pigment to that in the precursor was higher than that found for the same position of labeling in dopamelanin, as if any of the carbon atoms in the dopamine molecule was involved in other types of polymeric linkages. In other words, a considerable number of the units in dopamine melanin must be held together by C-N or C-O linkages, in contrast to dopamelanin, where only C-C linkages seem to be involved.

Analysis of the incorporation values led Swan to suggest that dopamine melanin was a copolymer containing a high proportion of uncyclized dopamine units and DHI units, along with some carboxylated pyrrolic units arising by peroxidative cleavage of the benzene ring of what were originally 1,7-linked indole units. The percentage composition of the proposed units and the positions involved in the polymeric linkages are indicated in Fig. 6.5.

Swan's conclusions are attractive in connecting a number of different data. However, they are based entirely upon the assumption that all measured radioactivity originated from a high-molecular-weight polymer rather than a mixture of oligomers of low molecular weight, as suggested by the pigment solubility and easy formation of derivatives. Yet, in spite of this and other doubts, a remarkable difference between dopamine melanin and dopamelanin is the presence in the former of a high proportion of uncyclized side chains, as indicated also by ^{13}C NMR spectroscopy (Peter and Förster, 1989). The reason for such a difference is probably due to the minor tendency of dopaminequinone with respect to dopaquinone to undergo intramolecular cyclization (Young and Babbitt, 1983; Jimenez *et al.*, 1984). This implies that at high substrate concentrations, like those used to obtain dopamine melanin, there may be competition for reaction at the 6 position of dopaminequinone between intramolecular cyclization and intermolecular addition of the amino group of another molecule of dopamine.

There would also be the possibility that dopaminequinone could react with the hydroxyl ion to give 6-hydroxydopamine. In support of this, Senoh and Witkop (1959a,b) reported that dopamine can be converted into 6-hydroxydopamine by the rat, as well as *in vitro*, and that the latter readily undergoes oxidation to give the *p*-quinonoid aminochrome (Fig. 6.6). This,

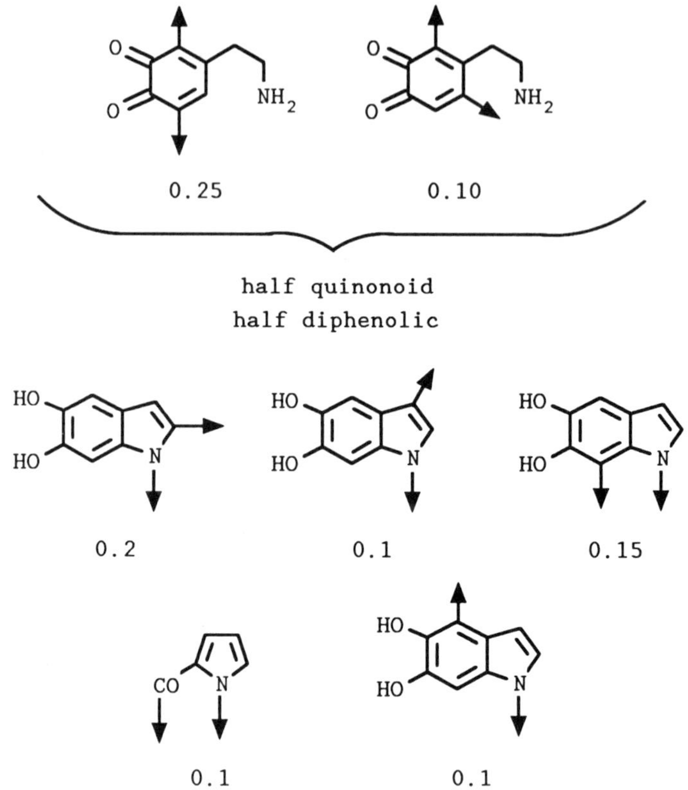

FIG 6.5 Percentage composition of the units present in dopamine melanin according to Swan (1976). Arrows indicate the position of attachment of the units in the pigment polymer.

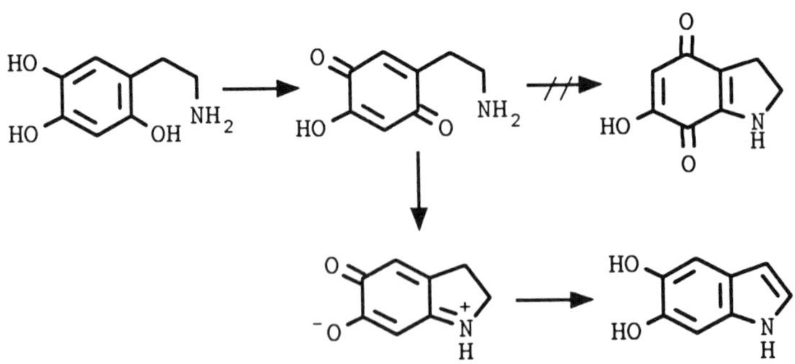

FIG 6.6 Oxidative cyclization of 6-hydroxydopamine.

however, was not isolated and its formation seems unlikely in light of a previous work by Harley-Mason (1953), later confirmed by Swan (1976), showing that oxidative cyclization of 6-hydroxydopamine proceeds as expected to give dopachrome and then DHI.

B. Adrenalin Black

The early stages of adrenalin oxidation have long been known and involve formation of adrenochrome (6.**1**) and its rearrangement to adrenolutin (6.**2**) (Heacock, 1965). The subsequent fate of these intermediates was until recently very little understood. In 1945 Cohen suggested that oxidation of adrenochrome leads to a dark red pigment, "oxoadrenochrome," which was tentatively identified as the *o*-quinone of adrenolutin (6.**3**). This structure was

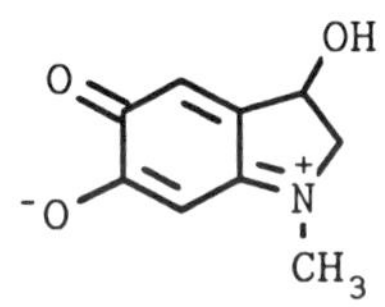

6.**1**

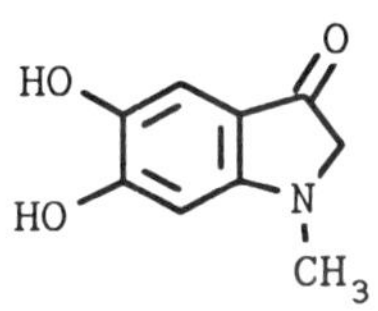

6.**2**

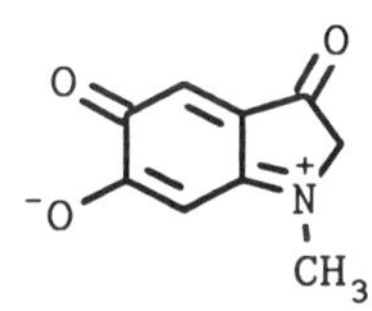

6.**3**

later questioned by Harley-Mason (1950), who failed to obtain oxoadrenochrome by oxidation of adrenolutin: The only recognizable product was a greenish-black pigment of indigoid nature. Moreover, he found that when adrenochrome was treated with acids or was left in aqueous solution under nitrogen for 24 hr, it changed nonoxidatively into a dark pigment termed "adrenalin black." This was an amorphous material insoluble in organic solvents, except pyridine, but soluble in sodium hydroxide solution, from which it could be reprecipitated with acids.

Based upon chemical and spectrophotometric experiments and partly on theoretical grounds, Bu'Lock (1961) suggested that formation of adrenalin black proceeded via the initial acid-catalyzed dehydration of adrenochrome to give 1-methyl-5,6-indolequinone, which subsequently polymerized through the 3 and 4 positions of the indole ring. In contrast with this view, a reexamination of the structure of adrenalin black (Corradini *et al.*, 1988) showed that the pigment consists of a complex mixture of oligomers of 5,6-dihydroxy-1-methylindole, some of which (present in somewhat greater amounts) have been isolated and characterized as the dimers 6.**4** and 6.**5** and the trimer 6.**6**. The structure of these products clearly indicates that pigment formation proceeds by a mechanism which involves the preliminary conversion of adrenochrome to 5,6-dihydroxy-1-methylindole, followed by acid-catalyzed enamine–imine type dimerization of the indole through the 2 and 3 positions.

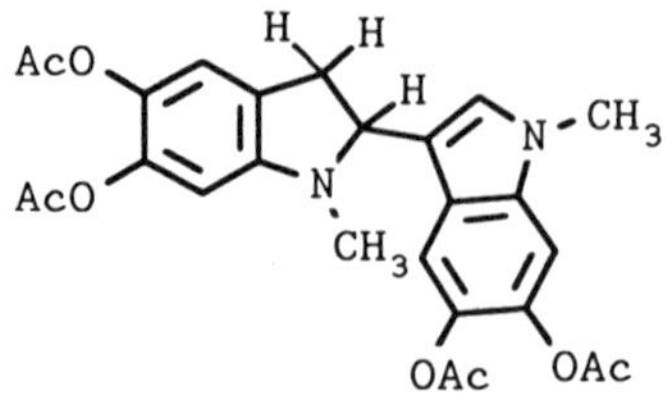

6.4

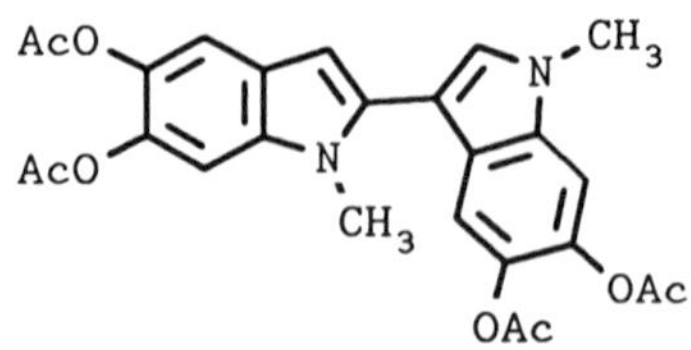

6.5

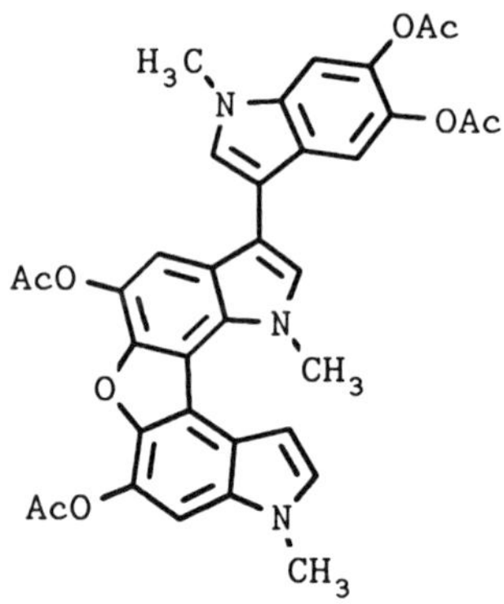

6.6

In an extension of this study (d'Ischia *et al.*, 1988) it was found that at physiological pH values, under anaerobic conditions, the major product of adrenochrome rearrangement is not adrenolutin, as generally believed, but a yellow dimeric product (6.**7**) consisting of an adrenolutin moiety covalently linked to the angular 9 position of adrenochrome. When air was allowed into the solution, the reaction took a more complex course, due to the formation, besides the dimer, of two additional products arising from autooxidation of adrenolutin. These were identified as 5,6-dihydroxy-1-methylisatin (6.**8**) and the corresponding 4,4′-dimer (6.**9**).

The product distribution of adrenochrome rearrangement was markedly dependent on the initial concentration of the aminochrome, the nature of the buffer and, notably the pH of the medium (Palumbo *et al.*, 1989). In particular, at pH higher than 8 the reaction was mainly directed toward the formation of adrenolutin, whereas at pH around or below neutrality both the dimer and 5,6-dihydroxy-1-methylindole prevailed. Transition metal ions of common occurrence in biological systems also affect adrenochrome decomposition, to an extent and with modalities which depend on the nature of the metal. Thus, redox-inactive or oxidizing cations, like Zn^{2+} and Cu^{2+}, markedly

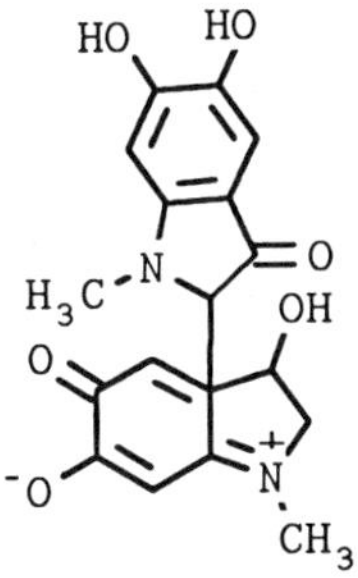

6.7

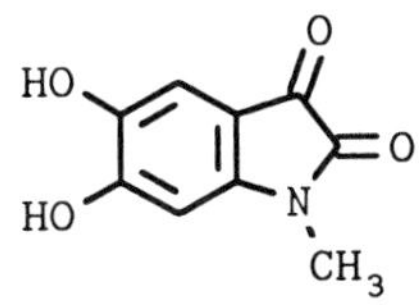

6.8

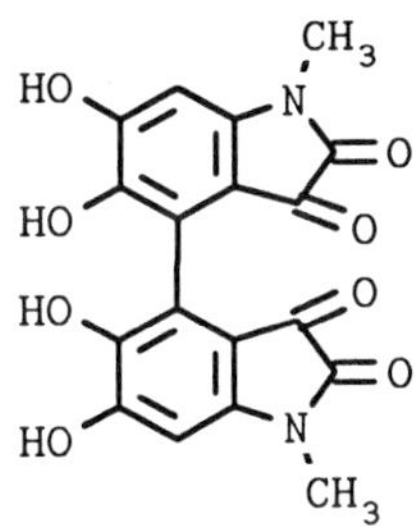

6.9

hasten the rate of rearrangement, inducing the formation of adrenolutin. Mild reductants like Fe^{2+} ions, on the other hand, cause adrenochrome to follow an entirely different decomposition pathway, involving reduction to leucoadrenochrome and its subsequent dehydration to 5,6-dihydroxy-1-methylindole. The results of these studies point to a profound difference in the oxidative pathways of dopamine and adrenalin, which is evidently due to the presence of the β-hydroxyl group affecting the reactivity of the resulting indole metabolites.

VIII. FUNCTIONAL SIGNIFICANCE OF NEUROMELANIN

Although the role of neuromelanin is not known, studies of the physicochemical properties of the pigment have given rise to several hypotheses that may or not be mutually related (Lacy, 1981; Youdim *et al.*, 1989; Rabey and Hefti, 1990). One view is that neuromelanin could function as a device for the inactivation of harmfully excited molecules into innocuous vibrational energy (McGinness and Proctor, 1973). Other authors have postulated that neuromelanin could exert a protective role by acting as a sink for free radicals (Commoner *et al.*, 1954) or as a redox buffer against reducing or oxidizing toxic conditions (Gan *et al.*, 1976, 1977). Another property of neuromelanin, which is typical of eumelanins (see Chapter 4), is the ability to bind a number of foreign substances and to retain them for a very long time (Ings, 1984). This has led Lindquist *et al.* (1987) to suggest that neuromelanin may protect the surrounding cells by keeping potentially harmful substances bound and thereafter slowly releasing them in low, nontoxic concentrations.

An interesting theory, proposed by Forrest (1974), envisages neuromelanin as serving a biocybernetic function, whereby its lack or depletion causes the *substantia nigra* to revert from its neuroendocrine role to its original motoric role. Moreover, it would inhibit the bioelectric activity of the neuron, producing a decrease in the cell's discharge frequency by controlling the electron flow from the cell cytoplasm into the original axon cylinder. Which of these theories has the greatest physiological relevance is at present difficult to assess, as too little is known on the origin and metabolism of neuromelanin.

IX. NEUROMELANIN AND PARKINSON'S DISEASE

Though degeneration of the *substantia nigra* in the *striatum* is undoubtedly associated with Parkinson's disease, as carefully documented by Hirsch *et al.* (1988), there is a question as to whether or not neuromelanin may have a part to play in the induction of parkinsonism and as to whether it may be either a cause or a consequence of the degradation of the dopaminergic neurons. Marsden (1983) argues that neuromelanin is not directly related to the pathogenesis of parkinsonism and that the Lewy body, an inclusion in cells of the dopaminergic system involved in the disease, may hold the key.

An alternate view stems from the aforementioned ability of neuromelanin to bind foreign substances, some of which may be a threat to the pigment cell, as suggested by certain forms of chemically induced parkinsonism (Ings, 1984). Typically, chronic exposure to manganese dust in miners is known to affect the nervous system with symptoms similar to those in Parkinson's disease (Lyden *et al.*, 1984). The damage seems to especially hit the melanin-containing cells of the *substantia nigra*, presumably by a mechanism involving metal-catalyzed autooxidation of dopamine with formation of cytotoxic quinones (Donaldsson *et al.*, 1982; Graham, 1984). Another neurotoxin with a high affinity for neuromelanin is 1-methyl-4-phenyl-1,2,3,6-tetrahydropyridine (MPTP), a meperidinelike narcotic drug which readily penetrates the blood–brain barrier giving rise to severe parkinsonian symptoms (Lindquist *et al.*, 1987; D'Amato *et al.*, 1986; Kopin *et al.*, 1986). Once in the brain, MPTP is converted in a two-step process (Fig. 6.7) to the 1-methyl-4-phenylpyridinium ion (MPP^+) which is believed to be the actual substance responsible for the neuronal destruction (Castagnoli *et al.*, 1985; Langstron *et al.*, 1987).

Poirer *et al.* (1985) have postulated that a radical intermediate in the initial oxidation of MPTP may potentiate the autooxidation of dopamine to dopaminequinone. However, a pulse radiolysis study of the decay of one-electron oxidized MPTP in the presence and in the absence of dopamine gave a

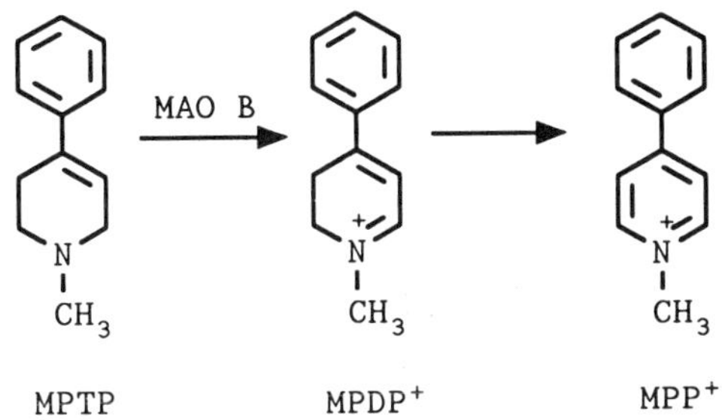

FIG 6.7 Bioactivation of MPTP.

rate-constant limit of $< 2 \times 10^7 M^{-1}s^{-1}$ for the reaction, suggesting that the proposed pathway is unlikely to be of importance in the development of Parkinson's disease (Chacon *et al.*, 1987; Land, 1988).

The degree of MPTP neurotoxicity seems to be related to the amount of neuromelanin present in the brain. Primates, including man, have much higher sensitivity to MPTP than laboratory animals, for example, mice (Heikkila *et al.*, 1984). Many other compounds, like MPTP and manganese, have melanin affinity and may cause lesions in pigmented neurons secondary to their accumulation on the neuromelanin granules. Among these, of particular interest is the pesticide paraquat, which is structurally similar to MPP^+. In a recent epidemiological investigation, the highest prevalence of Parkinson's disease was found in rural agricultural areas of high pesticide use (Barbeau *et al.*, 1986).

Other substances capable of inducing selective degeneration of catecholamine neurons in animals are 6-hydroxydopamine and 6-aminodopamine (Graham *et al.*, 1978; Morrison and Cohen, 1983). The selective toxicity of these molecules has been ascribed to the affinity for the uptake mechanisms of catecholaminergic neurons, coupled with their marked tendency to undergo autooxidation with formation of free radicals, particularly the hydroxy radical (Cohen, 1983). The superoxide ion radicals have also been implicated in the mechanism of cytotoxicity, but a recent pulse radiolysis study does not seem to support this notion (Kalyanaraman *et al.*, 1988).

It is well known that oxygen toxicity is dependent on normal functioning of antioxidant systems such as GSH, GSH peroxidase, SOD, and ascorbate, present in relatively high concentrations in the basal ganglia. A deficiency in any of these would inevitably make the dopamine neurons much more sensitive to chemical insults. Thus, ultimately, the balance between production and disposition of free radicals may be an important factor in the pathogenesis of parkinsonism.

CHAPTER 7

PHEOMELANINS AND TRICHOCHROMES

I. INTRODUCTION

The history of investigations pertaining to the discovery of the sulfur-containing pheomelanins as chemical entities distinct from eumelanins was outlined earlier in Chapter 1. We shall now consider the chemistry of these pigments which characterize certain types of reddish hair and feathers, wherein they usually occur together with the biogenetically related trichochromes. As the latter are well-defined molecular entities, a detailed account of what is known about them would be desirable before discussing the more complex and as yet little understood nature of the high-molecular-weight pheomelanins. An excellent review of the early studies carried out at Naples University on both trichochromes and pheomelanins has been written by Thomson (1974).

II. TRICHOCHROMES

A. STRUCTURE AND PROPERTIES

The basic structural unit of trichochromes is the 1,4-benzothiazine ring system, which can exist in two tautomeric forms, that is, 2*H*- and 4*H*-1,4-benzothiazine (Fig. 7.1). Because of its marked instability, the parent compound has not yet been isolated, although it has been detected by ^{1}H NMR spectroscopy as a transient species existing predominantly as the 2*H*-tautomer (Chioccara *et al.*, 1977). For a general review on the chemistry of 1,4-benzothiazines see Brown and Davidson (1985).

The presence of this heterocyclic system in trichochromes was first suggested on biogenetic grounds and then confirmed by extensive model studies which culminated with the synthesis of the hitherto unknown pigment chromophore, $\Delta^{2,2'}$-bi(2*H*-1,4-benzothiazine) (Prota *et al.*, 1967). This gave an important lead to interpretations of the structural work, involving inevitably extensive degradation of the benzothiazine rings (Nicolaus *et al.*, 1969; Prota

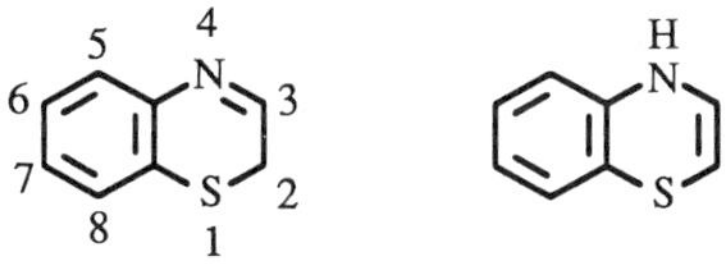

2H *4H*

FIG 7.1 Tautomeric forms of 1,4-benzothiazine.

et al., 1969; Prota *et al.*, 1971). The trichochromes so far identified are shown in Fig. 7.2.

The most commonly encountered trichochromes B and C were first isolated from the red feathers of New Hampshire hens and were later found in "carrot" red human hair, but though they are intensely colored they constitute only a minor fraction of the total pheomelanic pigments in hair. Apparently, trichochromes E and F do not occur in red human hair, and it is not yet clear whether the small amounts found in red feathers as well as in other sources

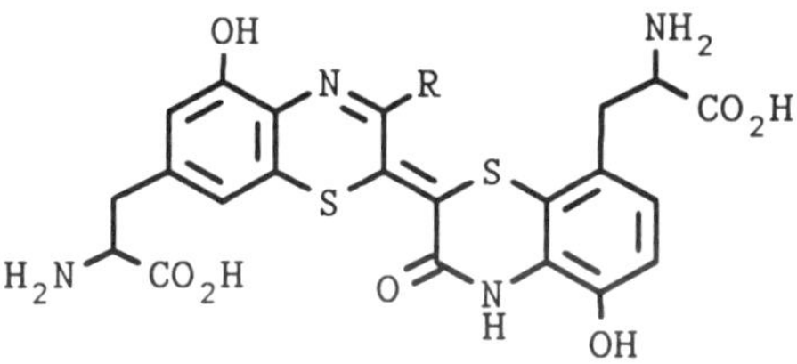

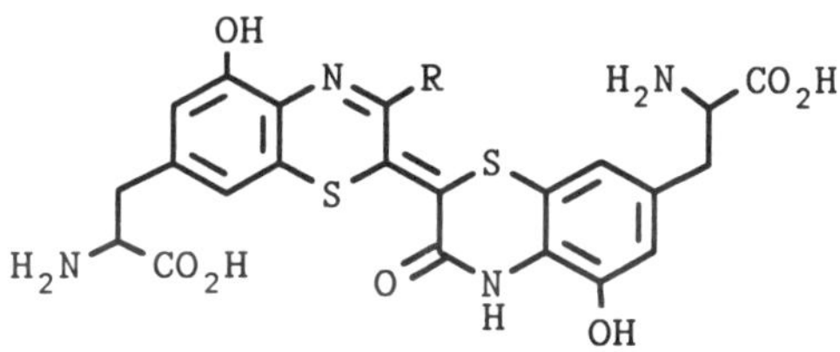

R=CO_2H : Trichochrome B
R=H : Decarboxytrichochrome B

R=CO_2H : Trichochrome C
R=H : Decarboxytrichochrome C

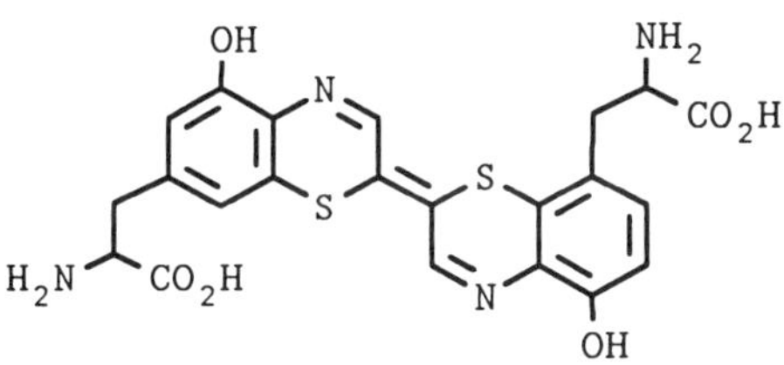

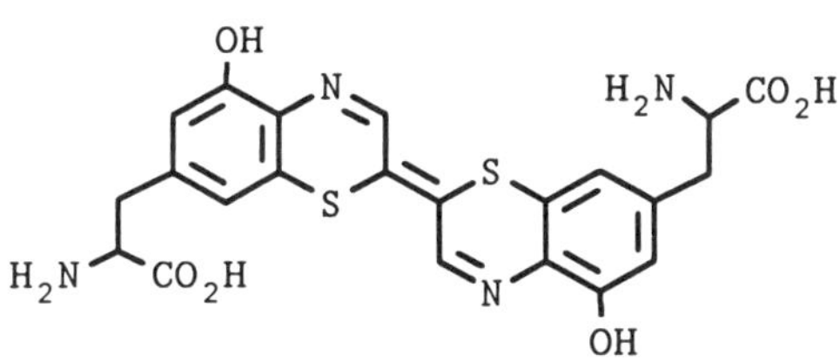

Trichochrome E

Trichochrome F

FIG 7.2 Structure of trichochromes.

arise as artifacts of oxidation of biosynthetic intermediates during the isolation procedure which involves both acid and alkali treatments (Prota, 1972; Agrup *et al.*, 1978a).

Trichochromes have, generally speaking, uncommon properties, for example, insolubility in organic solvents and in water at neutral pHs, difficult crystallization, and a strong tendency toward aggregation and adsorption. The yellow-orange trichochromes B and C are also relatively unstable in acids. On brief heating in 0.1 *M* HCl, they undergo selective decarboxylation of the acidic group at the C-3 position to give new pigments (decarboxytrichochrome B and C) corresponding to Rothman and Flesch's trichosiderins which can also be obtained directly from red hair or feathers by extraction with boiling dilute acid.

Unlike the parent trichochromes, the decarboxylated pigments exhibit pH-dependent colors, being reversibly red-purple in acid and yellow-orange in alkali (Table 7.1). Such a behavior results from protonation of the conjugated system –S–C=C–C=N–, which gives rise to a mesomeric ammonium–thionium cation accounting for the large bathochromic shift in acid media. In trichochromes B and C the presence of the carboxyl group at C-3 diminishes the basicity on the adjacent nitrogen so that the formation of the mesomeric cation does not normally occur. However, in acetic acid the thiazine nitrogen becomes more basic and protonation does occur with hydrogen chloride, the visible absorption maximum shifting to about 550 nm.

It is of interest that the basic chromophore of trichochromes contains the essential features of an indigoid dye, as defined by Wille and Lütkke (1971), namely two donor and two acceptor groups arranged to form a double cross-conjugated system (Fig. 7.3).

The parent $\Delta^{2,2'}$-bi(2*H*-1,4-benzothiazine) skeleton of trichochromes B and C, that is, 7.**1**, has been obtained by Kaul (1974) by condensation of

TABLE 7.1
VISIBLE ABSORPTION SPECTRA OF TRICHOCHROMES

Pigment	λ_{max}, nm (log ε) in 0.1*N* NaOH	λ_{max}, nm (log ε) in 2*N* HCl
Trichochrome B	454 (3.98)	450 sh
Trichochrome C	452 (4.13)	460 sh
Trichochrome E	482	587
Trichochrome F	482 (4.33)	587 (4.42)
Decarboxytrichochrome B	462 (4.19)	525 (4.16)
Decarboxytrichochrome C	458 (4.21)	525 (4.18)

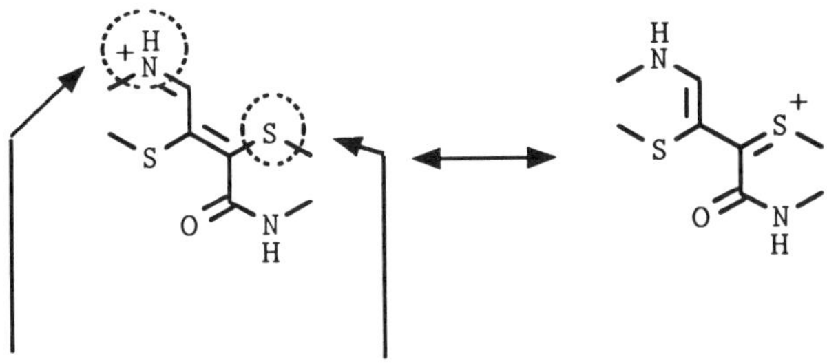

FIG 7.3 Indigoid-type chromophore of trichochromes.

o-aminothiophenol with mucochloric acid. A similar type of synthesis also has been utilized to obtain the new benzothiazinone dye 7.**2** using 2,3-dichloromaleic anhydride in place of mucochloric acid.

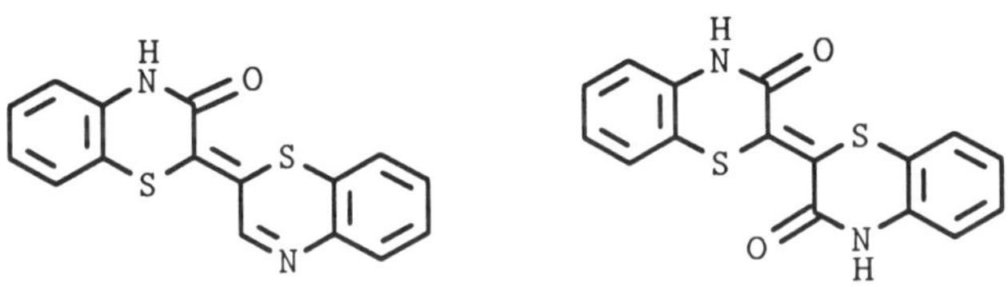

7.1 7.2

B. DEGRADATION

An interesting reaction of trichochromes is the reductive cleavage of the chromophore with hydroiodic acid and red phosphorus which was extensively used for structural studies. In the case of the symmetrical substituted trichochrome C, this reaction gave mainly the benzothiazinone amino acid (7.**3**) and AHP (7.**4**), along with the benzothiazole amino acid (7.**5**) in smaller amounts, arising evidently by ring contraction of the unsubstituted benzothiazine moiety. This interpretation is supported by the synthesis of the model compound 7.**6**, which exhibits an absorption spectrum (λ_{max} = 410 nm) different from that characteristic of the $\Delta^{2,2'}$-bi(2*H*-1,4-benzothiazine) chromophore of trichochrome C (Nicolaus *et al.*, 1969).

Likewise, degradation of the isomeric decarboxy-trichochrome B with hydroiodic acid and red phosphorus leads to the same benzothiazinone 7.**3** and a mixture of the isomeric aminohydroxyphenylalanines (AHP) 7.**4** and 7.**7**. Considering that 7.**4** is a secondary product derived by degradation of the amide 7.**3** by hydroiodic acid, the formation of 7.**5** is evidently due to the presence of a 1,4-benzothiazine unit substituted at position 8 in trichochrome B.

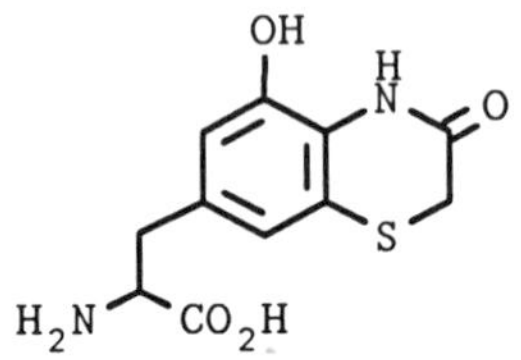

7.**3**

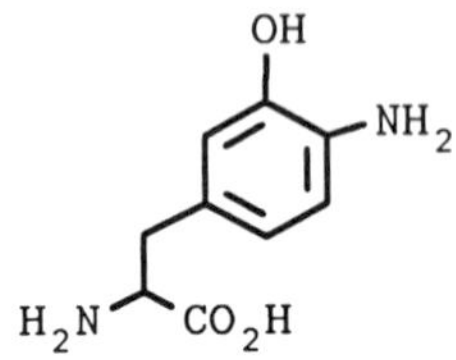

7.**4**

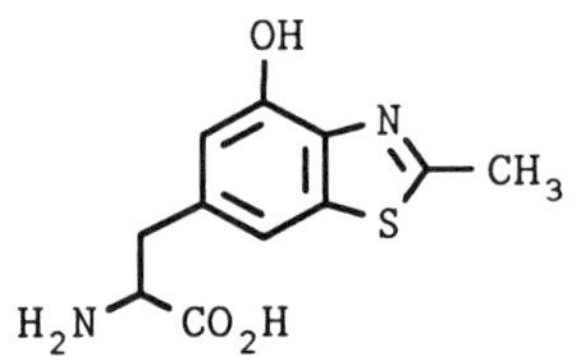

7.**5**

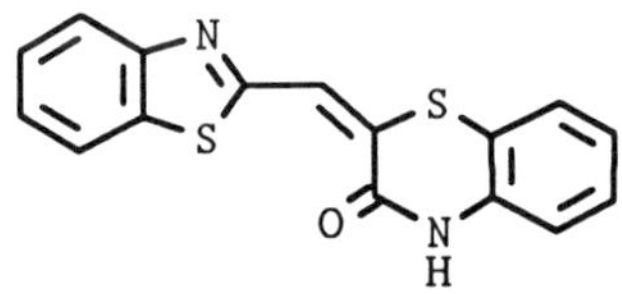

7.**6**

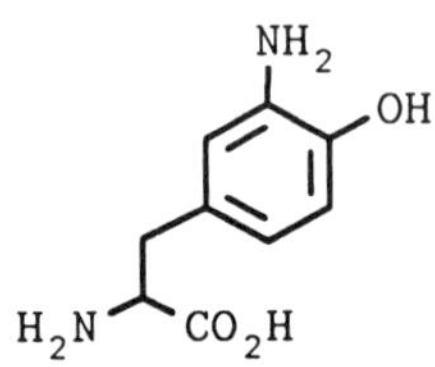

7.**7**

A similar difference in the orientation of the alanine side chains is observed between the two violet trichochromes E and F, which are characterized by a $\Delta^{2,2'}$-bi(2*H*-1,4-benzothiazine) chromophore unsubstituted at positions 3 and 3′. Consequently, on protonation they exhibit a larger bathochromic shift than that of decarboxy trichochrome B and C. On heating with hydroiodic acid trichochrome F gives mainly 4-AHP, meaning that the alanine side chains attached to the 1,4-benzothiazine chromophore are both at position 7. This is fully in accord with the ^{1}H NMR spectrum, in which the two azomethine protons appear as a singlet at δ 8.21 ppm. On the other hand, degradation of trichochrome E under similar conditions gives a mixture of the two isomeric

AHPs 7.4 and 7.7, consistent with the presence of the alanine residues at positions 7 and 8 of the bibenzothiazine chromophore.

C. Synthesis and Stereochemistry

No good synthesis of trichochromes is available, although they can be obtained in small amounts by aerial oxidation of the appropriate cysdopa isomer(s), followed by acidification of the solution after a short time (Prota *et al.*, 1968a; Prota *et al.*, 1976b). The pigments thus formed can be precipitated by lowering the pH of the solution and purified by ion exchange and Sephadex chromatography.

As yet the stereochemistry of the $\Delta^{2,2'}$-bi(2*H*-1,4-benzothiazine) skeleton in the trichochromes has not been defined, although indirect evidence indicates that the isolated pigments may well be an equilibrium mixture of both geometric isomers, the *trans* form predominating in the dark (Prota, 1988a). This view is supported by the behavior of the parent ring system which can be conveniently prepared by air oxidation of the 2*H*-1,4-benzothiazine, generated *in situ* by hydrolysis of 1-(*o*-aminophenylthio)-2,2-diethoxyethane (Prota *et al.*, 1974). This procedure yields only the yellow *trans* isomer (7.**8a**) which in solution exhibits a pronounced photochromism, being converted into the red unstable *cis* (7.**8b**) form on brief irradiation with sunlight (Fig. 7.4). In

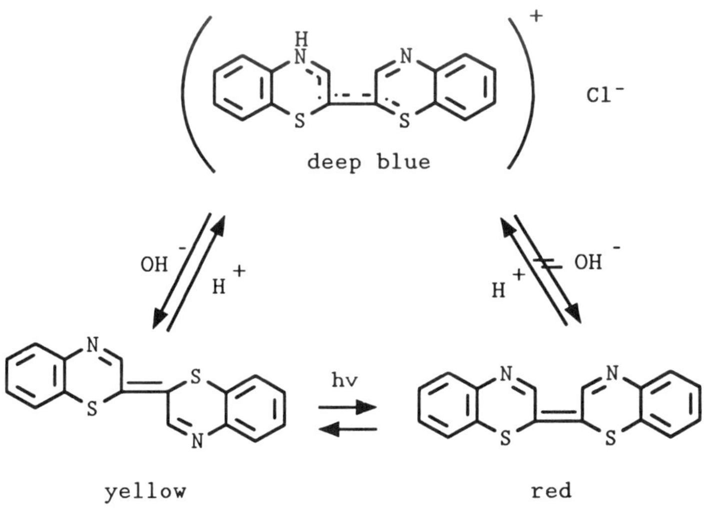

7.**8a** 7.**8b**

FIG 7.4 Interconversion of the *cis* and *trans* forms of $\Delta^{2,2'}$-bi-(2*H*-1,4-benzothiazine).

the dark the process reverses slowly at room temperature but rapidly above 60°C to regenerate the *trans* isomer. Alternatively, the red isomer can be transformed into the *trans* isomer by protonation with dilute acid to give the deep blue mesomeric cation from which 7.**8a** is obtained by neutralization.

In another study (Giordano *et al.*, 1971), the diasteroisomers of the 3-aryl-substituted $\Delta^{2,2'}$-bi(2*H*-1,4-benzothiazines) were found to readily interconvert in solution at room temperature with the red *cis* isomer, predominating over the yellow *trans* isomer (Fig. 7.5). However, X-ray analysis of the structures indicated that the dominance of the *cis* isomer is probably due to Van der Waals forces, which maintain the two benzene rings at the 3,3′ positions in an almost fixed conformation. As the temperature increases, thermal motion destabilizes the *cis* configuration and the less strained isomer becomes favored. The easy interconversion seems to take place between quasiplanar higher energy states involving resonance between conjugated double bonds. This gives a partial single-bond character to the central double bond and facilitates rotation about it.

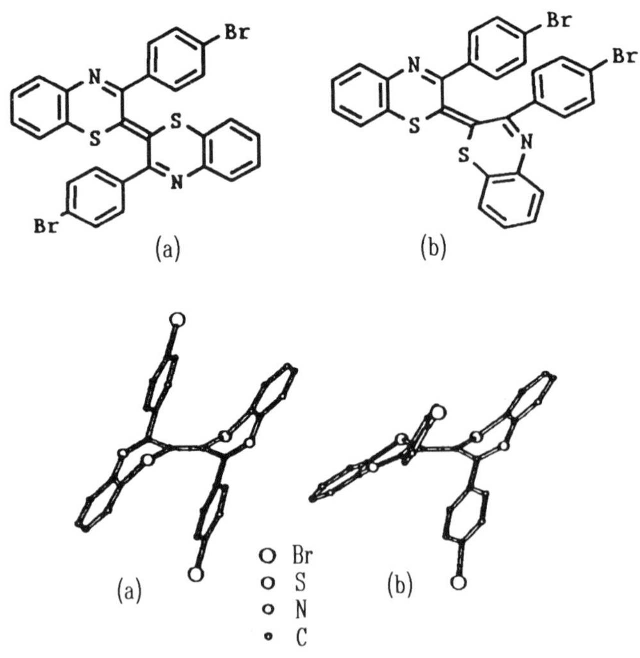

FIG 7.5 Structures and perspective drawings of *trans* (a) and *cis* (b) $\Delta^{2,2'}$-bi-(3-*p*-bromophenyl-2*H*-1,4-benzothiazine).

III. PHEOMELANINS

A. General Properties

Like the trichochromes, pheomelanins appear to be zwitterionic in character, being soluble both in dilute alkali and in certain acidic solvents such as formic acid or ethylene chloridrin. This behavior is due to the presence of alanine side chains attached to the subunits forming the pigment backbone, as indicated by early degradative studies (Minale *et al.*, 1967). The presence of phenolic groups has also been established and their content determined by analysis of the methoxy groups (8%) after treatment with diazomethane (Fattorusso *et al.*, 1968). Not much is known about the redox state of pheomelanins, though they are generally regarded as containing structural units at different oxidation levels. This, however, awaits confirmation.

Since natural pheomelanins are generally associated with protein, the possibility exists that they are chromoproteins. In support of this view, Fattorusso *et al.* (1968) found that the pigment-bound protein differs from feather keratin in the absence of cysteine from the 16 amino acids obtained in hydrolysis. Moreover, protein appeared tightly bound to the pigment chromophore, suggesting a covalent link between them. Similar results were obtained by Deibel (1983), who used isoelectric focusing to obtain more homogeneous pheomelanin fractions from red hair and feathers.

Different estimates of the apparent molecular weight of pheomelanins based on gel filtration have been reported ranging from 5000 up to 50,000 (Minale *et al.*, 1967). It must be cautioned that the material used in these studies is inhomogeneous from the molecular viewpoint and this, coupled with the severe tailing and irreversible binding of the pigment fractions to the chromatographic gels, makes the interpretation of the results uncertain. A more accurate determination of the molecular weight of pheomelanins has been carried out by Fattorusso *et al.* (1968), using a derivative of the protein-free pigment obtained by treatment with 2,4-dinitrofluorobenzene and subsequent methylation with diazomethane. Examination by vapor pressure osmometry gave an average molecular weight of 2365, corresponding to a value of ca. 1800 for the parent pigment.

B. Structural Characterization

Despite an immense amount of experimental work, little is known about the general structure of pheomelanins. The most extensively studied are the gallopheomelanins from the red feathers of New Hampshire hens (*Gallus gallus*), but chemical evidence (Fattorusso *et al.*, 1970b) indicates that the analogous pigments found in other red feathers, as well as in mammalian hair, are structurally similar, differing probably only in molecular size and protein content. In each case, these pheomelanins appear to be irregular polymers or mixtures of polymers which lack any well-defined spectral or other physical characteristics. Thus, despite the theoretical advantage of dealing with alkali-soluble material, the structural problem is similar to that presented by the eumelanins and it has been therefore tackled in the same way by the classical chemical approach. In early studies at Naples, the crude gallopheomelanins obtained by alkaline extraction were separated into four main fractions (1–4) by dialysis and chromatography on Sephadex (Prota and Nicolaus, 1967b) or Biogel P columns (Minale *et al.*, 1967). On analysis, all were found to contain nitrogen and sulfur in a ratio of approximately 1:2 (Table 7.2), consistent with their biogenetic relationship with trichochromes. The most abundant fraction, gallopheomelanin-1, was protein free and was therefore preferably used for degradative studies (Minale *et al.*, 1967; 1969; Fattorusso *et al.*, 1968, 1969b).

On alkali fusion, the pigment gave a mixture of simple catechols and *o*-aminophenols, for example, 4-amino-3-hydroxybenzoic acid and 3-amino-4-hydroxybenzoic acid, while oxidation with hydrogen peroxide gave cysteic acid in about 10% yield and small amounts of glycine and aspartic acid. Significantly, PTCA, the characteristic degradation product of eumelanins, was absent, implying the absence of DHI units in the pigment backbone.

Larger fragments, chiefly AHP (7.**4**) and the benzothiazole 7.**9** with smaller amounts of the corresponding isomers 7.**7** and 7.**10**, were obtained by heating the pigment in boiling hydroiodic acid or hydrochloric acid. The structure of all these amino acids, derived from chemical and spectral analysis, has been confirmed by stepwise syntheses (Patil and Chedekel, 1984). Other minor degradation products of gallopheomelanins with HI include the benzothiazole amino acid 7.**5** and, notably, the isoquinolines 7.**11** and 7.**12** which were

TABLE 7.2
ELEMENTAL COMPOSITION OF GALLOPHEOMELANINS[a]

Pigment fraction	C	H	N	S (%)	N/S
Gallopheomelanin-1	49.0	3.8	9.0	9.9	2.1
Gallopheomelanin-2[b]	53.5	6.5	14.0	3.0	11.
Gallopheomelanin-3[b]	51.5	4.7	11.8	7.4	3.6
Gallopheomelanin-4[c]	49.5	3.8	8.0	10.2	1.8

[a]From Minale *et al.* (1967).
[b]Isolated as a protein complex.
[c]Containing some protein.

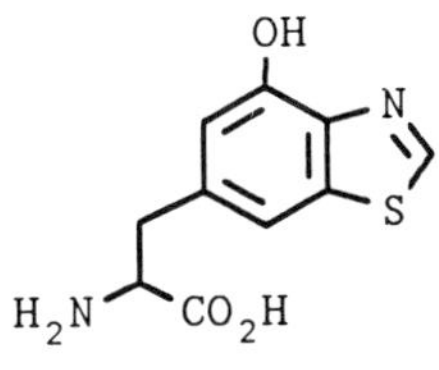

7.**9**

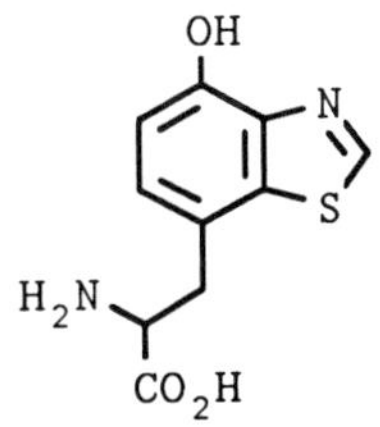

7.**10**

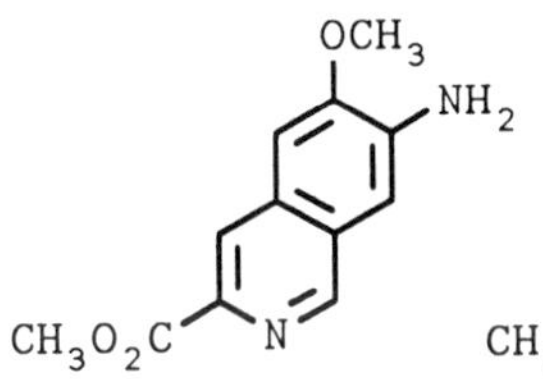

7.**11**

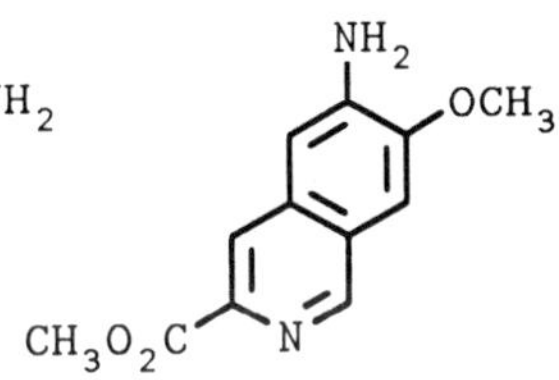

7.**12**

isolated after methylation of the crude reaction mixture with diazomethane (Fattorusso *et al.*, 1970b). Complementary information on the structure of gallopheomelanin-1 was obtained by oxidation with potassium permanganate, affording besides the thiazole acids 7.**13** and 7.**14**, which were derived apparently by degradation of benzothiazole units, a small amount (0.1%) of a large fragment identified as the polyacid 7.**15** (Minale *et al.*, 1969, 1970). This was envisaged as arising by oxidative degradation of a segment of the pigment backbone containing benzothiazole and tetrahydroisoquinoline ring systems tentatively linked as in the partial structure 7.**16**. Moreover, after consideration of the other fragments obtained, such as the benzothiazinone and cysteic acid, it was suggested that 1,4-benzothiazine units at various oxidative levels can also partake in the building up of the gallopheomelanin polymer and that some uncyclized cysteine residues are present as well. This has led to the view

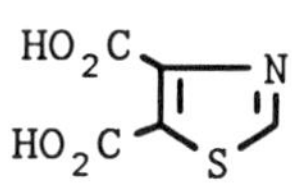

7.**13**

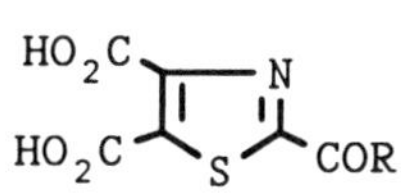

R=OH
R=NH_2

7.**14**

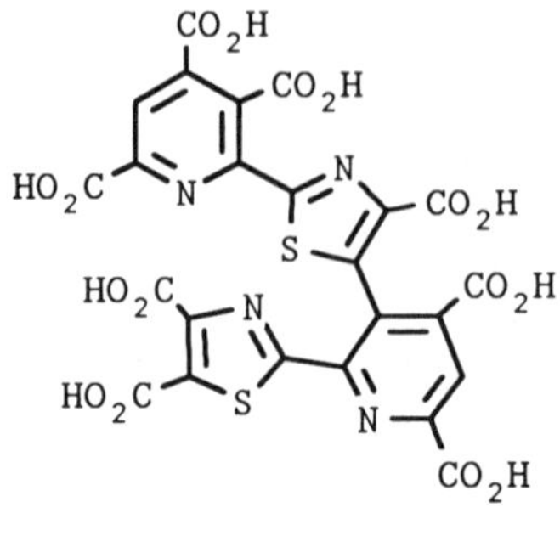

7.**15**

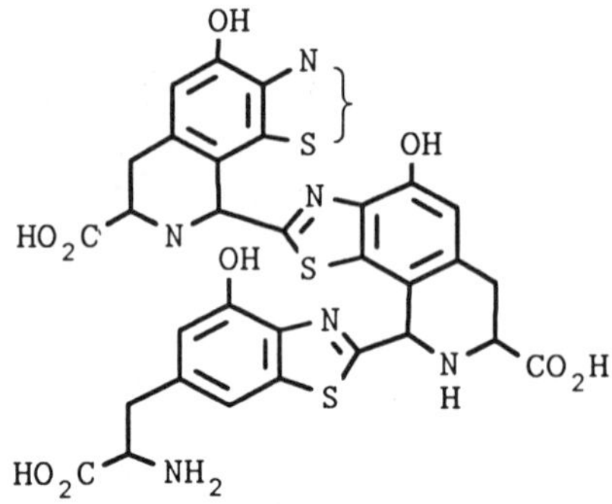

7.**16**

that gallopheomelanin-1 might be a very complex mixture of similar polymers in which a range of variation is possible depending upon the number and relative proportion of various types of structural units involved.

While the heterogeneous character of pheomelanins is consistent with the biosynthesis of these pigments involving different cysdopa precursors, the proposed structures and mode of linking of the units in the polymer(s) are difficult to accept without question. This stems primarily from the harsh degradative methods used, resulting probably in the production of artifacts which may not be related to units actually present in the intact polymer(s). In connection, it is noteworthy that a number of the degradation products isolated from gallopheomelanins, for example, the amino acids 7.**9** and 7.**10**, are also obtained by degradation of the trichochromes under similar conditions (Prota, 1988a). As the latter contain only 1,4-benzothiazinyl units, it is likely that the benzothiazole fragments isolated by degradation of gallopheomelanins are artifacts arising probably by acid-promoted ring contraction of similar benzothiazine units present in the pigment polymer. In line with this view, Ito and Fujita (1985) have reported that permanganate oxidation of trichochrome F yields 2,3-thiazoledicarboxylic acid, which is another characteristic fragment of gallopheomelanins. Against the proposed partial structure of pheomelanins are also the results of biosynthetic studies discussed in Section IV,B indicating that the alanine side chain of dopa is incorporated predominantly intact into the pigment backbone (Chedekel *et al.*, 1987). In light of these findings, the proposed general structure of gallopheomelanins appears unlikely and should be examined anew.

IV. BIOSYNTHETIC STUDIES

As in the case of eumelanins, knowledge of the biogenesis of pheomelanins preceded the structural studies and was acquired in an early experiment in which dopa was enzymically oxidized in the presence of cysteine. Under such conditions, a reddish-brown pigment was obtained which proved to be chemically similar to the pheomelanins extracted from hen feathers (Prota and Nicolaus, 1967a). Moreover, when the oxidation was stopped after a short time by acidification, the pale yellow solution changed to violet in a few minutes and it was possible to isolate small amounts of two pigments, identical in all respects to trichochrome E and F. Hence, it was suggested that pheomelanins as well as trichochromes could be formed *in vivo* by a deviation of the eumelanin pathway involving addition of cysteine to dopaquinone enzymically generated by oxidation of tyrosine.

This biogenetic route was later confirmed by a cytochemical study of pheomelanin formation in the feathers papillae of New Hampshire chick embryos (Misuraca *et al.*, 1969). As expected, both [3-^{14}C]cysteine and [2-^{14}C]tyrosine were specifically incorporated into pigment-producing cells, while in the presence of a tyrosinase inhibitor the incorporation was negligible. These initial studies provided the necessary background for the design of appropriate experiments aimed at understanding the chemistry underlying formation of pheomelanins and trichochromes.

A. SYNTHESIS OF CYSDOPAS

At the time these studies were performed, little was definitively known regarding the reactions of cysteine with dopaquinone or other *o*-quinones. Studies of the oxygen consumptions and spectroscopic characteristics of these reactions when the quinones from catechol or 4-methylcatechol were employed indicated that thiols reacted in a fashion similar to amines to give a 1,4 adduct, for example, 7.**17** and 7.**18**, respectively (Mason, 1955). Likewise, tyrosinase-catalyzed oxidation of dopa in the presence of cysteine and GSH was reported to give the analogous adducts 7.**19** and 7.**20** (Bouchilloux and Kodja, 1960). However, none of these compounds had been isolated and carefully analyzed to prove that the proposed structures were correct.

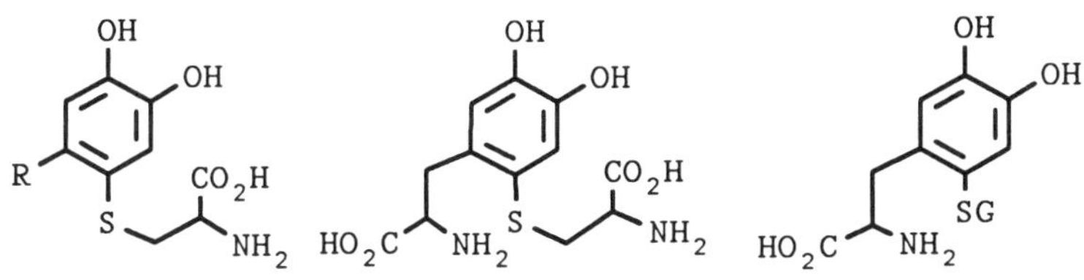

R=H, 7.**17**
R=CH$_3$, 7.**18**

7.**19**

7.**20**

In a reexamination of these early studies, Prota *et al.* (1967) found that reaction of cysteine with 4-methyl-*o*-quinone proceeds by 1,6 addition to give as a major product cystein-*S*-ylmethylcatechol 7.**21**, identified by chemical and spectral analysis. Confirming evidence for the attachment of the cysteine residue at the indicated position was provided by the marked tendency of the adduct to undergo aerial oxidation at alkaline pH to give the cyclic amide 7.**22** via the labile acid 7.**23**.

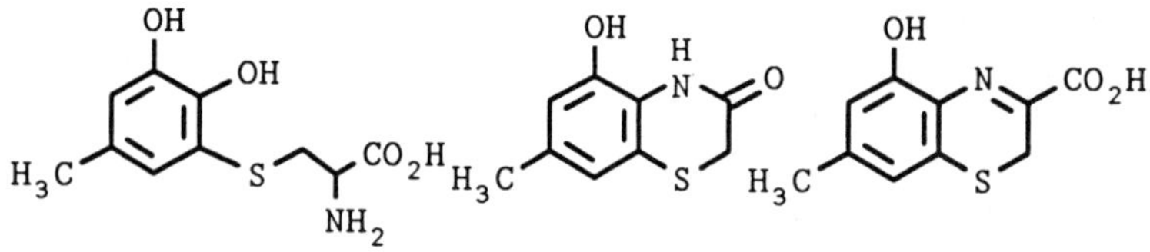

7.**21** 7.**22** 7.**23**

This and other features of the oxidative chemistry of 7.**21** strongly suggested that a similar adduct, that is, 5-cysdopa, could be the key intermediate in the biosynthesis of pheomelanins. The putative adduct was then synthesized (Prota *et al.*, 1968b) by an unambiguous route (Fig. 7.6) involving addition of cysteine ethyl ester to *N*-acetyldopaquinone ethyl ester, followed

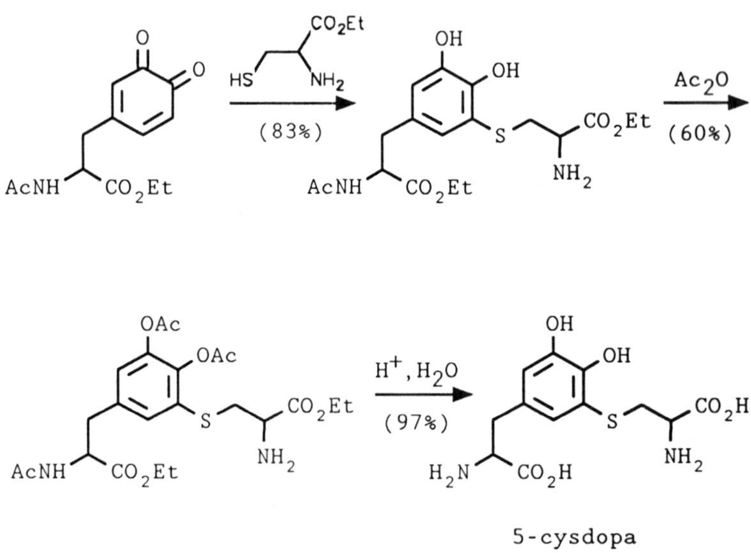

FIG 7.6 Synthesis of 5-cysdopa (Prota *et al.*, 1968b).

by purification of the resultant adduct as the acetyl derivative. Subsequent hydrolysis with acid afforded the free amino acid in an overall yield of ca. 48%. Later, Fattorusso *et al.* (1969a) found that, in addition to the 5-cysdopa, the reaction between cysteine ethyl ester and *N*-acetyldopaquinone ethyl ester leads also to a small amount (ca. 5%) of another adduct, which was isolated and identified as the 2-cysdopa isomer.

In further experiments, the formation of cysdopas was studied under biomimetic conditions by reaction of cysteine with dopaquinone generated *in situ* by tyrosinase-catalyzed oxidation of dopa (Ito and Prota, 1977a). Evidence was thus obtained that the reaction gives, in addition to 2- and 5-cysdopa, a small amount of the 6-cysdopa isomer, as well as a diadduct which was identified as 2,5-dicystein-*S,S*-yldopa (dicysdopa) (Fig. 7.7), previously found by Ito and Nicol (1977) in the *tapetum lucidum* of the gar *Lepisosteus.*

The yield of these products, as estimated by automatic amino acid analysis, is in the order 14, 74, 1, and 5%, which is consistent with the relative contribution of these intermediates in the biosynthesis of the natural pigments. Thus, for example, trichochrome C, which may be derived from two molecules of the major 5-cysdopa adduct, is normally found in red hair and feathers in greater amounts than trichochrome B, whose formation requires, besides 5-cysdopa, the less abundant 2-cysdopa isomer. Likewise, from the structure of the degradation products of gallopheomelanins it is clear that the major building block of these pigments is 5-cysdopa. However, 2-cysdopa and 2,5-dicysdopa also contribute to some extent, which accounts for the heterogeneous character of sulfur-containing pheomelanins.

B. Conversion of Cysdopas to Pheomelanins

Both 5- and 2-cysdopas have been detected during the early stages of the original enzymic oxidation of dopa and cysteine, mimicking the process of pheomelanogenesis *in vitro* (Fattorusso *et al.*, 1969a). In separate experiments, it was found that cysteinyldopas are not good substrates of tyrosinase; however, if a catalytic amount of dopa is added to the incubation mixture, they are smoothly oxidized to eventually give reddish-brown pigments, known as 2-cysdopapheomelanin and 5-cysdopapheomelanin, respectively (Fattorusso *et al.*, 1969a). On analysis, these proved to be similar to the synthetic pheomelanin obtained by direct enzymic oxidation of dopa and cysteine, except that they are more homogenous from the molecular viewpoint, as confirmed by subsequent structural studies (Deibel and Chedekel, 1984).

Regarding the conversion of cysdopas to pheomelanins, *in vitro* biosynthetic experiments (Prota *et al.*, 1970a,b) carried out mainly on the 5-isomer

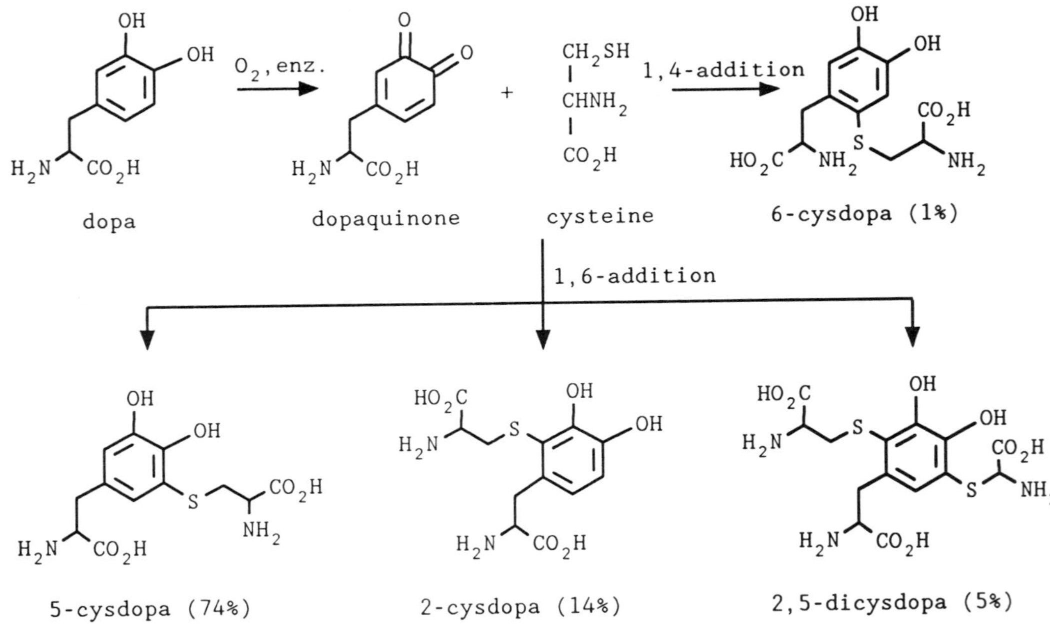

Fig 7.7 Formation and yields of isomeric cysdopas and 2,5-dicysdopa by reaction of cysteine with enzymatically generated dopaquinone.

and the model compound 7.**21** suggest that the reaction involves the oxidative ring closure of the cysteine side chain leading to the unstable *o*-quinoneimine intermediate 7.**24**; this can either give rise to the dihydrobenzothiazine 7.**25** by redox exchange with the starting material or undergo rearrangement with or without concomitant decarboxylation to give the 2*H*-1,4-benzothiazines 7.**26** or 7.**27** (Fig. 7.8). Which of these products prevails depends upon various parameters, such as pH, oxygen tension, and notably the presence of metal ions (see Chapter 8, Section II,B).

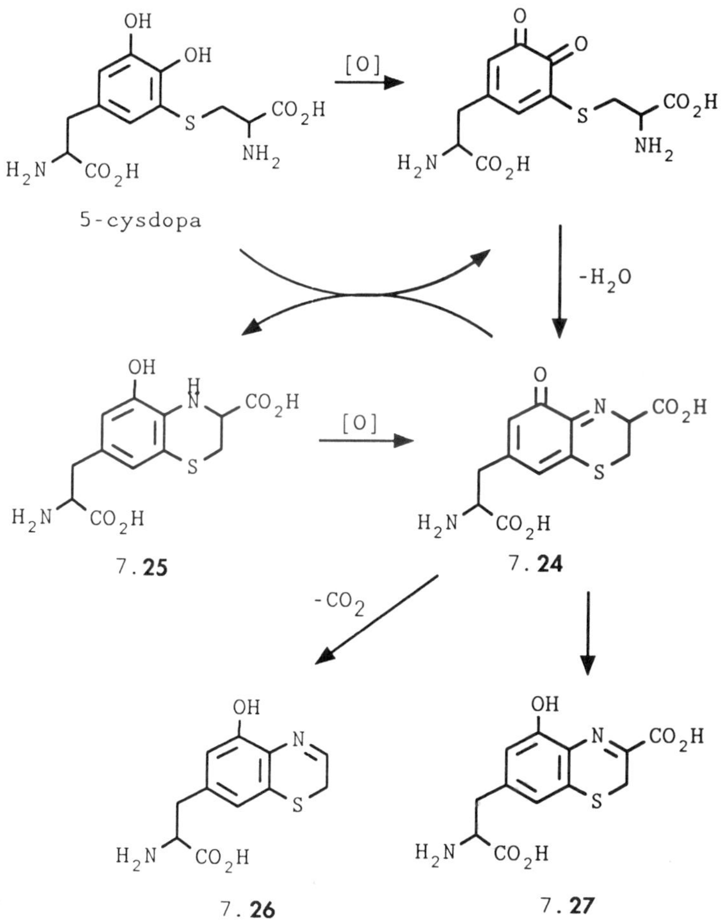

FIG 7.8 Oxidative cyclization of 5-cysdopa to 1,4-benzothiazines.

There is evidence which suggests that 2-cysdopa and 2,5-dicysdopa behave similarly, but little is definitively known about the later stages of pheomelanogenesis *in vitro* beyond the benzothiazine stage. Investigation is complicated by the instability of the various benzothiazine intermediates involved and by the fact that many concurrent reactions can take place. Using ^{13}C-, ^{14}C-, ^{35}S- and tritium-labeled cysdopa precursors, Deibel and Chedekel (1982, 1984) presented evidence suggesting that during polymerization a small percentage of the benzylic carbon of the alanine side chain undergoes oxidation to an sp^2 olefinic or aromatic-type carbon, while the majority exhibits no change in oxidation state or attached functionality. Moreover, the ^{13}C NMR spectrum of the pheomelanin pigment prepared from oxidation of 5-cysdopa labeled on the carboxyl group of the cysteine side chain indicates that the incorporated carbon is still in the form of carboxylic acid moieties (Chedekel *et al.*, 1987). However, apart from these and other structural details, the mode of polymerization of the postulated benzothiazine intermediate(s) remained undetermined. In an attempt to fill this gap, Costantini *et al.* (1990) recently reexamined the tyrosinase-catalyzed oxidation of 5-cysdopa in phosphate buffer at pH 6.8. Periodic HPLC analyses of the mixture showed the rapid and almost quantitative conversion to 7.**25**, after which the reaction proceeds smoothly to give eventually, besides a considerable amount of chromatographically ill-defined polymeric material, a number of oligomeric products, two of which, present in relatively greater amounts, were isolated and identified as two diastereoisomers corresponding to the gross structure 7.**28**.

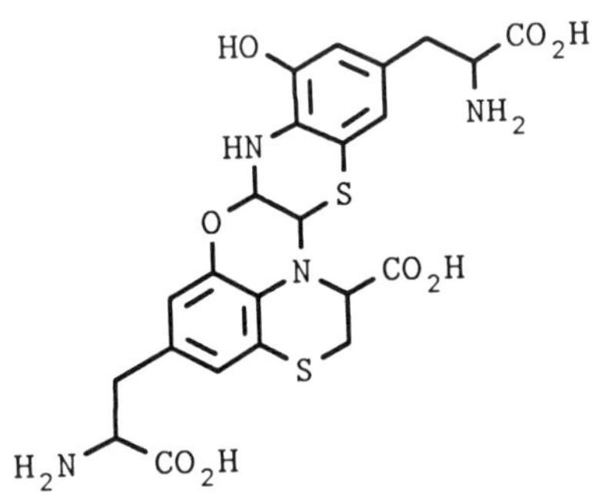

7.**28**

Mechanistically, compounds 7.**28** can be envisaged as arising from cycloaddition of the quinonimine 7.**24**, formed by oxidation of the dihydrobenzothiazine 7.**25**, to the 4*H*-1,4-benzothiazine 7.**29**, derived from decarboxylation of 7.**24**. Alternatively, the formation of 7.**28** (Fig. 7.9) could

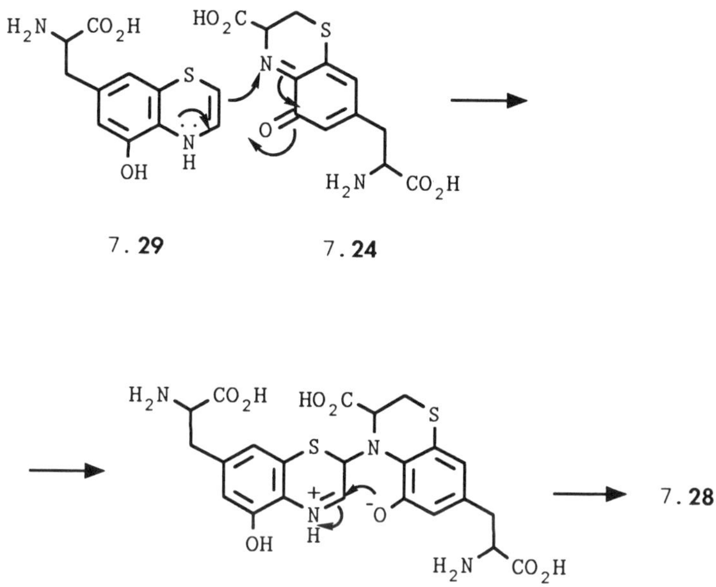

FIG 7.9 Proposed mechanism of formation of the diastereoisomeric dimers 7.**28** (Costantini *et al.*, 1990).

proceed by coupling of semiquinonimine radicals arising from 7.**26** and 7.**27**. The facile generation of such radicals has been demonstrated by EPR spectroscopy (Sealy *et al.*, 1982a). Which of these two mechanisms is actually operative is difficult to assess due to the intrinsic complexity of the chemistry of benzothiazines and the relatively low yield of 7.**28**.

A characteristic feature of the chemistry of 7.**28** is the marked instability to acids: When an aqueous solution is treated with hydrochloric acid a deep violet color develops in few minutes, and trichochrome F could be identified in the mixture. The formation of the trichochrome involves probably acid-catalyzed ring opening of 7.**28** to give the 1,4-benzothiazine 7.**26**, which, under acidic conditions, is highly unstable and readily undergoes oxidative coupling at the 2 positon.

In their extensive studies on pheomelanogenesis *in vitro* Fattorusso *et al.* (1969a) observed that when the enzymic oxidation of 5-cysdopa is stopped in the early stages by acidification of the mixture, a small amount of trichochrome F is formed. At the time, this behavior was ascribed to an acid-unstable benzothiazine intermediate. However, in light of these findings, it is more likely that formation of trichochrome F is due to the acid decomposition of the isomeric dimers 7.**28**.

Compounds 7.**28** are the first products beyond the benzothiazine stage that have been isolated from the biomimetic oxidation of 5-cysdopa. Although it is not clear whether they play a role in the biosynthesis of pheomelanins or merely represent a side branch of the pathway, their marked instability to acids is of interest in relation to the possibility that trichochrome and F (and E) may be artifactually derived from 7.**28** (or the analogous dimer from 2-cysdopa) during the extraction and work-up procedures.

CHAPTER 8

Pigment Cell Metabolism: Enzymatic and Chemical Control

I. BEYOND THE TYROSINASE CONCEPT

Traditionally, studies of the biosynthesis of melanins have been guided by the Raper–Mason scheme, in which all the steps following the tyrosinase-catalyzed formation of dopaquinone are envisaged as proceeding more or less spontaneously, without any further enzymatic assistance. Accordingly, pigmentary variations of skin and hair have mainly been ascribed to differences in the rate of synthesis and/or activation of tyrosinase, regarded as the major controlling factor of melanogenesis.

There are, however, a number of biological observations suggesting that tyrosinase, though essential for melanogenesis to occur, is not the sole requirement. At the cellular level, this is seen in the continent melanocytes, for example, of the adult eye, where the enzyme is definitely active (Dryja *et al.*, 1978) and yet no melanin is formed. Even more striking is the inverse relationship between the extent of melanization and the level of tyrosinase activity in different subcellular compartments of epidermal melanocytes (Imokawa and Mishima, 1981; Mishima and Imokawa, 1985). Typically, the highest activity is found in coated vesicles (CV) and premelanosomes (PMS), whereas the lowest is detected in melanosomes (MS) that are heavily pigmented (Fig. 8.1).

Why pigment formation does not occur whenever tyrosinase is present has always been a puzzling question. One proposed explanation is that tyrosinase is functionally active in CV and PMS but that melanization is interrupted, perhaps by an indole blocking factor, until entry of the enzyme-inhibitor complex into melanosomes (Pawelek *et al.*, 1980; Chakraborty *et al.*, 1989). Others have proposed that melanogenesis could be regulated by posttranslation inducers which might affect the glycosylation of the enzyme or its transport and subsequent binding to the melanosome membrane (Imokawa and Mishima, 1984; Fuller *et al.*, 1987). Such a mechanism is thought to operate in mouse hair follicle melanocytes where a reduction in tyrosinase activity

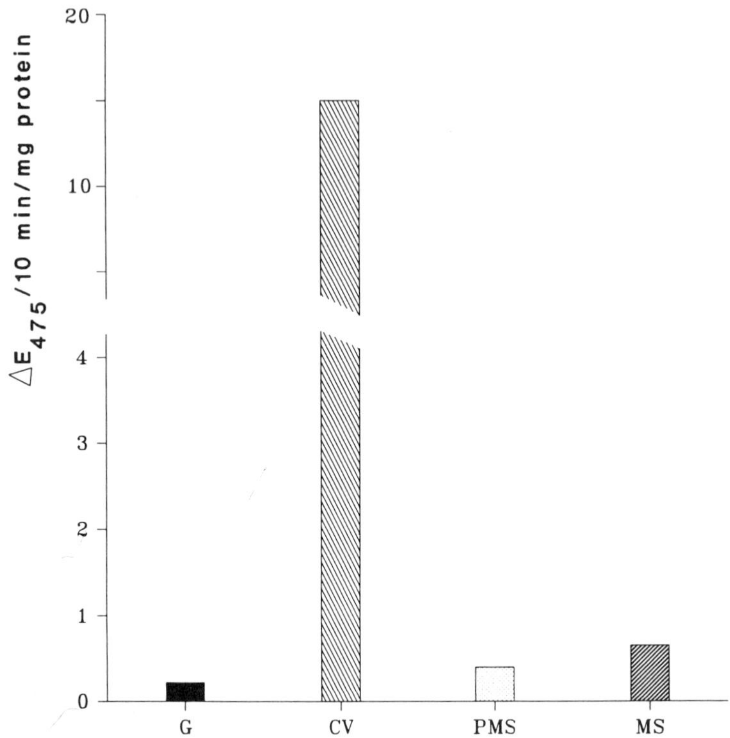

FIG 8.1 Distribution of tyrosinase in subcellular compartments of melanocytes (Imokawa and Mishima, 1981). G, Golgi apparatus; CV, coated vesicles; PMS, premelanosomes; MS, melanosomes.

seems associated with a decreased glycosylation of the enzyme (Burchill *et al.*, 1989).

Another suggestion is that tyrosinase at extramelanosomal sites is loosely bound to an inhibitor which is removed during purification (Chen and Chavin, 1975) or is diluted or inactivated upon entry of the enzyme into the melanosomal compartment (Hearing and Jimenez, 1987; Martinez *et al.*, 1987). This view would be in keeping with several reports dealing with inhibitors of tyrosinase activity, one of which has recently been isolated from unpigmented melanoma cells (Kamayama *et al.*, 1989). It has alternatively been proposed that stimulation of melanogenesis in melanocytes involves activation of pre-existing tyrosinase by MSH and other agents capable of raising the intracellular levels of cAMP (reviewed by Moellmann *et al.*, 1988).

Yet, in spite of these persistent efforts, it is clear that regulation of melanogenesis goes far beyond a mere control of tyrosinase activity. This is because once dopaquinone is generated its metabolic fate is ultimately dependent upon the sulfhydryl content within melanocytes. Thus, by implication, enzymes which influence the redox state of the glutathione system can indirectly affect the level of melanization, even if tyrosinase is normally expressed (Prota, 1980a, 1989a). In addition, evidence has been accumulating in recent years indicating the existence of a number of other enzymes and co-factors which can play a critical role in the later stages of melanogenesis. To gain perspective, in the following sections the nature and role of each of these regulatory factors will be dealt with separately; we will then see how these different pieces of information, derived from *in vivo* and *in vitro* studies, may be fitted together in a meaningful way to account for the biosynthesis of eumelanins, pheomelanins, and related metabolites.

II. CONTROLLING FACTORS ACTING DISTAL TO TYROSINASE

A. Sulfhydryl Compounds

1. Addition to Dopaquinone

The involvement of cysteine in the biosynthesis of pheomelanins and trichochromes was discussed earlier. In more recent years, emphasis has been placed on the regulatory role of sulfhydryl compounds, especially GSH, in eumelanin pigmentation. Credit for the first relevant investigation in this area goes to Flesch and Rothman (1948), who described the presence in aqueous extracts of human skin of "a dialysable, water soluble, –SH containing compound," capable of inhibiting the tyrosinase-catalyzed conversion of tyrosine or dopa to melanin. Later, Halprin and Ohkawara (1967) supplied circumstantial evidence that the inhibitory SH-containing substance postulated Flesch Rothman was GSH. They went on to show that GSH and GSH reductase levels were significantly lower in Negro skin than in white Caucasian skin.

At that time, little was definitely known about the chemistry of melanogenesis, and the observed reciprocal relationship between skin pigmentation and cutaneous level of GSH was generally explained in terms of the ability of SH groups to inhibit tyrosinase. A possible physiological role of GSH and cysteine as regulators of tyrosinase activity has been reconsidered recently (Jara *et al.*, 1988b). However, by analogy with the behavior of cysteine, a covalent binding of GSH to dopaquinone now appears more likely. This is supported by model experiments showing that tyrosinase-catalyzed oxidation of dopa in the presence of GSH proceeds in a similar fashion as with cysteine, to give the corresponding adducts, 5-, 2-, and 6-GS-dopa, as well as a diadduct tentatively identified as 2,5-diGS-dopa (Ito *et al.*, 1985a).

The yields of the glutathionyl monoadducts formed under these conditions are 76, 12, and 5% for the 5-, 2-, and 6-isomers, in that order. This pattern is evidently similar to that of the cysdopa isomers formed by addition of cysteine to dopaquinone (Ito and Prota, 1977a). Such a correlation is of interest in relation to the possibility that cysdopas may be formed *in vivo* via hydrolysis of the corresponding GS-dopas rather than by direct addition of cysteine to dopaquinone (Fig. 8.2).

The enzymes necessary for this transformation, γ-glutamyl transpeptidase (γ-GTP) and peptidase, are of common occurrence in biological systems (Meister and Anderson, 1983) and have also been detected in malignant melanoma cells (Agrup *et al.*, 1975; Hu and Buxman, 1981). Moreover, Mishima and co-workers (Mojamdar *et al.*, 1982, 1983) have found that the level of γ-GTP is higher in melanotic than in amelanotic cells. They also obtained evidence that the intracellular distribution of γ-GPT parallels that of tyrosinase, the highest concentration occurring in coated vesicles rather than melanosomes (Chakraborthy *et al.*, 1989). However, the question of which of the two possible pathways prevails *in vivo* has not yet been settled.

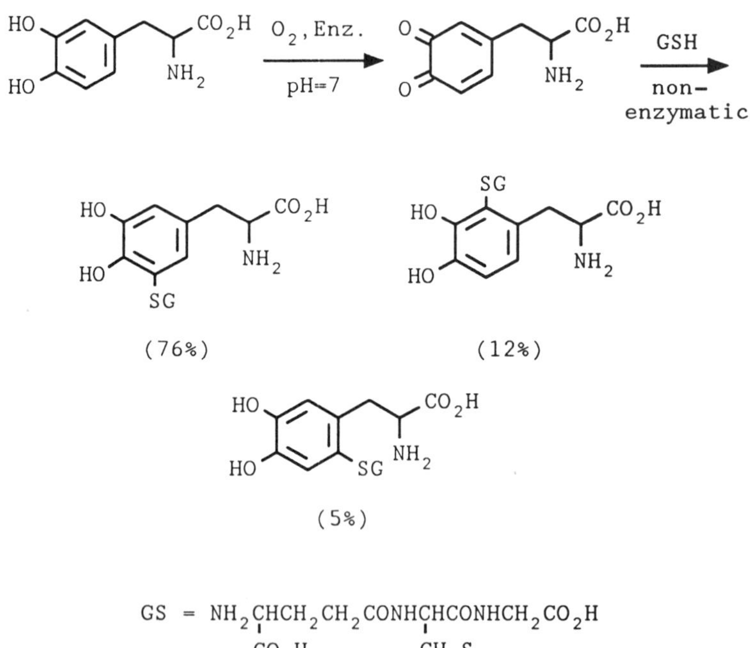

FIG 8.2 Reaction of glutathione with enzymatically generated dopaquinone.

According to Rorsman *et al.* (1988), though occasionally found in melanoma tissues, 5-GS-dopa may not be a normal metabolite of melanocytes, for it is probably produced when dopa oxidation occurs outside tyrosinase-containing cells. This is supported by the demonstration of 5-GS-dopa in several organs of albino rats after injection with large doses of dopa (Fehling *et al.*, 1981). Under such conditions some degree of nonspecific oxidation of dopa seems to take place, leading to GS-dopa(s), presumably by the action of methemoglobin or other oxidizing systems capable of generating dopaquinone, for example, peroxidase/H_2O_2.

In other studies at Lund, the concentration of 5-cysdopa in cultured melanoma cells was shown to be more than 100 times that of 5-GS-dopa (Carstam *et al.*, 1987; Karg *et al.*, 1991a). Moreover, 5-γ-glutamylcystein-*S*-yldopa was not found in the cells, indicating that GSH synthesis and dopaquinone formation occur at different sites in the melanocytes. When combined, these results were taken to mean that formation of cysdopas in melanocytes proceeds via direct addition of cysteine to dopaquinone rather than by hydrolysis of GS-dopas. However, while this is likely to be the case in cell culture systems, *in vivo* the tissue concentration of GSH should normally be such that it favors the formation of GS-dopas (Prota, 1989b).

Another aspect of the metabolism of GS-dopas that has been considered is their oxidative behavior under nonhydrolytic conditions. From a priori consideration of the structure of GS-dopas, it is apparent that the direct conversion of these metabolites to pheomelanins by way of 1,4-benzothiazine intermediates is precluded. Therefore, any oxidative transformation of GS-dopa(s) should invariably involve the intramolecular cyclization of the alanine side chain via glutathionyldopaquinone(s) structurally analogous to dopaquinone. These would then undergo ring closure to give sulfur-containing eumelanins via oxidative polymerization of glutathionyl-DHICA intermediates, as outlined in Fig. 8.3.

In support of the proposed pathway, Carstam *et al.* (1987) reported spectroscopic evidence that on oxidation GS-dopa is converted to eumelaninlike pigments through a series of colored intermediates similar to those observed in the enzymic oxidation of dopa. In another study, Ito *et al.* (1988) found that both cysteine and GSH can be incorporated in synthetic melanins, the former as part of benzothiazine units in pheomelanin-type polymers, and the latter covalently linked to 5,6-dihydroxyindole units in a eumelanin-type polymer.

2. *Addition to Other Melanogenic Precursors*

In the course of a survey of the reaction of sulfhydryl compounds with melanogenic quinones, Bouchilloux and Kodja (1960) found that the oxidation of DHI to melanin in the presence of tyrosinase is inhibited by an excess of GSH. This was attributed to the addition of GSH to 5,6-indolequinone to give a colorless

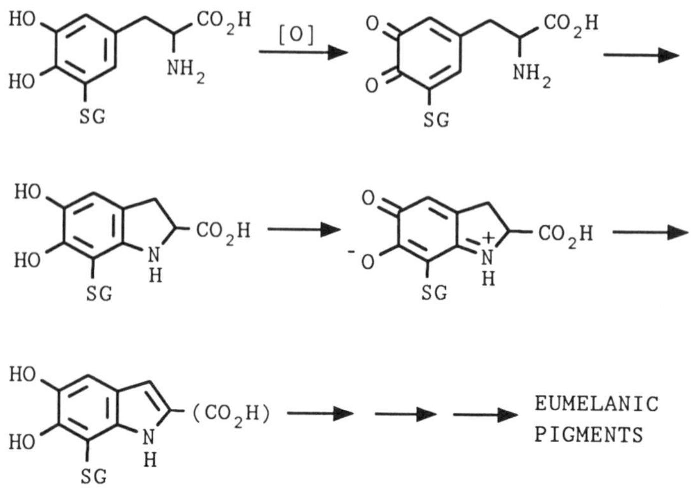

FIG 8.3 Oxidative cyclization of 5-GS-dopa.

adduct. A similar conclusion was later drawn by Mason and Peterson (1965) on the basis of spectrophotometric analysis of the incubation mixture.

In a reexamination of these studies, d'Ischia *et al.* (1987a) succeeded in isolating the observed adduct, which was identified as 4-glutathion-*S*-yl-5,6-dihydroxyindole by analysis of its ^1H and ^{13}C NMR spectra. Tyrosinase-catalyzed oxidation of DHICA in the presence of GSH proceeded in a similar fashion to give the corresponding 4-glutathion-*S*-yl adduct (Prota *et al.*, 1988).

Another potential target of sulfhydryl compounds that has been considered is dopachrome. In the case of GSH, the reaction gives, besides the known rearrangement products (DHI and DHICA), a colorless adduct identified as 8.**1** (d'Ischia *et al.*, 1987b). This probably arises by direct addition of GSH to the electrophilic 4 position of dopachrome to give an indoline intermediate, followed by reoxidation and rearrangement of the resultant 4-glutathion-*S*-yldopachrome. Replacement of GSH with cysteine results in the formation of the new pyrrolebenzothiazine adduct (8.**2**) which rapidly autooxidizes to give complex mixtures of yellowish-brown oligomeric materials.

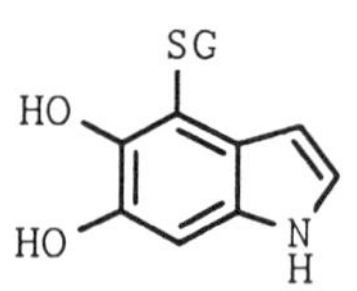

8.**1**

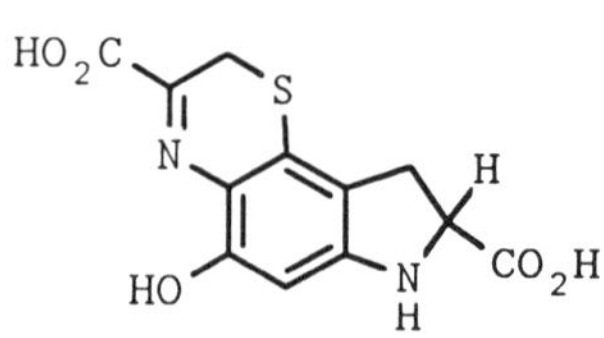

8.**2**

These results indicate the possibility that sulfhydryl compounds could also play a regulatory role at various stages of the eumelanin pathway distal to dopaquinone formation. However, direct evidence for this notion is not yet available.

B. Metal Ions

In addition to sulfhydryl compounds, certain metal ions, especially Cu^{2+}, Zn^{2+}, and Fe^{2+}, have long been known to be involved in melanin pigmentation (Lerner and Fitzpatrick, 1950). Typically, copper-deficient diets invariably result in depigmentation of hair in rats, cats, rabbits, and cattle. Addition of traces of copper to the deficient diet restores pigmentation. Zinc and iron are ineffective by themselves in reversing the depigmentary process, although their administration in combination with copper is sometimes more effective than copper alone. Moreover, very high levels of these metals and others, for example, manganese, have been found in a variety of melanized structures, including the choroid of the eyes (Bowness *et al.*, 1952), red or black hair (Flesch, 1968; Dorea and Pereira, 1983), pigmented moles (Molokhia and Portnoy, 1973), and isolated melanosomes from Harding-Passey, horse, and human melanomas (Horcicko *et al.*, 1973). In a recent study (Shibata *et al.*, 1990), the localization and distribution of metal ions within melanocytes have been examined in more detail using inductively coupled plasma mass spectrometry (ICP-MS). The most abundant metal ion was found to be zinc, followed by iron and copper, the latter in much lower amounts. The relative quantities of each of the metal ions in subcellular compartments of melanocytes from Green's hamster melanoma are shown in Fig. 8.4.

It may be noted that the metal content is lower in melanosomes than in premelanosomes, where little or no melanin is present. This would suggest that the accumulation of the metal ions in these structures is not merely the result of the binding properties of melanins, but is rather an intrinsic biochemical feature of melanin-producing cells that is likely to be under genetic control. This aspect, however, has not yet been investigated.

At the chemical level, the possibility that metal ions could catalyze some steps in the biosynthesis of melanins had long been considered (Nicolaus, 1968; Swan, 1974). The first relevant observation was reported by Palumbo *et al.* (1983) in the course of a study on the ability of 5-cysdopa to form complexes with various metal ions, including zinc, copper, and iron. In the presence of oxygen, these complexes were shown to undergo rapid nonenzymic oxidation, to give one of the postulated benzothiazine intermediates in the biosynthesis of pheomelanins. The proposed sequence of reactions, shown in Fig. 8.5, is basically similar to the enzymic oxidation of cysdopa, except for the

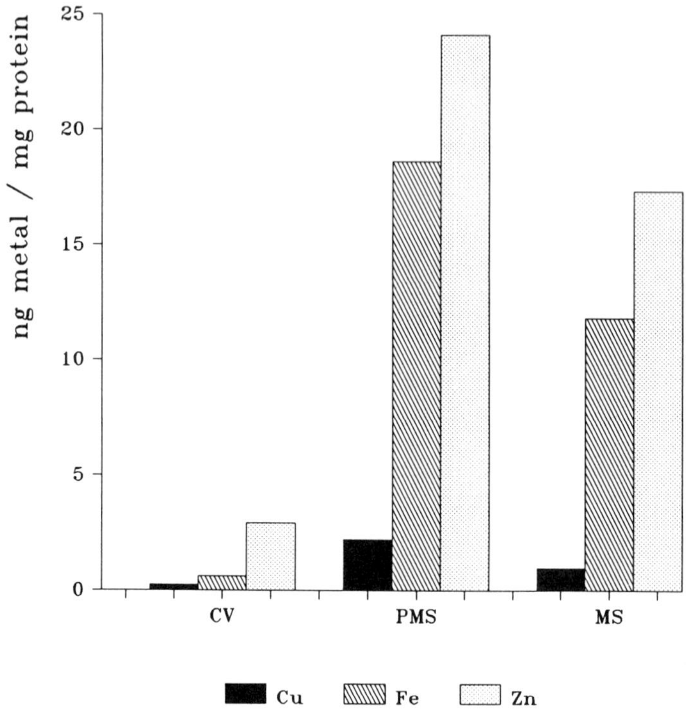

FIG 8.4 Distribution of metal ions in subcellular compartments of mammalian melanocytes (Shibata *et al.*, 1990). CV, coated vesicles; PMS, premelanosomes; MS, melanosomes.

rearrangement of the quinoneimine intermediate which, in the presence of metal ions, proceeds without decarboxylation. The rate of formation of the benzothiazine product is dependent upon the nature of the metal ion and is markedly accelerated by the action of the peroxidase/H_2O_2 system, consistent with the intermediacy of cysteinyldopaquinone. Notably, in the absence of metal ions oxidation of cysdopa with peroxidase gives mainly the dihydrobenzothiazine acid in 72% yield (Ito and Fujita, 1981).

A possible role of metal ions in the rearrangement of dopachrome was suggested by Bu'Lock and Harley-Mason (1951b). They reported that in the presence of zinc acetate dopachrome, generated *in situ* by chemical oxidation of dopa with potassium ferricyanide, rearranges very rapidly to give DHI. Since these studies, Zn^{2+} has often been included in the Raper–Mason scheme as a natural promoting factor for the conversion of dopachrome to DHI. However, the concentration of zinc acetate used in these experiments was exceedingly

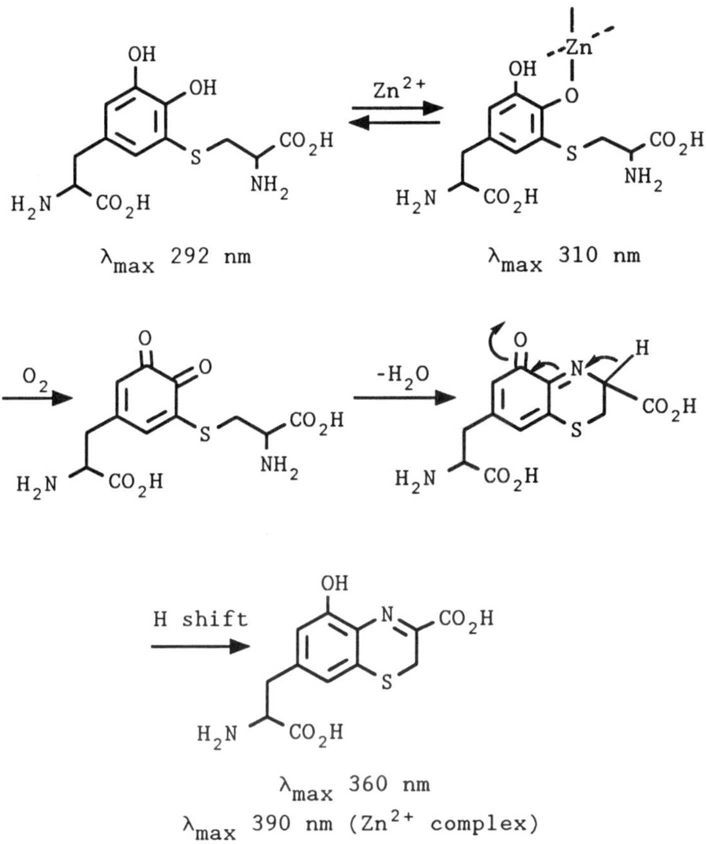

FIG 8.5 Metal-catalyzed autooxidation of 5-cysdopa (Palumbo *et al.*, 1983).

high (about 20% by weight) and, as admitted by the authors, in only one experiment was a 30% yield of DHI obtained. In fact, when monitored spectrophotometrically, addition of zinc acetate to dopachrome results in the formation of a purple pigment with a broad absorption maximum at 540–560 nm, typical of melanochrome (Palumbo *et al.*, 1987a). The reaction rate is dependent upon the concentration and nature of the metal added as well as the pH and type of buffer used. Under comparable conditions, copper proved to be by far the most active, inducing an almost instantaneous conversion of dopachrome to melanochrome (Fig. 8.6). Ferrous and ferric ions were less effective than copper but definitively more active than zinc.

In further studies (Prota, 1988b) it was found that the enhanced rate of conversion of dopachrome to melanochrome was due to the ability of the metal ions to catalyze each of the two sequential reactions involved, that is, the rearrangement of dopachrome to 5,6-dihydroxyindole(s) and their subsequent oxidative polymerization to melanochrome. To distinguish the two effects, the rearrangement of dopachrome was then investigated in the absence of oxygen, whereby oxidation of the resultant products to melanochrome was prevented. Typically, in the absence of metal ions, the rearrangement of dopachrome in phosphate buffer at pH 6.8 proceeded at very low rate, with 50% decomposition being reached only after 1 hr. Addition of copper at very low concentration resulted in a dramatic enhancement of the reaction rate, as evidenced by the rapid decay of the dopachrome chromophore at 475 nm (Fig. 8.6).

Among the other metal ions tested, Ni^{2+} and Co^{2+} also affected significantly the rate of dopachrome rearrangement, while Mn^{2+}, Ca^{2+}, Fe^{3+} and, notably, Zn^{2+} had a smaller yet detectable effect. Kinetic studies indicated that the metal-promoted rearrangement is first order with respect to both dopachrome and metal concentration and is progressively slowed down in the presence of increasing concentrations of EDTA, consistent with a mechanism involving a direct 1:1 dopachrome–metal ion interaction in the transition state (Palumbo *et al.*, 1987a).

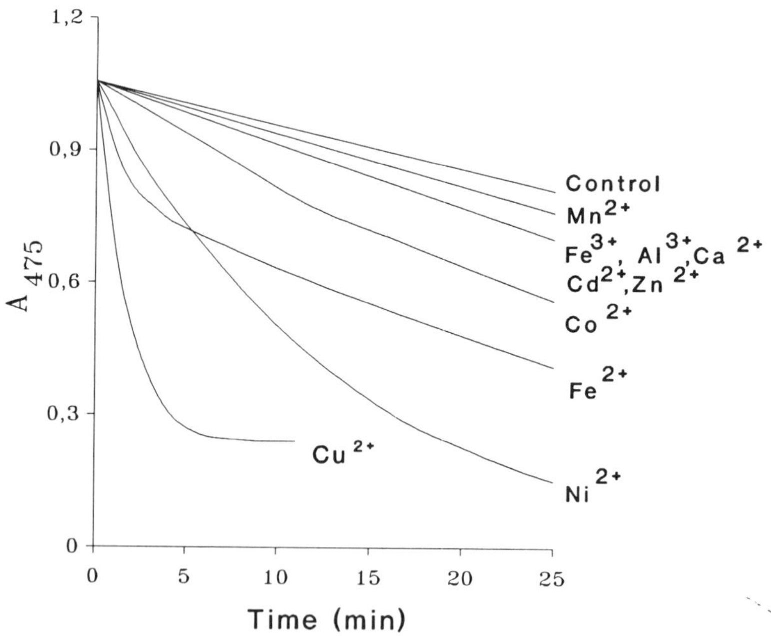

FIG 8.6 Effect of metal ions (2.0×10^{-4} *M*) on the rearrangement of dopachrome (4.0×10^{-4} *M*).

HPLC analysis of the incubation mixture after decomposition of dopachrome showed that metal ions affect not only the kinetics of the rearrangement but also the ratio of formation of DHI/DHICA. This is shown in Table 8.1 for Cu^{2+}, Ni^{2+}, and Co^{2+} at various concentrations. Again, it may be noted that the ratio of formation of DHI/DHICA depends upon the nature and concentration of the metal ion. As a rule, the more active is the metal on the kinetics of the reaction, the more it induces the nondecarboxylative rearrangement of dopachrome to DHICA.

Similar results on the effect of metal ions on dopachrome rearrangement have been reported by Jara *et al.* (1990), who used both spectrophotometric and radiometric assays to evaluate the rate of formation of DHICA versus DHI and their extent of incorporation into the pigment polymer. A major consequence proposed by these findings would be that melanins prepared in the presence of metal ions must contain a high proportion of DHICA-derived units. This has been substantiated by chemical analysis of a series of enzymic dopamelanins prepared in the presence of Cu^{2+}, Co^{2+}, Ni^{2+}, and Zn^{2+} (Palumbo *et al.*, 1988). When subjected to acid decarboxylation and permanganate oxidation these melanins released significantly higher quantities of carbon dioxide and PTCA with respect to a control pigment prepared in the absence of metals.

Turning back to the rearrangement of dopachrome, Costantini *et al.* (1991) have recently reported kinetic and isotopic experiments showing that in the absence of metal ions the reaction proceeds by a mechanism involving as the

TABLE 8.1

EFFECT OF METAL IONS ON THE MODE OF REARRANGEMENT OF DOPACHROME[a]

Metal ($\times 10^{-4}$ *M*)	Yield of DHI (%)	Yield of DHICA (%)	Ratio DHICA/DHI
—	97	1.4	0.01
Cu^{2+} (2.0)	26	66	2.5
Cu^{2+} (1.0)	41	53	1.3
Cu^{2+} (0.5)	48	50	1.0
Cu^{2+} (0.2)	56	39	0.7
Cu^{2+} (0.1)	71	23	0.3
Ni^{2+} (4.0)	16	84	5.2
Ni^{2+} (2.0)	25	70	2.8
Ni^{2+} (1.0)	45	55	1.2
Ni^{2+} (0.5)	60	40	0.6
Co^{2+} (4.0)	38	55	1.4
Co^{2+} (2.0)	60	34	0.6
Co^{2+} (1.0)	72	28	0.4

[a]4.0×10^{-4} *M*.

rate-determining step abstraction of the proton at position 3 to give a quinonemethide intermediate (Fig. 8.7) which then undergoes decarboxylation to DHI. The same pathway was inferred by the isolation of the stable quinonemethide by rearrangement of α-methyldopachrome methyl ester, in which both decarboxylation and deprotonation are blocked (Cheng *et al.*, 1982; Crescenzi *et al.*, 1990; Sugumaran and Semensi, 1990).

The mechanism of the metal-catalyzed rearrangement of dopachrome to DHICA has not yet been elucidated. However, from consideration of the above pathway, it is likely that the nondecarboxylative route results from an increased acidity of the H-2 proton due to metal coordination at C-5 and C-6 oxygens, whereby removal of this proton prevails over carboxyl loss, even at neutral pH.

C. Dopachrome Tautomerase

1. Occurrence and Nomenclature

The story of this enzyme begins in 1980 when Pawelek and his associates at Yale (Korner and Pawelek, 1980; Pawelek *et al.*, 1980) reported the occurrence in melanoma cell extracts of a number of regulatory factors capable of affecting the enzymic formation of dopamelanin after the dopaquinone stage. These included a "dopachrome conversion factor" (DCF) acting on the kinetics and mode of rearrangement of dopachrome, an "indole blocking factor" (IBF) inhibiting the conversion of DHI to melanin, and an "indole conversion factor (ICF) favoring the oxidation of DHI to melanochrome. Part of this latter activity was ascribed to tyrosinase (Korner and Pawelek, 1982; Murray *et al.*, 1983), and later the other two factors were shown to be associated with the activity of a single enzyme, named dopachrome oxidoreductase (DCOR), found in both normal and malignant melanocytes (Barber *et al.*, 1984, 1985;

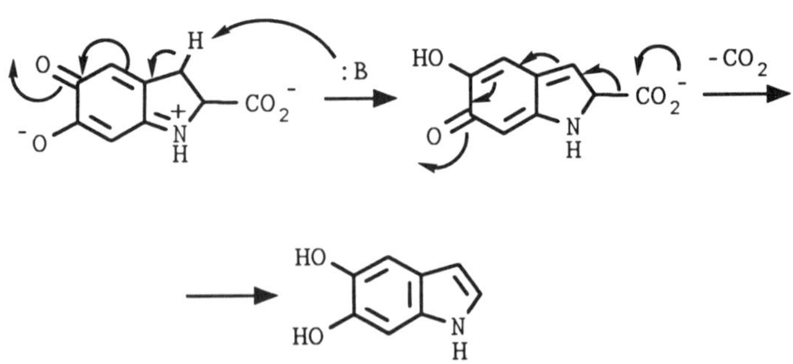

FIG 8.7 Mechanism of rearrangement of dopachrome.

Lamoreaux *et al.*, 1986). Early studies on the activity of this enzyme were carried out on crude or partially purified homogenates from melanoma cells and the interpretation of the observed effects on melanogenesis was not entirely unambiguous with respect to such parameters as the nature and molecular size of the active components, presence of protein and metal contaminants, and product identification. Thus, for example, the reaction catalyzed by DCF or DCOR was originally thought to involve the rearrangement of dopachrome to DHICA followed by a spontaneous decarboxylation to DHI (Pawelek and Korner, 1982; Barber *et al.*, 1984). However, at physiological pH DHICA does not undergo decarboxylation, implying that DHI, if formed, must derive directly from the decarboxylative rearrangement of dopachrome (Napolitano *et al.*, 1985a). In fact, when [^{14}C]carboxyl-labeled dopachrome was used as substrate for DCF, Korner and Pawelek (1980) observed that the rate of decoloration of dopachrome did not correlate with the $^{14}CO_2$ liberation. This discrepancy was reconciled by Korner and Gettins (1985), who found that the colorless product arising from the rearrangement of dopachrome catalyzed by DCF is DHICA, identified by mass spectrometry and ^{1}H NMR. This has been since confirmed by other workers using various analytical techniques such as HPLC (Leonard *et al.*, 1988; Pawelek, 1990) and UV and radiometric analysis (Aroca *et al.*, 1990). There are grounds to believe that DCOR and DCF probably correspond to the same enzyme insofar as both are capable of catalyzing a tautomeric shift of dopachrome with formation of the more stable isomer DHICA. Accordingly, this enzymatic activity has been renamed dopachrome isomerase (Pawelek, 1990) or, more appropriately, dopachrome tautomerase (DT), as suggested by Aroca *et al.* (1990). In the IUB classification, DT has therefore been given the entry EC 5.3.2.3.

2. Distribution and Structural Features

Dopachrome tautomerase activity has been detected in both melanotic and amelanotic melanoma cell lines as well as in extracts of melanocyte-containing tissues from mice, hamsters, and humans (Pawelek, 1991). There are, however, significant differences in the properties of the enzyme preparations from various sources, and even from the same source but from different laboratories. Korner and Pawelek (1980) reported that in crude cells lysates DCF activity was associated with a membrane-bound heat-labile glycoprotein which co-migrated with tyrosinase through several purification steps. However, they found that one particular treatment, that is, salt elution from a QAE-Sephadex resin, converted DCF activity to a heat-stable form with a molecular weight or less than 1000.

On the other hand, the DCOR activity originally described by Barber *et al.* (1984) was a protease-sensitive protein with an estimated molecular weight of 34 kDa. Moreover, DCOR was unaffected by tyrosinase inhibitors such as DDC and PTU. More recently the same group (Townsend *et al.*, 1990), using an improved isolation procedure, obtained a purified protein fraction exhibiting DCOR activity with an approximate molecular weight of 100 kDa. In another study by Aroca *et al.* (1990), dopachrome tautomerase from B16 melanoma was reported to have an oligomeric structure of about 300 kDa, estimated by gel filtration, the molecular weight of the monomer being probably 46 kDa, as determined by SDS–PAGE electrophoresis. The activity of this enzyme preparation was competitively inhibited by DHICA and other indoles, including L-tryptophan. Pawelek *et al.* (1991), however, failed to observe inhibition of DCF by tryptophan even at concentrations as high as 40 m*M*. They also reported experimental evidence suggesting that DCF exists probably as a complex with MSH, tyrosinase, and other melanosomal proteins.

One explanation that has been offered to account for these discrepancies is the marked sensitivity of the enzyme to proteolysis (Barber *et al.*, 1984; Aroca *et al.*, 1990) and/or the degree of glycosylation (Pawelek, 1991), though the existence in melanocytes of multiple factors acting on the rearrangement of dopachrome cannot be ruled out at present without further investigation.

Given the difficulties encountered with the isolation of dopachrome tautomerase by conventional techniques, another approach (Tsukamoto *et al.*, 1992) has been to identify the gene encoding for the enzyme by careful analysis of the melanosomal-specific proteins from melanoma cell extracts. One of these proteins, termed TRP2, which putatively maps the slaty locus, was found to share significant sequence homology with tyrosinase, including a perfect conservation of the critical domains, such as putative copper and iron binding regions, the cysteine and tryptophan residues (Jackson *et al.*, 1992). The size of the encoded TRP2 protein (75–80 kDa) is somewhat larger than the product of the albino locus. By use of immunoaffinity chromatography, the protein has been purified and shown to possess DT activity. These results provide a solid background to look into the mode of action of the enzyme and its physiological role in melanin pigmentation.

3. Dopachrome Tautomerase and Metal Ions

Since both DT and metal ions are present in melanocytes, efforts have been made to discriminate their effects on the kinetics and mode of rearrangement of dopachrome (Leonard *et al.*, 1988; Jara *et al.*, 1990). In a vis-a-vis comparison using a partially purified DT preparation, Palumbo *et al.* (1991b) reported a number of basic differences between the modes of action of the enzyme and metal ions, which permit a clearcut distinction when both factors are present in biological extracts. In particular, they found that the enzyme-promoted reaction is highly stereospecific for L-dopachrome, is unaffected by EDTA, and has an optimal pH of about 6.7. By contrast, the kinetics of the metal (Cu^{2+})-catalyzed rearrangement are not dependent on the stereochemistry of the substrate, are affected by EDTA, and are not influenced by the pH of the medium in the range between 5 and 8. Moreover, the ratio of formation of DHICA/DHI determined by HPLC is significantly higher in the enzyme-catalyzed than in the metal-catalyzed reaction (Table 8.2).

A comparison of the relative activities of the two factors is difficult due to their entirely different chemical natures, that is, a small metal ion and an apparently large and inhomogenous protein fraction. However, as is usually the case, the enzyme appears to be far more active than the metal ion in catalyzing the formation of DHICA from the rearrangement of dopachrome. Assuming a molecular weight for the enzyme of 50–100 K, Pawelek (1991) has estimated that DT is 10^5 to 10^6 times more potent than Cu^{2+} on a mole-per-mole basis in catalyzing the reaction. This has raised some doubts about the relative importance of metal ions in the regulatory control of the rearrangement of dopachrome. However, the roles of DT and metal ions do not seem to be mutually exclusive; indeed, they could act in a complementary fashion in the overall metabolism of tyrosine in mammalian melanocytes. It has also been suggested that DT is a metalloprotein that could utilize iron for its catalytic action (Pawelek, 1991). In this connection, Chakraborthy *et al.* (1989) have reported biochemical evidence showing that the subcellular distribution of DT in B16 mouse and hamster melanoma cells parallels that of tyrosinase, that is, very high in coated vesicles, low in premelanosomes, and absent in melanosomes, which is exactly the opposite of what we have seen for metal ions. From a functional viewpoint, another difference is that the activity of DT is mainly directed to convert dopachrome into DHICA, whereas metal ions have additional catalytic activities, for example, pseudocatalase or peroxidase activity, that may play a role in the oxidative polymerization of 5,6-dihydroxyindoles. Yet, a definite assessment of the actual role of metal ions in melanogenesis must await further investigation.

TABLE 8.2
EFFECT OF CUPRIC IONS AND DCT ON THE MODE OF REARRANGEMENT OF DOPACHROME[a]

Factor	Yield of DHI (%)	Yield of DHICA (%)	Ratio DHICA/DHI
Cu^{2+} (0.1×10^{-4} *M*)	71	23	0.3
Cu^{2+} (0.5×10^{-4} *M*)	43	55	1.3
Cu^{2+} (2.0×10^{-4} *M*)	26	66	2.5
DCT (3 mU)	52	61	1.2
DCT (6 mU)	26	75	2.9
DCT (12 mU)	7.5	90	12

[a] 2.0×10^{-4} *M*.

As a concluding remark, it is of interest that an enzyme acting on dopachrome has been described in insects (Aso *et al.*, 1989; Sugumaran, 1990). This has been named dopaquinoneimine conversion factor and has the remarkable ability to catalyze the decarboxylative rearrangement of dopachrome to DHI. The enzyme from *Manduca* hemolymph exhibits a narrow substrate specificity and attacks only the L-aminochromes such as L-dopachrome, L-dopachrome methyl ester, and α-methyl-L-dopachrome, but not the D-isomers. It did not act on dopaminochrome, indicating the requirement of the carboxyl group for the reaction to occur (Sugumaran and Semensi, 1991). Whether this enzyme is involved in melanogenesis is currently under evaluation, but it is noteworthy that different enzymes catalyzing the rearrangement of dopachrome and related aminochromes may occur in organisms at various levels on the evolutionary scale.

D. Peroxidase

Peroxidases are heme-containing proteins of widespread occurrence in plant and animal tissues. The best studied of this group of enzymes is horseradish peroxidase; it has a molecular weight of 40K, of which 18% is carbohydrate, and a ferriprotoporphyrin as the prosthetic group. Animal peroxidases contain an unidentified green hemin and occur mainly in membrane-bound forms. Functionally, however, they all behave similarly in utilizing hydrogen peroxide to oxidize various phenolic substrates, including monophenols, *ortho*- and *para*-diphenols, enediols, and aromatic amines. In all these reactions, a true peroxidase acts as a dehydrogenating agent rather than as an hydroxylating agent. In contrast with this notion, in 1970 Okun and his associates reported histochemical evidence that oxidation of tyrosine to melanin in mammalian melanocytes could be mediated by peroxidase. In further biochemical studies using isolated mieloperoxidase or horseradish peroxidase the same group (Patel *et al.*, 1971; 1974; Okun *et al.*, 1973) found that these two enzymes were able to convert tyrosine to dopachrome or melanin in the presence of dopa or dihydroxyfumarate as cofactor. They therefore concluded that "epidermal melanocytes having peroxidase activity could form melanin from tyrosine whereas epidermal melanocytes containing tyrosinase activity but not peroxidase could form melanin from dopa but not from tyrosine. Tyrosinase may therefore function primarily as a dopaoxidase, as conceived by Bloch" (Okun *et al.*, 1970).

Okun's interpretation of the enzymatic assistance in melanogenesis did not pass unnoticed and stimulated considerable activity in various laboratories. Thus, a few years later, Smith and Swan (1976) repeated the *in vitro* experiments of Okun to show that the alleged formation of dopachrome by peroxidatic oxidation of tyrosine was actually a misinterpretation of the spectral changes occurring during the reaction. On the other hand, Hearing and Ekel (1975) and Mufson (1975) obtained additional evidence that a true tyrosinase capable of hydroxylating tyrosine to dopa is definitively present in mammalian melanocytes. This series of papers apparently settled the question and the peroxidase hypothesis of melanogenesis was then relegated to history.

In a reexamination of these early studies at Naples, attention was focused on the possibility that peroxidase could act on another critical step of melanogenesis, that is the oxidative polymerization of DHI to melanin. This has been traditionally envisaged as the third site of intervention of tyrosinase (Korner and Pawelek, 1982; Hearing *et al.*, 1982). However, considering the marked tendency of 5,6-dihydroxyindoles to undergo oxidation, one would expect that a high specificity of the assisting enzyme, if any, is not required. And indeed, as we have seen in Chapter 5, peroxidase is more effective by far than tyrosinase in catalyzing the oxidative polymerization of DHI. This has been further substantiated by the results of a kinetic study (d'Ischia *et al.*, 1991) showing that under comparable conditions oxidation of DHI with the peroxidase/H_2O_2 system proceeds almost instantaneously, whereas with tyrosinase the reaction is rather slow (Fig. 8.8).

Even more dramatic is the difference in the relative effectiveness of the two enzymes when DHICA is used as substrate. The resulting data (Fig. 8.8) indicate a very poor catalytic effect of tyrosinase, in marked contrast with the peroxidase system, which again induces a fast and complete conversion of DHICA to melanin.

These results, besides supporting the involvement of peroxidase in melanogenesis, also provide an explanation for the incorporation of DHICA into the pigment polymer, which would otherwise be difficult to reconcile with the low oxidizability of this substrate by tyrosinase. Preliminary experiments from the author's laboratory strongly suggest that peroxidase may also play a critical role in the biosynthesis of pheomelanins, acting in particular on the oxidative polymerization of cysdopas. Like DHICA, cysdopas are very poor substrates of tyrosinase unless a suitable redox cycler such as dopa is present in the incubation mixture.

The biosynthetic implications derived from these *in vitro* studies have recently been integrated by a reexamination of the distribution of peroxidase in different subcellular compartments of melanocytes from hamster melanoma tissues (Shibata *et al.*, 1990). Interestingly, the highest levels of peroxidase activity were found associated with stage II and stage III premelanosomes, wherein pigment deposition begins to appear. Moreover, no significant peroxidase activity was detected histochemically in amelanotic melanoma cells, consistent with the putative role of this enzyme in melanogenesis.

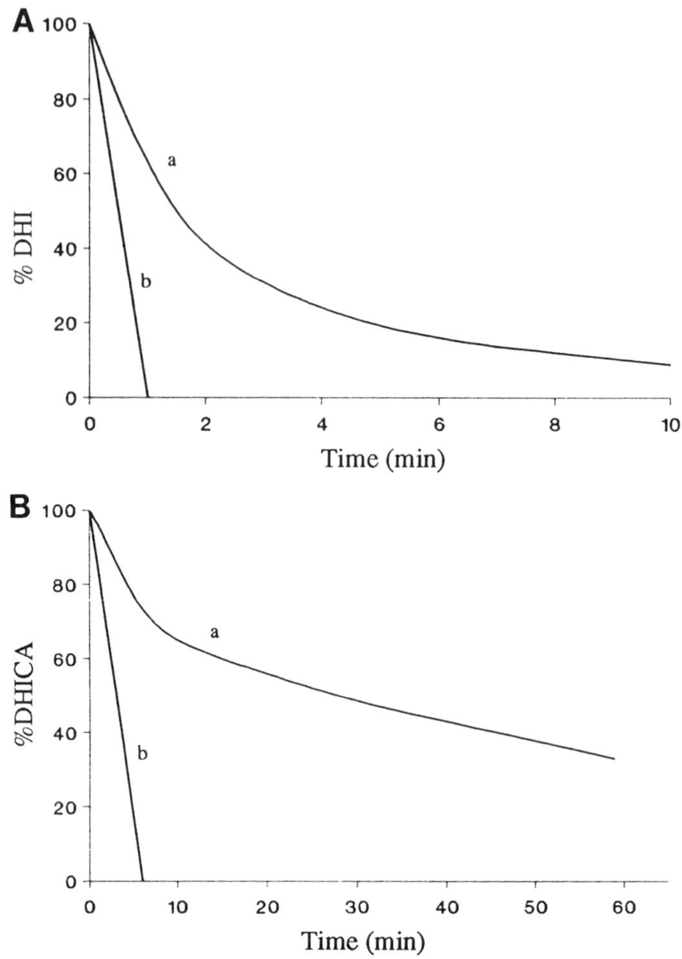

FIG 8.8 Comparison of the rate of oxidation of DHI (A) and DHICA (B) catalyzed by tyrosinase (a) and peroxidase-hydrogen peroxide (b), as followed by HPLC monitoring of substrate decay.

III. OVERVIEW OF METABOLIC PATHWAYS

It may be helpful at this point to provide the reader with an updated "working scheme" of melanogenesis (Fig. 8.9), highlighting the major known biosynthetic intermediates and possible control points.

As pointed out earlier, the primary requirement for melanogenesis is the presence of tyrosinase, which catalyzes the conversion of tyrosine to dopa-

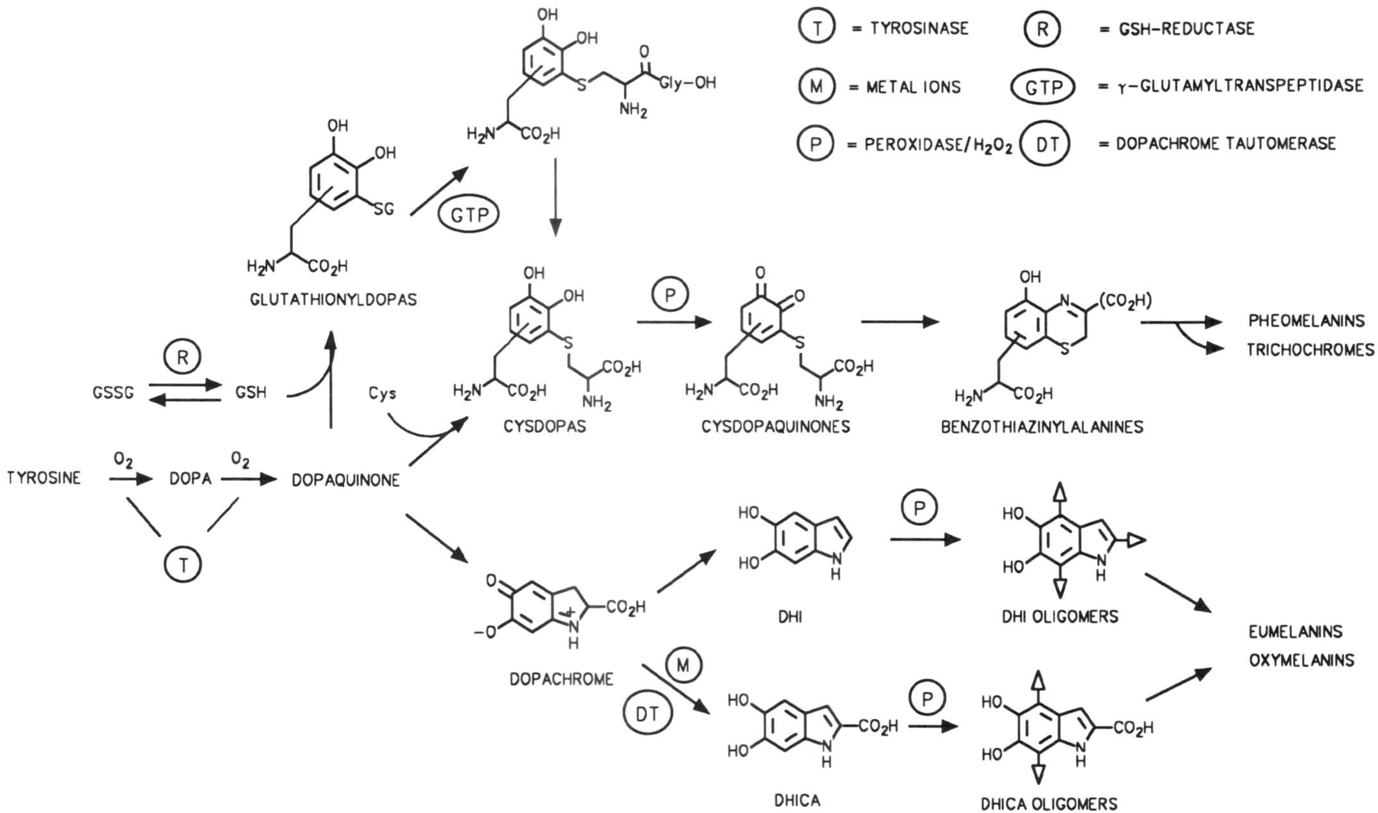

FIG 8.9 Comprehensive scheme of metabolic pathways leading to melanins and related metabolites.

quinone. The subsequent fate of this unstable intermediate is the result of its intrinsic reactivity coupled with the biochemical milieu of the subcellular compartments in melanocytes, which is genetically controlled.

Under conditions of low sulfhydryl content, as in eumelanin-forming melanocytes, most of the dopaquinone is converted to dopachrome, which may undergo enzymatic or metal-catalyzed rearrangement to give DHI and/or DHICA. The subsequent polymerization of these two indoles to melanin is probably associated with a depletion of the antioxidant defenses at the melanosome site, leading to an increase in the intracellular level of reactive oxygen species. As a result, some hydrogen peroxide is eventually generated which, in combination with peroxidase, can efficiently bring about the oxidative polymerization of both DHI and DHICA.

Hydrogen peroxide is also responsible for the subsequent oxidative breakdown of part of the 5,6-dihydroxyindole units in the pigment polymers, accounting for the increased molecular heterogeneity of natural eumelanins. However, whether this side reaction occurs within melanocytes or after transfer of the pigment granules to keratinocytes is not clear. As indicated in the proposed scheme, a more extensive attack of the eumelanin polymer(s) by hydrogen peroxide could reasonably explain the formation of oxymelanins, but this remains to be confirmed at both the chemical and the biological level.

The cytophysiology of pheomelanin synthesis has been most extensively investigated in the formation of the yellow (agouti) band during hair growth in mice (Quevedo *et al.*, 1981; Takeuchi, 1985). An interesting observation, supported by more recent studies (Burchill *et al.*, 1986; Movaghar, 1989; Granholm *et al.*, 1990) is that tyrosinase activity during pheomelanogenesis is lower than in eumelanogenesis. The mechanisms that regulate tyrosinase synthesis and processing are also different in eumelanin- and pheomelanin-forming melanocytes (Burchill, 1989). Moreover, the switching from eumelanogenesis to pheomelanogenesis is accompanied by a marked overall decrease in the melanin content of hair, as indicated by the ratio AHP/PTCA (Ito and Fujita, 1985). These associations were taken to mean that the level of tyrosinase in melanocytes may be a critical factor in the switch mechanism leading from eumelanin to pheomelanin synthesis. Such a view, however, seems to be in contrast with an earlier report by King *et al.* (1978), showing that the enzyme levels in the human follicular melanocytes from red-haired subjects are higher that those from dark-haired subjects. Similar results have been recently confirmed by Burchill *et al.* (1991) upon comparing tyrosinase expression in some twenty three red- and dark-haired subjects. Thus, if it is conceded that pheomelanogenesis in mouse hair follicles follows the same pathway so that which occurs in the human counterpart, it would seem that the switching mechanism is not primarily associated with a specific level or change of tyrosinase activity but rather with an increased availability of sulfhydryl compounds in the melanogenic compartments of the pigment-producing cells, as previously suggested by Prota (1980a).

Direct support for this view is provided by comparative studies (Benedetto *et al.*, 1981, 1982) of the levels of glutathione and related enzymatic activities in tortoise shell guinea pig skin of different colors (black, yellow, red, and white) as well as in the skin of pure black (*a/a*) and yellow (A^y/a) mice. As expected, the lowest levels of glutathione reductase activity were found to be associated with eumelanin-type pigmentation, whereas the highest ones were found in the skin with pheomelanin-producing melanocytes. Moreover, analysis of the nonprotein thiol compounds revealed that the GSH level is lower in black skin than in yellow skin of the agouti mice. These results are in keeping with early studies by Cleffman (1963, 1964) showing that melanocytes producing pheomelanins give a very strong cytochemical reaction for free SH groups, suggesting that a cyclic change in the level of sulfhydryl compounds was responsible for the switching mechanism leading from eumelanin to pheomelanin synthesis during hair growth in the agouti mouse. In addition, he found that in tissue culture pheomelanin-forming melanocytes from agouti mouse follicles produce only black pigment but revert to pheomelanin production when sufficient glutathione or cysteine are added to the nutrient medium. The amount of sulfhydryl compound required for this reversal depends on the genetic makeup of the strain.

While the effect of cysteine on the synthesizing activity of melanocytes does not involve any additional enzymatic system, the participation of GSH in the biosynthesis of pheomelanins is more complex, since it requires the intervention of hydrolytic enzymes to release the bound cysteinyl residue after the coupling of the thiol group to the quinone. This implies that, in addition to serving as a possible substrate for the synthesis of cysdopas, GSH could play a regulatory role by converting part of the dopaquinone formed in melanocytes into GS-dopas, which, as such, cannot give rise to pheomelanins unless they are transformed to cysdopas. Hence, in the absence of this latter step, the formation of GS-dopas would only sidetrack the dopaquinone without producing either pheomelanin or eumelanin.

At last, we are beginning to perceive the pigmentary system as a complex machine where the various cogwheels are synchronously activated and/or inhibited in response to constitutive factors or contingent conditions. There are still many aspects to be explained, at both the chemical and the biological level. We need to learn more about the relative contribution of GSH and cysteine in melanogenesis, and what other enzymes, in addition to those mentioned in the scheme, control the redox state of the various subcellular compartments in melanocytes. Reports have appeared suggesting that superoxide dismutase (Carraro and Pathak, 1988; Sichel *et al.*, 1987) and catalase (Halaban and Moellmann, 1990; Schallreuter *et al.*, 1991) are involved in melanin pigmentation. In addition, the thioredoxin reductase/thioredoxin system, a major antioxidant catalyst in the human epidermis, has been increasingly implicated in the regulation of tyrosinase activity and the control of free radicals at the surface of melanocytes (Schallreuter and Wood, 1989). Further work along this line will significantly contribute to our understanding of the chemistry of melanogenesis.

IV. INTERMEDIARY PRODUCTS OF MELANOCYTE ACTIVITY

Most of the foregoing conclusions regarding the biosynthesis of melanins have been derived from studies of the melanization processes under biomimetic conditions, which may not necessarily reflect tyrosine metabolism in melanocytes. Perhaps the most convincing evidence that a close similarity exists between the *in vitro* and *in vivo* pathways comes from parallel studies on the nature and origin of urinary melanogens. This is a collective name for a series of phenolic metabolites responsible for the spontaneous darkening of the urine from patients with metastasizing melanoma (Duchon and Pechan, 1963). Until now, more than 20 different urinary melanogens have been identified, including the postulated melanin precursors and their metabolites. Several reviews are available on the occurrence, detection, and clinical significance of intermediary products of tyrosine metabolism in melanocytes: A most comprehensive one is by Rorsman and his associates (1983) who, during the last 20 years, have made important contributions to the field. Other more recent reviews have been written by Duchon (1987) and Rorsman and Pavel (1990).

A. Dopa and Catecholamines

Dopa is the first product of the tyrosinase-catalyzed oxidation of tyrosine and it may also be formed in melanocytes by reduction of dopaquinone, for example, by a redox reaction with dopachrome (see Chapter 1). Thus, a certain amount of dopa can be expected to be constantly present in melanocytes and partly excreted, to be metabolized in the body by decarboxylation, transamination, *O*-methylation, and conjugation with sulfate or glucuronic acid (Fig. 8.10). This has been documented by several studies showing that an increased excretion of dopa and related metabolites, for example, dopamine, dopac, and homovanillic acid (HVA), does occur in the urine of patients with melanoma metastasis (Voorhees, 1970; Hinterberger *et al.*, 1973). In such patients increased levels of 3-methoxytyrosine may also be present in the urine in amounts greater than those of free dopamine (Rorsman *et al.*, 1983).

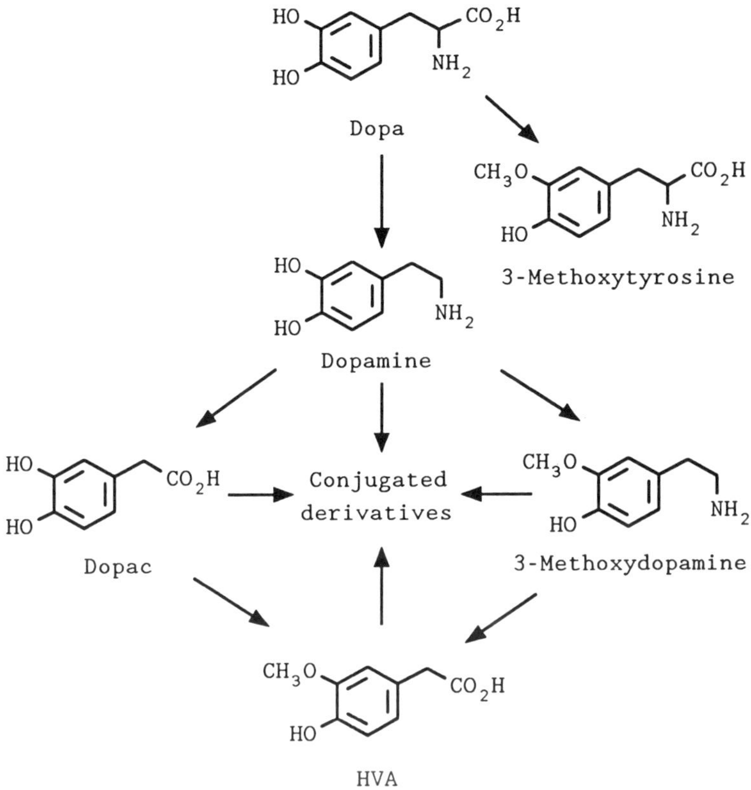

FIG 8.10 Nonoxidative metabolism of dopa and related catecholamines.

An interesting metabolite of dopac has been recently isolated from urine and identified as 5-cystein-*S*-yl-3,4-dihydroxyphenylacetic acid, arising by nucleophilic addition of cysteine or glutathione to the dopac *o*-quinone (Carstam *et al.*, 1990).

Goodall and Alton (1972) found that dopa metabolites such as dopamine, dopac, and HVA are to some extent conjugated with glucuronic acid or sulfate, and hydrolysis of these acid-sensitive derivatives therefore appears desirable before analysis of urine and serum (Hansson *et al.*, 1979b). A different approach, taken by Ito *et al.* (1985b), is based upon determination of the "total" excretion of acidic metabolites of catecholamines in urine after reductive hydrolysis with HI to convert methoxyphenylacetic acids and their conjugates to a single product, dopac, which is then directly measured by HPLC with electrochemical detection.

It may be noted that the nervous system and the adrenals contain the enzyme tyrosine hydroxylase which is also capable of converting tyrosine to dopa, but this is rapidly decarboxylated to dopamine by the enzyme dopa-decarboxylase. In contrast, it seems that melanocytes do not contain dopa-decarboxylase, so that dopamine and related catabolites should be formed outside these cells, especially in the kidney (Rorsman *et al.*, 1983). A few reports, however, have appeared indicating that malignant melanoma cells are capable of synthesizing catecholamines both in culture (Semple *et al.*, 1982; McEwan and Parsons, 1987) and *in vivo* (Faray *et al.*, 1981). This is an interesting metabolic difference between normal and malignant melanocytes which may have a practical implication in the control of melanoma growth.

B. 5,6-Dihydroxyindoles

Historically, DHI and related metabolites were the first urinary melanogens to be intensively studied and used for screening of metastisizing melanoma diseases (Duchon *et al.*, 1981). In the early clinical studies, the Thormählen reaction (sodium nitroferricyanide) was used, which gives a blue color (λ_{max}, 620 nm) with indole derivatives unsubstituted at the 2 and 3 positions (Leonhardi, 1953). Two types of melanogens were then found in melanoma urine: The Thormählen-positive and Thormählen-negative ones. In 1954 Leonhardi showed that the Thormählen-positive melanogens could be separated by paper chromatography into three major apparently homogenous spots, which he designated A, B, and C. These were very soon recognized as derivatives of DHI (Leonhardi, 1955), but it was several years before the structure of the individual components could be definitively established. This is because the Thormählen-positive melanogens occur in the form of sulfates and glucuronates which are unstable to acid and are also difficult to separate on a milligram scale by chromatography. Yet, using different approaches and analytical techniques, it was eventually possible to identify the main Thormählen-positive indoles corresponding to Leonhardi's A, B, and C compounds (Fig. 8.11).

Compound A is a mixture of glucuronides of 5-hydroxy-6-methoxyindole and 6-hydroxy-5-methoxyindole in an approximate ratio of 5:1 (Pavel *et al.*, 1981; Matous *et al.*, 1981). The 6-methoxy derivative, which is relatively more abundant in melanoma urine, has also been detected in normal urine. Compound B is also a mixture of the same isomeric methoxyindoles which are conjugated in the free OH position with sulfate, while compound C appears to be the 6-*O*-sulfate of the parent indole (Pavel, 1989).

Regarding the nature of the Thormählen-negative indoles, in the 1960s Duchon and his collaborators in Prague (Duchon *et al.*, 1966; Duchon and

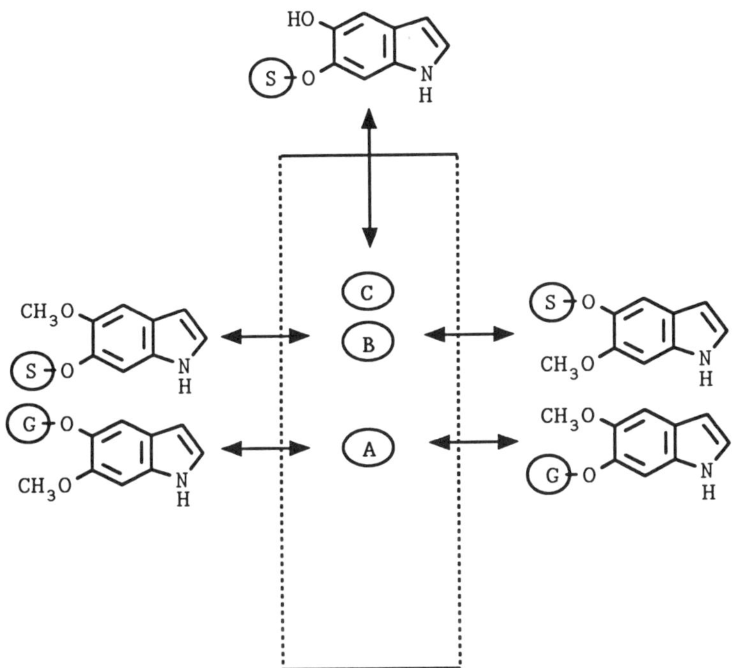

FIG 8.11 TLC separation of the Thormählen-positive melanogens on cellulose (*n*-butanol-pyridine-water, 1:1:1) Courtesy of Dr. S. Pavel.

Matous, 1967) first reported the occurrence in melanoma urine of two metabolites of DHICA, identified as the 5-methoxy and 6-methoxy derivatives, 5-hydroxy-6-methoxyindole-2-carboxylic acid (5H6MICA) and 6-hydroxy-5-methoxyindole-2-carboxylic acid (6H5MICA) (Fig. 8.12). The identification of these compounds was confirmed by Badinand *et al.* (1969); later they were also found by gas chromatographic–mass spectrometric analysis of urine from melanoma patients as well as from hamsters with transplantable melanotic melanoma (Pavel *et al.*, 1978; Wakamatsu and Ito, 1990). Small amounts of the dimethylated compound, 5,6-dimethoxyindole-2-carboxylic acid (DMICA), may be found in melanoma urine (Pavel, 1989). More recently, the two methylated isomers of DHICA, along with smaller amounts of the nonmethylated indole (DHICA) have been detected in normal urine by HPLC analysis using fluorimetric detection (Hansson, 1988).

Biogenetically, the indolic melanogens are likely to be formed in melanocytes via enzymic and/or metal-catalyzed rearrangement of dopachrome and subsequent methylation and/or conjugation with glucuronic and sulfuric acids. However, whether these additional steps occur in melanocytes or some-

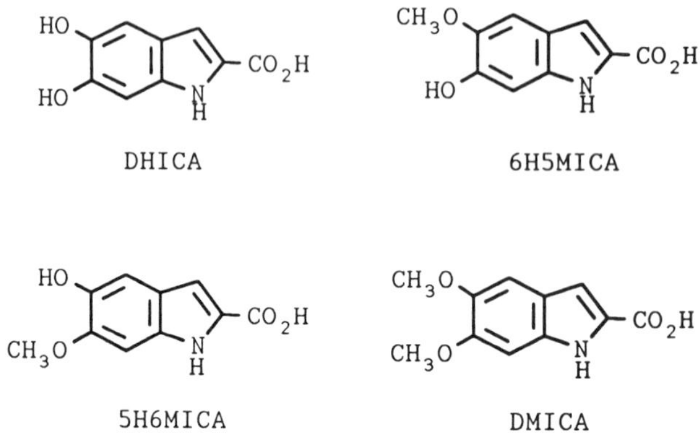

FIG 8.12 Structure of the Thormählen-negative indole melanogens.

where else in other tissues is not yet clear. Axelrod and Lerner (1963) found that *O*-methylation of dihydroxyindoles is catalyzed by catechol-*O*-methyltransferase (COMT) present in hamster melanoma and suggested that this enzyme may play a regulatory role in melanin pigmentation. This has been substantiated by the identification of COMT and various methylated indoles, that is, 5-hydroxy-6-methoxyindole (5H6MI), 6-hydroxy-5-methoxyindole (6H5MI), and 6H5MICA in the supernatant of melanoma cell cultures (Smit *et al.*, 1990). The determinations of COMT activity in subcellular fractions have shown that the enzyme is localized mainly in the cytoplasm, partly integrated in the structures forming endoplasmic reticulum. However, a more recent study on the distribution of COMT activity in melanotic and amelanotic hamster melanoma does not seem to support a regulatory role of this enzyme in melanogenesis (T. Shibata *et al.*, unpublished).

1. Analysis of Indole Melanogens

In recent years improved analytical techniques have been used for identification and quantitation of 5,6-dihydroxyindoles in blood and urine. Among these, the quantitative mass fragmentometric method, making use of deuterim-labeled internal standards, provides a valuable tool for analysis of the whole pattern of 5,6-dihydroxyindole metabolites (Pavel, 1989). However, not all indole metabolites are suitable for analytical purposes. This holds particularly true for DHI and DHICA, which are easily oxidized during collection, storage, and processing of urine samples. The monomethylated indoles are relatively more stable and 5H6MI was initially used by Pavel *et al.* (1986) as a specific melanocyte metabolite for monitoring increased melanogenesis in patients with pigmented malignant melanomas.

Recently, however, the 2-carboxylated compound, 5H6MICA, proved to be a more suitable marker of melanocyte activity (Westerhof *et al.*, 1987). A quantitative method based upon HPLC with fluorimetric detection has been developed which permits a rapid analysis of this metabolite in urines or tumor extracts by direct injection of the samples into the chromatographic system (Hansson, 1988). Another method developed by Wakamatsu *et al.* (1988) involves HPLC analysis with electrochemical detection of 5H6MICA, its isomer 6H5MICA, and their sulfates, referred to as total indoles. Their excretion value seems to provide a more complete insight into the production, circulation, and excretion of melanin-related metabolites. In clinical studies, however, quantitation of the sole 5H6MICA can be used for such a purpose as well, because it constitutes half of the total indoles in urine and its percentage ratios are markedly constant.

C. Cysdopas

The definition of the mechanism of formation of the sulfur-containing pheomelanins created an important background for the discovery of cysdopas as excretion products of normal and pathological melanocytes. The major one of these metabolites, 5-cysdopa, was first detected in melanoma tissues and in the urine from patients with metastatic pigmented melanoma (Bjorklund *et al.*, 1972). Later it was found also in the urine, serum, and epidermal tissues (skin and hair) of healthy subjects (Rorsman *et al.*, 1979). The two other isomers of 5-cysdopa, namely, 2- and 6-cysdopa, have also been found in melanoma urine, as has the diadduct, 2,5-dicysdopa (Prota *et al.*, 1977b). This latter had been previously found in *tapetum lucidum* of catfish (Ito and Nicol, 1977).

The ratios between isomers of cysdopa in the urine are similar in control and melanoma patients and correspond to the quantities produced in model experiments on incubation of tyrosine and cysteine with tyrosinase (Ito and Prota, 1977a). This remarkable fact is evidence that the ratio of formation of cysdopas *in vivo* is solely dependent upon the positional reactivity of dopaquinone toward cysteine which, as mentioned earlier, follows the order 5- >> 2- >> 6-cysdopa. Thus determination of the major isomer indirectly provides information on the whole pattern of the metabolites. At one time it was thought that the urinary level of this metabolite was related to the ability of melanocytes to form pheomelanins rather than eumelanins. In fact, as we now know, 5-cysdopa is normally found in the urine of both eumelanic and pheomelanic subjects, and there is no substantial difference in the normal mean values between people of different race or complexion (Morishima and Hanawa, 1981; Wirestrand *et al.*, 1985; Nimmo *et al.*, 1988). Notably, 5-cysdopa has also been found, albeit in smaller amounts, in plasma and urine from several types of human and mouse albinos, implying that it is a nonspecific, that is, nontyrosinase-dependent, oxidation product of dopa in melanocytes (Acquaron *et al.*, 1981; Carstam *et al.*, 1985; Nimmo *et al.*, 1985).

Studies of the distribution of dopa and 5-cysdopa in different compartments of murine melanoma cells have revealed that the ratio of 5-cysdopa to dopa is higher in the soluble fraction than in the large granule fraction containing melanosomes (Rorsman *et al.*, 1983; Chakraborty *et al.*, 1991). This would suggest that a large portion of 5-cysdopa is formed in the cytoplasm of the melanocytes, from which it is excreted into the surrounding tissue fluid, thus passing into circulation (Rorsman *et al.*, 1983). The same authors have shown that 5-cysdopa is not decarboxylated in the kidney or other organs, and this partially explains why the ratio of dopa to 5-cysdopa is high in serum but very low in urine. Moreover, no conjugation of 5-cysdopa with glucuronic or sulfuric acid has been observed. These and other properties, including a high renal plasma clearance, make 5-cysdopa a useful marker of oxidative metabolism of tyrosine in mammalian tissues.

1. Analysis of 5-Cysdopa

Several methods are available for detection and quantitative determination of 5-cysdopa in plasma, urine, and epidermal tissues. These include fluorimetry (Rorsman *et al.*, 1973), ion exchange chromatography in combination with colorimetry of reaction products (Banda *et al.*, 1977), and HPLC with electrochemical detection (Hansson *et al.*, 1979b; Kagedal and Petterson, 1985). The last is the one which is preferably used in clinical laboratories, and an automated method has been described by which about 25 determinations can be performed per day (Kagedal *et al.*, 1989). As an internal standard for analytical determinations, a diastereoisomer of 5-cysdopa, 5-D-cystein-*S*-yl-L-dopa, has been recommended because of the close similarity in chemical properties (Agrup *et al.*, 1983b).

The new methods for quantitating 5-cysdopa have proved most helpful in investigative studies of the production of metabolites by cultured melanocytes, in which both the intracellular concentration of 5-cysdopa and the excretion into the culture medium have been defined (Aubert *et al.*, 1975, 1988). In general, the quantity of 5-cysdopa in melanoma cells greatly exceeds that of dopa.

Determination of 5-cysdopa appears to be of no value in the diagnosis of primary melanoma, but it may serve in the follow-up of patients after surgical treatment and chemotherapy. Attempts have been made to increase the diagnostic value of 5-cysdopa analysis by giving dopa, but the results are not entirely unambiguous from the clinical viewpoint. Rorsman *et al.* (1983) reported that dopa loading can lead to high serum values of 5-cysdopa in normal subjects. Apparently under such conditions 5-cysdopa can be formed by nonspecific oxidation of dopa, catalyzed by methemoglobin (Agrup *et al.*, 1983a), for example, or other mechanisms (Mojamdar *et al.*, 1982).

The increased demand for 5-cysdopa for clinical studies has prompted several studies aimed at developing laboratory syntheses of this catechol amino acid. Some of these, involving addition of cysteine to enzymically or chemically generated dopaquinone, were described earlier (Chapter 7). For large-scale preparation, the multistep synthesis with *N*-acetyldopaquinone ethylester is preferable, as well as the newly developed electrochemical oxidation of dopa to dopaquinone in the presence of cysteine (Hansson, 1981). As an alternative, 5-cysdopa can been obtained in 64% yield by addition of cysteine to dopaquinone generated by oxidation with ceric ammonium nitrate in 2 *M* sulfuric acid (Chioccara and Novellino, 1986). Another approach is based upon electrophillic substitution of the cysteine sulfenyl cation, generated *in situ* by reaction of cysteine with aqueous HBr, on the activated catechol ring of dopa (Ito and Prota, 1977b). The reaction is of general application and has been used for the synthesis of structural analog of 5-cysdopa, including cysteinyl and cysteaminyl monophenol derivatives (Ito *et al.*, 1981).

D. Cytotoxicity of Melanin Precursors and Allied Substances

In 1963 Hochstein and Cohen suggested that catecholic and quinonoid species generated by oxidation of tyrosine during melanogenesis might have cytotoxic effects toward the cell in which they are produced, that is, the melanocyte. This concept gained increasing credit in the subsequent decades as a possible approach to the chemotherapy of disseminated melanomas (Graham *et al.*, 1978; Pawelek and Lerner, 1978; Riley, 1991). Putative melanocytoxic agents, such as dopa or DHI, would rely for their activity on the measurable differences in the oxidative metabolism rather than on more subtle differences between normal and malignant melanocytes. Furthermore, since the cytotoxicity would be directed primarily against the hyperactive melanocyte, it was envisaged that these substances would be relatively nontoxic to other cells.

Initial experiments were focused on L-dopa and its methyl ester which, as expected, exhibited selective cytotoxicity to pigmented human and murine melanoma cells as compared to nonpigmented cells and fibroblasts (Wick, 1977; Wick *et al.*, 1977). Later, a variety of compounds structurally related to dopa, such as α-methyldopa, dopac, epinephrine, and 3,4-dihydroxybenzylamine, were shown to be highly inhibitory to the growth of melanin-producing melanoma cells (Parsons and Morrison, 1982; Kable and Parsons, 1988, 1989; Prezioso *et al.*, 1990; Karg *et al.*, 1991b). Side effects, however, have been reported for all these dopa analogs, and they include hepatic injury, cardiotoxicity, hemolysis, and photosensitivity.

As an aside to these studies, a wide range of substituted phenols were investigated as potential antimelanoma agents (Riley, 1981) and one of them, 4-hydroxyanisole (4-HA), has been used in clinical pilot studies with promising results (Morgan *et al.*, 1981; Riley, 1984). Interestingly, Frenk and Ott (1971) found that topical application of 10% 4-HA elicited depigmentation of eumelanin-forming melanocytes in the skin and hair of guinea pigs but had little effect on their pheomelanic hair follicles. Similar results were obtained in studies with Syrian hamsters and, more recently, with mice treated with monobenzylether of hydroquinone (Quevedo *et al.*, 1990). At present there is no obvious explanation for the resistance of pheomelanic follicular melanocytes to destruction by depigmenting phenols. For discussions and relevant observations see Riley (1974b) and Nordlund and Ackles (1985).

Another melanogenic precursor which has attracted a good deal of attention is 5-cysdopa. When tested on mice with B16 melanoma, this proved to be much more effective than dopa in inhibiting cell growth proliferation with no untoward effects on the host (Fujita *et al.*, 1980). Since this observation, a number of structurally related compounds have been synthesized and their cytotoxicity assayed *in vitro* and *in vivo* (Jimbow *et al.*, 1989). Among the new compounds tested, 4-cystein-*S*-ylphenol (4-CP) and 4-cysteamin-*S*-ylphenol (4-CAP) showed significant depigmenting and antimelanoma activity, the latter phenol being the more effective in prolonging the lifespan of melanoma-bearing mice. Evaluation of cytotoxic effects of 4-CAP on normal epidermal melanocytes by topical application revealed a decrease in a number of functioning melanocytes, a decrease in the number of melanosomes synthesized within melanocytes and transferred to keratinocytes, and degeneration and destruction of melanocytes (Yamada *et al.*, 1987). Whole-body autoradiographic studies on B16 melanoma-bearing mice indicated selective incorporation of these compounds into tumor tissues (Jimbow *et al.*, 1989). Four homolog of 4-CAP, namely *N*-Ac-4-CAP, α-Me-4-CAP, 4-homoCAP, and *N,N*-diMe-4-CAP, have also been synthesized and their cytotoxicity has been assayed using as the screening method the depigmenting potency in black follicles in mice (Fig. 8.13). Of the four compounds, *N*-Ac-4-CAP induced the most efficient depigmentation (98%) and was therefore proposed as a useful model for developing new active drugs against melanoma (Alena *et al.*, 1990).

The marked cytotoxicity of *para*-substituted phenols with a heteroatom, for example, O and S, directly linked to the aromatic ring recently led to extending the investigation to the nitrogen series. As test molecules, a number of structural analogs of *N*-*p*-hydroxyphenylglycine (HPG), including the methyl (Me-HPG) and dimethyl (di-Me-HPG) derivatives, as well as *N*-piperidinoethyl-*p*-aminophenol (PEAP) were used (Fig. 8.13). Preliminary *in vitro* assays

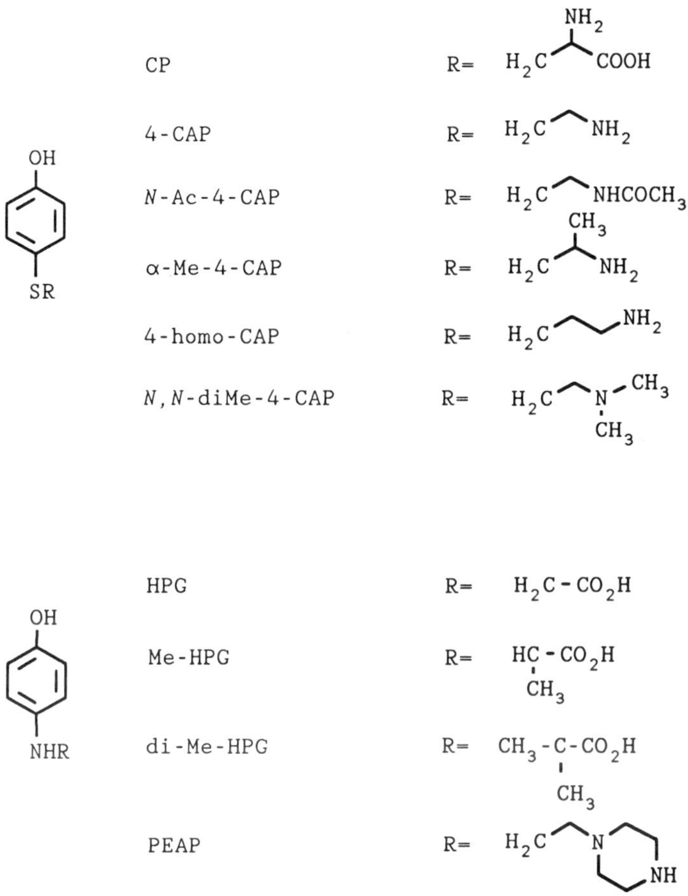

FIG 8.13 Structure of some phenolic compounds currently under investigation as antimelanoma agents.

on pigmented (HBL) and unpigmented (LND1) human melanoma cell lines showed in all cases significant inhibitory effects on both cells, with IC_{50} values on the order of 10^{-4} *M* (Mascagna *et al.*, 1992).

1. Mechanism of Action

The melanocytotoxic properties of melanin precursors and allied substances have been generally ascribed to the intracellular generation of highly reactive quinones capable of covalently reacting with crucial sulfhydryl-dependent enzymes, especially DNA polymerase (Wick, 1980; Yamada *et al.*, 1987; Land *et al.*, 1990). This theory assigns a central role to tyrosinase as the specific enzyme required for the metabolic activation of the phenolic substrates and subsequent formation of the potentially damaging *o*-quinones. However, though widely implicated as a targeting strategy for selective cytotoxicity, the role of tyrosinase is not unequivocal, since other enzymatic systems may be involved in the oxidative metabolism of phenols (for review see Kalyanaraman *et al.*, 1985). In support of this, Parsons (1985) found no correlation between dopa sensitivity and melanin or tyrosinase levels in human melanoma cells. On the other hand, Passi *et al.* (1987) presented evidence that phenolic agents have various toxic effects on cell functions unrelated to melanogenesis, including *inter alia* mitochondrial respiration and microsomal mixed-function oxidase. The same authors also drew attention to the fact that most cytotoxic effects were evaluated on cultured melanoma cells, under conditions where many of the most active cytotoxic phenols may be oxidized in the culture media (Picardo *et al.*, 1987). Therefore, their toxic action may partly result from extracellular oxidation with generation of toxic oxygen species rather than from intracellular effects justified by adequate membrane permeability. This would be in keeping with the observed ability of catalase to completely suppress the cytotoxic effect of cysdopa on melanoma cells in culture (Ito *et al.*, 1983). Likewise, dopac toxicity to cultured melanoma cells seems to be mediated by hydrogen peroxide since it is totally prevented by the addition of catalase to the oxidation medium (Karg *et al.*, 1991b). Similar arguments are reported by Kable and Parsons (1988), who showed that human tumor cells are extremely sensitive to extracellular active oxygen species during a relatively short period in the early S phase.

More recently, Inoue *et al.* (1990) have compared the effects of tyrosinase and MAO inhibitors on the cytotoxicity of 4-CAP and related compounds on melanoma cells. They found that the activity of 4-CAP and 4-homoCAP is mediated through MAO-dependent conversion to aldehydes, while that of α-Me-4-CAP is dependent on tyrosinase activity to some extent. In another study, however, Yamada *et al.* (1991) reported that 4-CAP is particularly effective in heavily melanized melanoma cells but has little or no inhibitory effect on amelanotic melanoma and nonmelanoma cells. Accordingly, it was suggested that 4-CAP may become toxic to melanoma cells only after oxidation by tyrosinase to the corresponding quinone.

It is clear from the preceding discussion that the molecular mechanisms of melanocyte destruction by phenolic agents are at present largely speculative. However, the current intensification of interest in the synthesis of new melanocytotoxic agents, coupled with our improved understanding of the chemistry of melanogenesis hold good promise for the future.

CHAPTER 9

GENETIC AND HORMONAL REGULATION OF MELANOGENESIS

I. PIGMENT-RELATED GENES

Genes affecting coat color were among the first to be studied for Mendelian inheritance in mice (Searle, 1968; Silvers, 1979). Some 130 determinants at more than 50 loci in the genomes have been described, which in the mutated states cause abnormalities in eye, skin, and hair pigmentation. At present it is not clear how many of these genes are structural genes coding for proteins which directly or indirectly form the phenotype or how many are regulatory genes which regulate the expression of the structural ones. Investigation is complicated by the fact that most, if not all, of the pigment-related genes have numerous pleiotropic effects on development that affect parameters other than pigmentation and further obscure their mode of action. It must be cautioned that the story of pigment genetics is mainly one concerned with hair color, which is biologically and biochemically distinct from skin color. Several lines of evidence indicate that a relatively small number of gene pairs, that is, from three to six, may account for the impressive array of human skin colors (Nicholls, 1973; Quevedo, *et al.*, 1985). This would support the notion that only a few genes may be directly involved in the regulation of the basic steps of melanogenesis (Prota and Searle, 1978).

Table 9.1 lists the main allelic series determining coat color in mice together with what is known about the mode of action of the gene concerned. The *albino* locus controls the expression of tyrosinase activity and is the key lesion in many types of albinism. The wild-type allele (*C*) is dominant over all other alleles and provides full tyrosinase activity. In contrast, the albino mice (*c/c*) have no melanin at all, either in the skin or in the eyes, and produce a characteristic pinkeye white phenotype. This mutation, however, does not interfere in other respects with the differentiation of the melanocyte (amelanotic melanocytes), which is identical to the normal pigmented melanocyte. It should be noted that albinism can also result from white spotting genes

TABLE 9.1
CHARACTERISTICS OF MAIN ALLELIC SERIES DETERMINING COAT COLOR IN MAMMALS

Series	Name	Main alleles	Effects and mode of action	Site of action
Albino	*C*	C^+, c^{ch}, c^b, c	Reduces intensity of pigmentation, first yellow and then black, until none is left in *cc* (albino).	Within melanocyte
Agouti	*A*	A^y, A^w, A^+, a^t, a, a^e	Controls regional distribution of black and yellow pigments in body and individual hairs, from all yellow (dominant in mouse) to all black.	Outside melanocyte
Extension	*E*	E^d, E^+, e^{br}, e	Extends black (dominant) or yellow (recessive) pigment in body as a whole, with e^{br}giving a black/yellow variegation.	Within melanocyte
Brown	*B*	B^{lt}, B^+, b	Affects eumelanins, changing black to brown in *bb*. May lighten eyes (*bb*) and underfur (B^{it}).	Within melanocyte
Dilute	*D*	D^+, d, d^l	Dilutes both black and yellow colors by clumping pigment granules. Dilute lethals (d^l) have convulsion.	Within melanocyte changing shape
Pinkeyed	*P*	P^+, p, p^s	Main effect on eumelanosomes, with dark colors much more diluted than light ones. Retinal pigmentation removed in *p* and p^s. p^s also causes male sterility.	Within melanocyte

such as piebald(s) and the mutant microphthalmia (*mi*), which prevent neural crest and optic cup melanocytes from reaching their destinations in the epidermis and retina. However, little is known about the molecular basis of these white spotting genes (Gordon, 1979; Searle, 1989).

The best known allele at the *B* locus is recessive brown (*b*), which acts on eumelanin pigmentation only, but a dominant allele (B^{lt}) which lightens the coat, especially the underfur, has been found in the mouse. Melanosomes of brown mice are rounded rather than ovoid but also have a rather disordered internal structure with coarsely granular pigment granules (Moyer, 1966). The pinkeye dilution series is also apparently involved in the differentiation and fine structure of melanosomes, although a more recent study would suggest some regulatory function at the level of tyrosinase processing and/or translocation in melanocytes (Halaban *et al.*, 1988a). The dilute coat color locus (*d*), which maps to chromosome 9 on mouse, is not correlated with abnormal melanin synthesis but rather with an aberrant morphology of the *d/d* melanocytes (Strobel *et al.*, 1988).

Of special interest from the biochemical viewpoint is the *agouti* locus (*A* on chromosome 2), which controls the overall type of pigmentation and determines whether pheomelanin or eumelanin is produced in the follicular environment (Quevedo *et al.*, 1981). The wild-type mouse expresses the character called agouti. The melanocytes located in the hair follicles produce eumelanin when hair starts to grow. They subsequently produce yellow pigment (pheomelanin) and finally produce eumelanin, thus producing a yellow band. The functional mutant gene at the agouti (*a*) locus is lethal yellow (A^y), which is lethal when homozygous, but heterozygotes have a uniform golden color attributed to pheomelanin. Another mutant gene is the nonagouti (*a*) that exhibits a black fur when homozygous.

As to the mode of action of the gene at the *A* locus, all the evidence is consistent with the original work of Silvers and Russell (1955) showing that the biochemical information comes from outside the melanocyte, presumably via an increase in the sulfhydryl level in the follicular environment. Another interesting feature of this locus follows from the work of Geschwind and co-workers (1972), who were able to induce eumelanin synthesis in pheomelanin-forming melanocytes of hair follicles by injecting the lethal mouse with α-MSH. A possible mechanism accounting for this effect has been suggested by Takeuchi and Yamamoto (1988), involving a complex interaction of the product at the *A* locus with the α-MSH–cAMP signal transduction system.

The extension (*E*) locus is also involved in the pheomelanin formation, since alleles are known to extend or restrict the amount of yellow pigment in the coat of mice (Searle, 1968), as well as in the feathers of domestic fowl (Brumbaugh and Moore, 1969). Homozygous mice for the bottom recessive allele *e* have a coat color similar to that of the heterozygous lethal yellow (A^y). In the extension series, however, gene action series is autonomous, as shown by Lamoreaux and Mayer (1975), and uninfluenced by α-MSH (Takeuchi and Yamamoto, 1988). Therefore it seems likely that these genes control distinct enzymic systems, the nature of which, however, remains elusive.

Little is known about the relationship between the yellow pheomelanin(s) expressed by the *agouti* or *extension* loci in mice and the reddish-brown pheomelanins typically found in red human hair and chicken feathers. The action of the mouse-equivalent agouti gene in man is unknown. Genetically controlled red hair occurs in humans and shares features of genes at the extension locus (Witkop *et al.*, 1988). Recently, a major gene for red hair has been assigned to chromosome 4 (Eiberg and Mohr, 1987), but several questions remain to be answered before the genetics underlying formation of pheomelanins can be resolved.

A. Culture and Cloning of Melanocytes

Malignant melanocytes were among the easiest to establish in culture and several cell lines have been derived from both human and animal melanoma (Bennett, 1989; Yaarand Gilchrest, 1991). Some of these lines have already been mentioned and include GR1, B16, and Cloudman, or S91, which proved particularly useful for studies on pigment cell metabolism and differentiation. Although short-term cultures of normal melanocytes were reported by Hu *et al.* (1957) and others, it was only in 1982 that Eisinger and Marko succeeded in obtaining long-term cultures, using as effective additives to the culture medium cholera toxin and the tetradecanoyl phorbol acetate (TPA). These two substances were clearly not natural mitogens for melanocytes, but they were capable of inducing a spontaneous immortalization of the normal cells that proliferated at rates similar to those of their malignant counterparts.

Following these results, a number of pure lines of normal murine melanocytes could then be established by different groups (Sato *et al.*, 1985; Bennett *et al.*, 1987; Tamura *et al.*, 1987). One of these lines was from black mice and was followed by lines from mice carrying albino (*c*) and other coat color genes. Immortalized melanocytes usually retain a high level of differentiated function in culture, including tyrosinase activity and pigment characteristics of the mice of origin. Thus, melanocyte lines from black (nonagouti, *a*/*a*) and brown (*b*/*b*) mice produce black and brown melanins, respectively. Likewise, *albino* melanocytes in culture maintain the ability to produce premelanosomes which are unpigmented. Immortalized melanocytes also tend to retain other features of the original cells, such as a diploid or near-diploid chromosome number and lack of tumorogenicity in syngenic or nude mice (Bennett, 1991).

The availability of a variety of such cultured melanocytes has been invaluable in cloning pigment-specific genes and identifying mutations which affect the properties of the encoded proteins. An excellent account of the different strategies used in various laboratories is given by Hearing and Jimenez (1989). Much of the research currently pursued revolves around the structure and functional activity of *albino* and *brown* genes, and a separate account of the progress made in this rapidly developing field is given below. Supplementary information on the subject can be obtained from the reviews of Hearing and Tsukamoto (1991) and Takeuchi (1992).

B. STRUCTURE OF THE TYROSINASE GENE

Until recently, it was not clear whether the *C* locus encoded the gene for tyrosinase or a *trans*-regulatory protein, which inhibits or activates the enzyme (Hearing, 1973; Pomerantz and Li, 1974). A first move to distinguish between these two alternatives was made by Yamamoto *et al.* (1987), who succeeded in cloning the mouse tyrosinase cDNA (*Tyrs33*). At the same time, the isolation of a candidate cDNA clone (*pmel34*) for human tyrosinase was reported by Kwon *et al.* (1987). In both studies, antityrosinase antibodies were used for screening the clones, coupled with other techniques to verify the structure of the encoded protein. The two cDNAs thus obtained, however, exhibited significant differences in two regions; in the mouse cDNA, *Tyrs33*, the region corresponding to the sequence from 1222 to 1422 in the human cDNA was missing. On the other hand, the human cDNA, *pmel34*, lacked the initiation codon and the 5′ noncoding sequence. These sequence errors were later corrected through additional cloning and sequencing of the full-length transcript (Kwon *et al.*, 1988; Terao *et al.*, 1989; Yamamoto *et al.*, 1989). Further studies by Ruppert *et al.* (1988) showed that the *albino* gene in mouse is composed of five exons separated by four large introns and spans a chromosomal region of approximately 70 kb. In humans the gene is >35 kb (Takeda *et al.*, 1989), while the processed mRNAs are 2.4 kb in length. This splicing is not always accurate, and misplaced mRNAs (10 to 40% of the total) are generated which translate into altered proteins (Shibahara *et al.*, 1988; Porter and Mintz, 1991). The physiological implications of alteration in the mRNA splicing of the gene transcript are still in question (Takeuchi, 1992). Since the altered proteins are not competent catalytically as tyrosinase, they may fulfill some other role in melanocytes, for example, as competitive inhibitors (Hearing and Tsukamoto, 1991), but this has not yet been demonstrated.

1. *Catalytic Function*

Several approaches have been used to show the ability of the *albino* locus gene to function catalytically as tyrosinase. Müller *et al.* (1988) reported that one of their DNA clones (*pmcTyr1*) containing the full complement of exons was able to confer tyrosinase enzymatic activity to tyrosinase-negative cell lines when tested by transient transfection. This result was corroborated by Yamamoto *et al.* (1989), who found that following transfection with a mouse tyrosinase cDNA cultured albino melanocytes derived from BALB/c mice acquired the ability to produce melanin. Moreover, when amelanotic melanoma cells were transfected with tyrosinase cDNA the cells expressed both tyrosine hydroxylase and dopa oxidase activity (Takeda *et al.*, 1989). Bouchard *et al.* (1989) also reported a case in which cultured mouse fibroblasts transfected

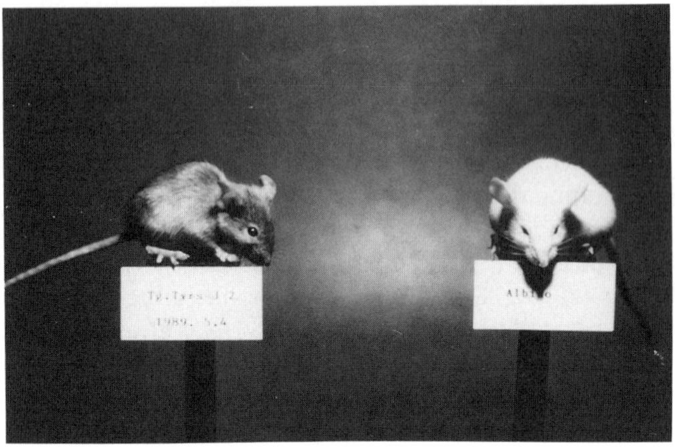

FIG 9.1 Expression of a functional tyrosinase "minigene" (mg-Tyrs-J) in albino melanocytes. Left: Transgenic mouse exhibiting black eyes and brown agouti coat color. Right: one of the nontransgenic litter mates. Courtesy of Professor T. Takeuchi.

with the human tyrosinase cDNA developed membrane-bound vesicles containing melanin.

In further studies, minigenes containing the tyrosinase coding sequence as well as a competent regulatory sequence were constructed and used to produce transgenic mice. In one such minigene, *mg-Tyrs-J*, in which the authentic genomic 5′ noncoding flanking sequence was fused to a mouse tyrosinase, cDNA was microinjected into fertilized eggs of albino BALB/c mice (Tanaka *et al.*, 1990). Of the 25 animals that developed, four mice exhibited pigmented hair and eyes (Fig. 9.1). Moreover, histological analysis of the transgenic mice thus generated revealed that melanogenesis was restricted to hair bulb and eyes. A similar rescue of the *albino* phenotype by injection of a functional tyrosinase gene into mice was achieved at almost the same time by the German group at Heidelberg (Beermann *et al.*, 1990). Five transgenic mice were obtained which transmitted the transgene to their offspring, which inherited the pigmented phenotype. The ability of the *albino* locus-encoded protein to function catalytically as tyrosinase has also been confirmed using antibodies generated against peptides which correspond to the amino and carboxyl termini predicted by the nucleic acid sequence (Jimenez *et al.*, 1989, 1991; Halaban and Moellmann, 1990).

2. Mutations

Knowledge of the structure of the tyrosinase gene has provided the necessary background for the characterization of the exact amino acid change which is responsible for the *albino* phenotype. It is a point mutation in the coding region at nucleotide 387 (G-C transversion) which results in the change of a conserved cysteine to a serine in the first cysteine-rich domain of the tyrosinase protein (Shibahara *et al.*, 1990). This single change in the protein structure causes the virtually quantitative loss of catalytic activity and consequent inability of the *albino* mouse melanocytes to produce melanins (Jackson and Bennett, 1990).

Known mutations at the *albino* locus, for example, chinchilla and Himalayan, are associated with significant alterations in the structure and catalytic activity of tyrosinase (Kidson and Fabian, 1981). Thus, in the Himalayan (c^h) there is an alteration in glycosylation which results in a temperature-sensitive phenotype. Kwon *et al.* (1989) have shown that the tyrosinase gene in Himalayan mouse (c^h/c^h) contains an A-G substitution at nucleotide 1259 resulting in a change of histidine to arginine at position 420. Since the levels of tyrosinase mRNA in $c(c^h)$ mice are equal to those in the wild-type (C) strains, it has been suggested that the reduced pigmentation in chinchilla mice is due to a weaker activity of a mutated protein (Halaban *et al.*, 1988a). This is supported by the observation that the enzyme isolated from chinchilla mice is more susceptible to proteolytic inactivation. Sequencing of the complete coding information of the genomic tyrosinase gene derived from $c(c^h)$ allele has revealed a single amino acid exchange at position 464, changing Ala (GCA) in the wild-type sequence to Thr (Beermann *et al.*, 1990).

Albinism in laboratory animals is equivalent to human tyrosinase-negative oculocutaneous albinism (OCA). However, a number of other kinds of human albinism have been described, including several tyrosinase-positive ones (Witkop *et al.*, 1978). These can be distinguished from one another on the basis of clinical, ultrastructural, and biochemical criteria and by their genetic characteristics as determined by matings of albinos of different genotypes (Witkop *et al.*, 1988). Thus, while expression of *albino* locus-encoded tyrosinase is necessary for pigment production, there are other control points that are critical for melanogenesis to occur, consistent with the biochemical pathway discussed in Chapter 8. Hopefully the next few years will be quite decisive in understanding the mechanism preventing pigment formation in tyrosinase-positive albinism.

Using genomic amplification of the five tyrosinase exons coupled with dideoxy sequencing, King *et al.* (1991a,b) have identified 12 different genetic lesions in individuals with type 1 OCA. These mutations cluster in a nonrandom fashion, occurring at various sites in either the structural or the promoter region of the gene. Giebel *et al.* (1990) have reported on two novel mutations for type 1B OCA (yellow), in which the defect is either within an intervening sequence of tyrosinase gene or distant to it. In another study by the same group (Spritz *et al.*, 1990), a mutation in the human gene has been identified which results in a temperature-sensitive tyrosinase that is analogous to the Himalayan mutation in mice. An interesting case of OCA in which a single base insertion in the exon II shifts the reading frame, thus introducing a premature termination, has been reported (Tomita *et al.*, 1989).

C. The Brown Locus Gene

One of the first cDNA clones (*pMT4*) studied as a candidate for mammalian tyrosinase (Shibahara *et al.*, 1986) proved to encode in fact a tyrosinase-related protein which mapped to the *brown* locus on chromosome 4 in mice (Jackson, 1988). The sequence nonetheless showed high homology to authenticated tyrosinase and was designated tyrosinase-related protein-1 or TRP-1. A human cDNA sequence for TRP-1 has also been reported which shows about 90% identity with that expressed by murine melanocytes (Cohen *et al.*, 1990).

The protein sequence of TRP-1 is about 40% homologous with that of tyrosinase and shares its known functional characteristics, including a leading signal sequence, a *trans*-membrane region, nearly all cysteine residues, and two potential-binding sites (Müller *et al.*, 1988; Jackson, 1988; Bennett, 1991). Thus, TRP-1 may be another pigmentary enzyme, located with tyrosinase on the various subcellular compartments of melanocytes.

The sequences of cDNAs for the murine brown mutant TRP-1 gene and an induced revertant to wild-type have been reported (Zdarsky *et al.*, 1990). Although there are several differences between the two sequences, only one of these is restored in the revertant, showing that it is the critical mutation. This is in the codon for a conserved cysteine residue, Cys-86 which, interestingly, falls very close to Cys-85 of the *albino* mutation when the two proteins are aligned by homology. Both cysteine residues lie within a region particularly highly conserved between the two proteins and presumably vital from the functional viewpoint. As with the basic *albino* mutation, all strains of brown mice so far studied have this single mutation, suggesting that all derive from the same original mutation (Jackson, 1988).

The function of brown protein is controversial and forms the subject of lively debate. Different groups have postulated that it may be a tyrosinase-related enzyme (Jimenez *et al.*, 1989, 1991), or dopachrome tautomerase (Jackson, 1988), or dihydroxyindole conversion factor (Kwon *et al.*, 1988), or melanosome specific catalase (Halaban and Moellmann, 1990). Although the latter possibility is very attractive in view of the critical role played by hydrogen peroxide in melanogenesis, the substrate(s) and catalytic activity of the brown protein have not yet been identified conclusively, and the question remains open at this time.

II. HORMONAL CONTROL

It was work with lower vertebrates, principally amphibians, which laid the foundations of our present understanding of hormonal control of pigmentation. In 1916 P. E. Smith at the University of California and B. M. Allen at the University of Kansas found that removal of the pituitary gland from animals such as tadpoles and frogs results in marked lightening of skin, whereas injection of pituitary extracts produces darkening of skin color. Moreover, when extracts from the pineal gland of cattle were added to the water in which tadpoles were swimming, the skin of the tadpoles became so transparent that their hearts and intestines were clearly visible (McCord and Allen, 1917). From the very early studies it was obvious that these changes were caused by the clumping and dispersion of the pigment granules in the melanophores under the influence of specific hormones present in the pituitary and pineal extracts. The term "intermedin" was initially coined by Zondek and Krohn (1932) for the darkening hormone, because its site of origin was the intermediate lobe of the pituitary gland. Later, however, the term melanocyte-stimulating hormone (MSH) was introduced by Lerner and his associates to embrace an action upon epidermal melanocytes, in keeping with their finding of a darkening of human skin in response to this hormone (Lerner and McGuire, 1961). The search for the melanosome aggregating agent was very challenging due to the minute size of the pineal gland and the almost infinitesimal quantities of the substance. In a reviewing paper, Lerner (1961) reports that it was only thanks to a generous supply of about 200,000 beef pineal glands from Armour & Co. that some 100 μg of the pure hormone could be isolated. By careful analysis, this was eventually identified as 5-methoxy-*N*-acetyltryptamine and named melatonin in order to reflect its paling effect on dermal melanophores (Fig. 9.2) (Lerner *et al.*, 1959).

Among the lower vertebrates, it is generally only on amphibian melanophores that the pineal gland appears to play a role in the control of color changes. In teleosts, such as bony fishes, aggregation of melanosomes within melanophores is mediated by the pituitary gland through the release of another hormone, named melanin-concentrating hormone (MCH), which like MSH is a peptide. The existence of such a hormone is of interest in relation to the bihumoral control theory of color changes, as elaborated by Hogben and Slome (1931), involving an antagonistic pituary hormone capable of

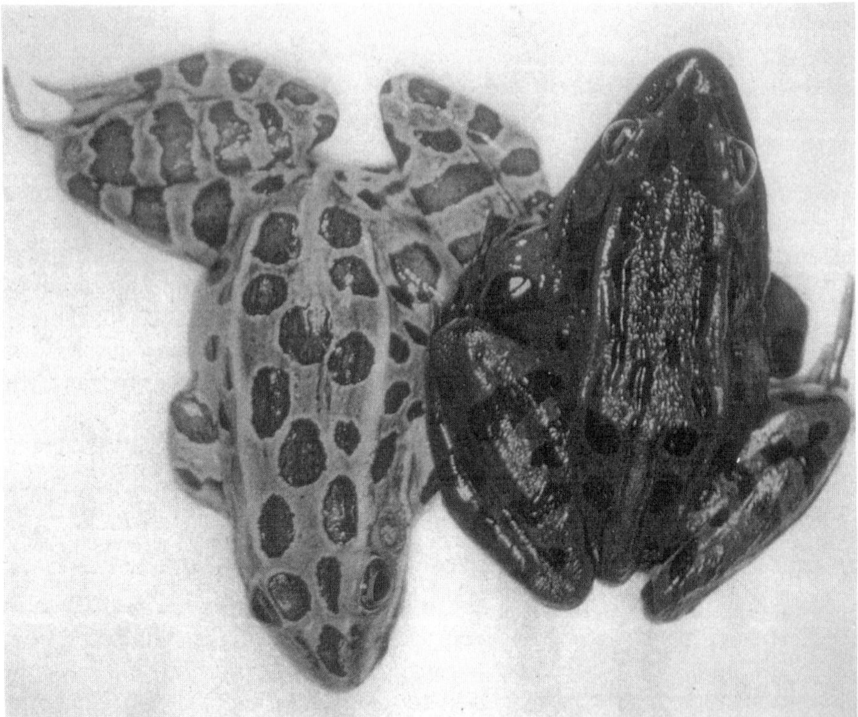

FIG 9.2 Effect of hormones on amphibian melanophores. The frog at left is lighter than normal after immersion in water containing melatonin, whereas the frog at right is much darker than normal after injection of MSH. Courtesy of Professor A. B. Lerner.

overriding the effects of MSH at the effector cell level. The evidence from amphibians on which the proposal was made had been misinterpreted, but it is noteworthy that MCH is responsible for the mechanism by which fish are able to adapt to light-colored backgrounds (Baker, 1991).

A. MELANOCYTE-STIMULATING HORMONE

MSH belongs to a family of peptide hormones, including adrenocorticotropin (ACTH), α-lipotropin (LH), endorphins, enkephalin, and others, all of which are derived from a common ancestor protein proopiomelanocortin (Eberle, 1988). Basically, two distinct forms of MSH are found in pituitary extracts: α-MSH and β-MSH. Both are small peptides containing the heptapeptide sequence Met-Glu-His-Phe-Arg-Trp-Gly-, present in ACTH and LH, which is essential for MSH biological activity (Fig. 9.3). Pure ACTH has inherent MSH activity, being about 1/40 as active as α-MSH on a molar basis.

MSH activity is usually determined by the *in vitro* frog skin assay which is based on the centrifugal dispersion of melanin granules in the dendritic processes of dermal melanophores, leading to skin darkening (Eberle, 1988). Lizard skin can also be used, but it is less sensitive. Under optimal conditions, the frog skin assay allows detection of minimal concentrations of α-MSH of 10^{-11} *M* and gives a dose–response curve in the range 2.5×10^{-11} to 4×10^{-10} *M*. It finds an interesting application for detection of human pregnancy in which there is an increase of circulating MSH-like activity in blood and hence in urine. This MSH-like activity may be of fetal origin, but its source and identity have not yet been established.

Under normal physiological conditions, adult humans do not seem to possess a circulating MSH. However, the α form is definitively present throughout the central nervous system, especially in the hypophysiotropic area of the hypothalamus. In both humans and animals there is evidence that α-MSH may play a role as a neuropeptide in a variety of pigment-unrelated physiological functions, such as increased motivation and learning ability, arousal, fetal growth, sexual excitement, and others (Hadley, 1988). α-MSH is synthesized and secreted by opiomelanotropinergic cells of the pair intermedia of the pituitary gland and by neurons in the central nervous system (O'Donohue and Dorsa, 1982). In species lacking the pairs, for example, chicken, synthesis of α-MSH takes place in the neurohypophysis. Several studies, reviewed by Kastin and Schally (1972), have demonstrated that the release of α-MSH in mammals as well as in amphibia is under the control of

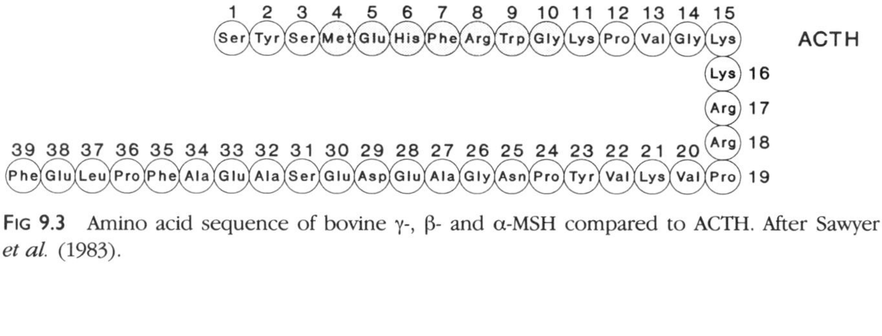

FIG 9.3 Amino acid sequence of bovine γ-, β- and α-MSH compared to ACTH. After Sawyer *et al.* (1983).

a specific inhibiting factor, MSH-release inhibiting factor (MIF), of hypothalamic origin. There is also evidence for a hypothalamic releasing factor (MRF). All these facts must be carefully considered before attempting to rationalize the effects of α-MSH on the melanocyte system. Indeed, as first pointed out by Lerner (1961), the main physiological role of this hormone may not be the control of pigmentation, though this phenomenon has led to its discovery.

1. Chemistry

Mammalian α-MSH is an acetyltridecapeptide, identical with the amino acid sequence 1–13 of ACTH, shown in Fig. 9.3. The deacetyl form is less potent and acetylation/deacetylation probably has a regulatory function in the control of α-MSH activity. In salmon only the deacetylated form has been found. Shark (*Squalus acanthias*) α-MSH is also deacetylated but contains Met at the C terminus of the tridecapetide chain (Hruby *et al.*, 1984).

As mentioned before, β-MSH contains the same messenger heptapeptide sequence, but with different flanking amino acids. Many species synthesize variants of β-MSH with 18 amino acid residues, for example, bovine, while human β-MSH has four additional residues (Ala-Glu-Lys-Lys-) at the N terminus. β-MSH has biological activity similar to that of α-MSH in the frog skin bioassay. However, current evidence suggests that α-MSH is the physiologically relevant systemic melanotropic hormone; β-MSH found in the pituitaries

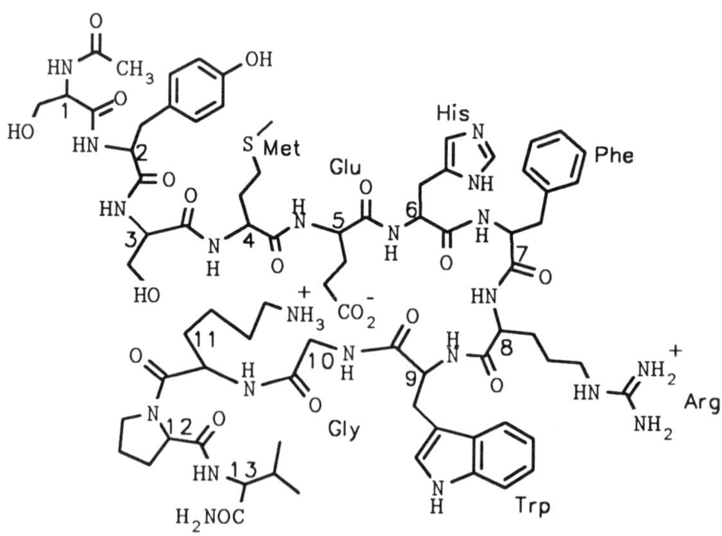

FIG 9.4 Pseudocyclic representation of α-MSH showing the proposed reverse-turn domain of the His-Phe-Arg-Trp sequence.

is probably a degradation product of β-lipotropin which has 91 amino acid residues and the same peptide core as ACTH (Hruby *et al.*, 1984).

Structure–activity studies of α-MSH suggest that the active molecule possesses a reverse-turn configuration in the central tetrapeptide sequence: His-Phe-Arg-Trp (Fig. 9.4). This is supported by the synthesis of a conformationally restricted α-MSH analogs, in which Met-4 and Gly-10 are replaced by Cys residues. Oxidative cross-linkage of these Cys residues maintains the molecule in a permanent reverse-turn configuration with a significant increase of the biological activity with respect to unmodified α-MSH (Sawyer *et al.*, 1983; Hruby *et al.*, 1984).

More recently, Hruby *et al.* (1988) have reported the synthesis of several analogs of α-MSH which are 100–1000 times more active than the native hormone. One such analog, having NLe and D-Phe at positions 4 and 7, respectively, is currently being evaluated as a "tanning hormone" for increasing skin pigmentation as well for treating vitiligo and other hypopigmentary disorders (reviewed by Hadley and Dawson, 1988).

2. Pigmentary Actions of MSH

As discussed by Hadley (1988), the biochemical events underlying MSH-induced melanosome dispersion within the chromatophores of cold-blooded vertebrates are analogous to those for smooth muscle relaxation. Both processes involve activation of membrane-bound adenylate cyclase and subsequent increases in the intracellular cAMP level, which results in a sequestering of Ca(II) from the cytosol to the plasma membrane of the cell. Absence of this divalent cation might then activate (directly or indirectly) melanofilaments, thus resulting in melanosome dispersion. In support of this, Moellmann *et al.* (1973) found that cytoclasin B, which causes a disappearance of microfilaments, reversibly inhibits melanosome dispersion in response to MSH. Furthermore, the divalent cation ionophore A23187 induces a marked reversible aggregation of melanosomes that is partially dependent upon the calcium in the medium (Novales, 1983; Lucas *et al.*, 1987; Miyashita and Moriya, 1990). This divalent cation seems specifically required for signal transduction between the α-MSH receptors and adenylate cyclase, but not for melanosome movements per se, because pigment dispersion is also observed in a Ca(II)-free medium in response to theophylline or to dibutyryl cAMP (Veseley and Hadley, 1979). A working model for the role of calcium and α-MSH in the control of melanosome movements within melanophores is shown in Fig. 9.5.

The subsequent events leading to enhanced melanogenesis are less well defined and are usually envisaged as resulting from the activation of the same cAMP-dependent protein kinase. The action of this kinase would be to phosphorylate one or more proteins which could initiate a so-called "cascade of

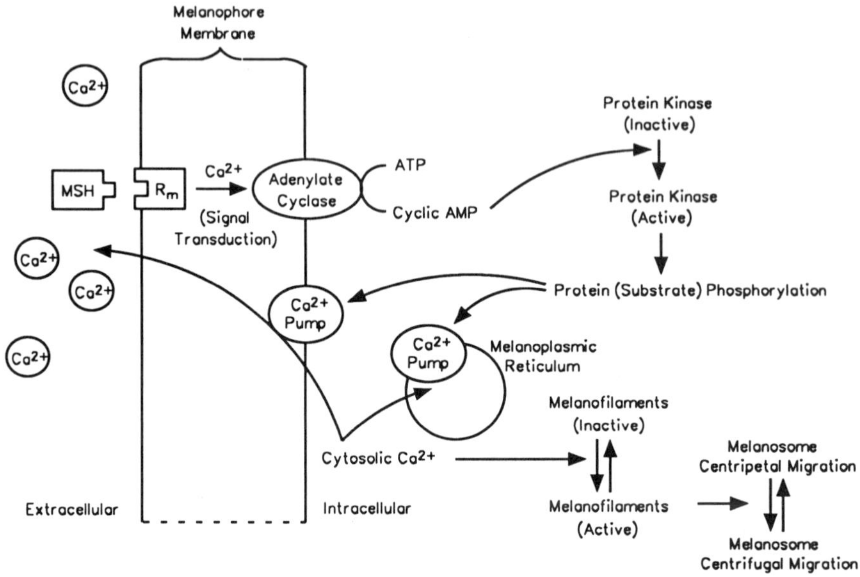

FIG 9.5 Hypothetical model for the role of MSH and Ca^{2+} ions in the control of melanosome movements within melanophores. After Sawyer *et al.* (1983).

events" leading eventually to an increase in the intracellular level of tyrosinase and consequent increase in melanogenesis (Pawelek, 1976; Fuller and Viskochil, 1979; Fuller *et al.*, 1987). Such a mechanism is supported by the observation that the effects of α-MSH on tyrosinase activity and melanogenesis can be mimicked by agents such as dibutyryl cAMP and phosphodiesterase inhibitors. However, most of this evidence is drawn from studies on melanoma cells and direct evidence from normal melanocytes is incomplete and species dependent.

The best documented effects of α-MSH are upon tyrosinase activation in hair follicle melanocytes from laboratory animals, especially mice (Weatherhead, 1988). However, the genetic constitution of the strain under investigation is important in determining whether it will respond to MSH or not. For instance, Geschwind *et al.* (1972) found that α-MSH has a pigmentary effect on the coat color of the agouti mouse, while it does not affect to any significant extent the hair color of yellow mice homozygous for the recessive gene *e*, even if they carry the A^y allele. Other groups have since confirmed that α-MSH has no effect on tyrosinase activity and melanogenesis on follicular melanocytes of nonagouti mice, suggesting that its action is dependent upon the expression of the *agouti* locus (Tamate and Takeuchi, 1984; Thody *et al.*, 1984; Burchill and Thody, 1986).

It has been suggested that although intracellular mechanisms that regulate tyrosinase are operational, melanocytes may not necessarily express their α-MSH receptors continuously (Pawelek, 1985). This could explain why hair follicular melanocytes in the Siberian hamster and mice only respond to α-MSH during the period of hair growth (Weatherhead and Logan, 1979; Thody *et al.*, 1984). Differences in responsiveness to MSH could also account for the coat darkening and increase of tyrosinase activity that occur in the hair follicular melanocytes of C_3H-HeA^{vy} mice during the period of eumelanin synthesis (Burchill *et al.*, 1986). In these mice the effect of α-MSH is reduced as the synthesis of eumelanin decreases and is eventually lost by the time the melanocytes produce mainly pheomelanin. This and other experiments (Burchill and Thody, 1986) would suggest that while cAMP-dependent mechanisms may regulate tyrosinase during the period of eumelanin synthesis, they do not operate in phoemelanin-forming melanocytes. If so, the failure of melanocytes to respond to α-MSH during pheomelanogenesis could be due to a change in responsiveness to cAMP rather than in receptor expression.

It is generally thought that α-MSH is also able to stimulate epidermal melanocytes and can therefore play a role as a pigmentary hormone in man. Experimental evidence supporting this view was first provided by Lerner and his co-workers (1954), who found that intramuscular injection into man of a crude extract of pig pituitary glands caused darkening of the skin and nevi. Some 7 years later, Lerner and McGuire (1961) repeated this experiment using pure α- and β-MSH and confirmed that these hormones were indeed able to darken human skin. Microscopic examinations revealed that dispersion of melanin granules into the dendritic processes preceded and initiated the onset of new melanin formation (Lerner *et al.*, 1966) by activation of tyrosinase. However, whether the observed effects are pathological rather than physiological is difficult to say due to the numerous effects of MSH on mammalian systems. In this connection, Friedmann and Thody (1986) found no correlation between skin pigmentation and circulating levels of MSH except when the levels are extremely high as in certain disorders of the pituitary adrenal axis. It has also been reported that MSH fails to induce any effect on human epidermal melanocytes in culture (Halaban *et al.*, 1983; Friedmann and Gilchrest, 1987). Whether this lack of effect is peculiar to man or whether it is a characteristic of epidermal melanocytes is not yet clear.

3. MSH Receptors

The voluminous pharmacological and physiological literature on the localization, "molecular anatomy," and methods of detection of MSH receptors in melanophores from cold-blooded vertebrates has been carefully reviewed by Eberle (1988). According to current ideas, similar receptors are localized on the cell membrane of melanocytes of skin and other tissues of homotherm vertebrates, including mammals. However, direct evidence in support of this notion is remarkably limited, as most of the work has been carried out on mouse melanoma cells, especially S91 Cloudman clones. These proved to be uniquely sensitive to α-MSH both *in vitro* and *in vivo* with a marked increase in tyrosinase activity and melanogenesis (Pawelek, 1985). More recently, clones of B16 melanoma cells have also been used for studying MSH receptors and their role in regulating responsiveness to the hormone (Eberle, 1988; Gerst *et al.*, 1988).

Evidence for a specific MSH binding to Cloudman S91 cells was first reported by Varga *et al.* (1974, 1976) using ^{125}I-conjugated α-MSH. They found that the label was not bound randomly over the surface of the melanocytes, but concentrated close to the nucleus. A similar concentration of bound labeled α-MSH was found in cultured guinea pig melanocytes (Fritsch and Varga, 1976). As only a low proportion (18 to 34%) of their cell populations bound the α-MSH, they surmised by analogy with findings in melanoma cells that α-MSH binding might be discontinuous with a maximal number of membrane receptors in the G-2 phase. Varga *et al.* (1976) have also shown that ^{125}I-labeled α-MSH is "internalized" to the Golgi region of melanocytes, but whether this is important for the stimulation of the cell is controversial (Weatherhead, 1988).

In more recent years, MSH receptors have been identified also in both normal and malignant human melanocytes by binding experiments with labeled α-MSH (Lejeune *et al.*, 1987; Ghanem *et al.*, 1988). However, the level of binding is much lower than that observed on murine melanoma cells, and the data relating to the hormone–cell interaction are contradictory. Legros *et al.* (1981) reported an increase of intracellular cAMP followed by stimulation of tyrosinase activity and inhibition of cell growth after treatment with α-MSH in the HM6A human melanoma cell line. On the other hand, Halaban *et al.* (1983) found tyrosinase activity and rate of cellular proliferation of normal and malignant human melanocytes to be insensitive to α-MSH. Likewise, Hadley and Dawson (1988) reported that a potent α-MSH analog, the Nle(4),D-Phe(7), was ineffective in stimulating cultured human melanocytes, as determined by the tyrosinase assay. More recently, using the human melanoma cell line IGR 1, Karg *et al.* (1991a) failed to observe any increase in adenylcyclase activity after incubation with 10^{-7} to 10^{-5} *M* α-MSH. It seems therefore that, though MSH is capable of stimulating some follicular and epidermal melanocytes, its pigmentary action is not as widespread as previously believed.

B. Melanin-Concentrating Hormone

The existence of a putative MCH causing the aggregation of melanin granules in melanophores was first suggested in catfish pituitary glands by Enami in

1955. However, it was only in 1983 that the active principle was isolated from chum salmon pituitaries by Kawauchi *et al.* (1983) and identified as a novel heptadecapeptide (Fig. 9.6). The proposed structure has been since confirmed in various laboratories by chemical synthesis (Okamoto *et al.*, 1984, Wilkes *et al.*, 1984; Eberle *et al.*, 1986). MCH cDNAs from a chum hypothalamic cDNA library have also been isolated and their genes characterized (Takayama *et al.*, 1989).

The purified peptide is active at nanomole concentrations on teleost melanophores, that is, it is 10^4 to 10^5 times more potent than catecholamines, which are well-known melanin aggregating agents in many teleost species. While the adrenal glands exert their influence on skin pigmentation indirectly through the pituitary, the actions of MCH are direct on the target cells and are unaffected either by Ca^{2+} or by α- or β-adrenoreceptor blocking agents (Baker, 1991). Melanosome aggregation in response to MCH is reversible, since removal of the peptide from the bathing medium results in a redispersal of the pigment granules within the melanophores. The melanosome aggregating activity can also be reversed if MSH is added to teleost skins lightened by MCH (Castrucci *et al.*, 1987).

MCH has been found in both the pituitary and the hypothalamus of several teleosts, the highest activity being usually in the neurointermediate lobes. Immunochemical and biochemical studies indicate that MCH is produced in the hypothalamus, from which it is transported to the neurohypophysis for storage and release (Kawauchi, 1989). Antigenically similar peptides have been detected in extracts of rat hypothalamus by radioimmunoassay and by an *in vitro* fish-scale bioassay after fractionation by HPLC and gel filtration (Kawauchi and Kawazoe, 1988). They appeared to be somewhat smaller in size and more hydrophobic than salmon MCH. However, more recent studies on the structure of rat (Vaughan *et al.*, 1989) and human (Presse *et al.*, 1990) MCH have shown the active peptide sequence has been highly conserved, the greatest changes involving the N-terminal region, where there are two substitutions and two additional residues.

MCH seems to be abundant in the mammalian brain, but its physiological significance is at present obscure (Baker, 1991). The persistence of MCH

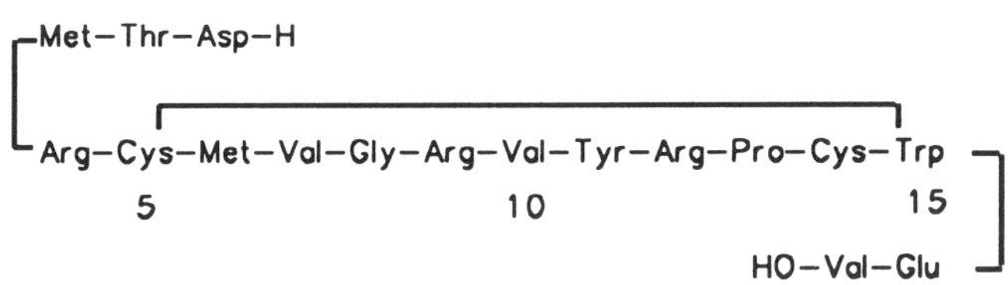

FIG 9.6 Primary structure of melanin-concentrating hormone.

hormone(s) in the mammmalian hypothalamo–neurohypophyseal system would suggest a role in the posterior pituitary function, distinct from mediation of color change seen in lower vertebrates, for example, as a neuromodulator or neurotransmitter. In support of this view, Baker *et al.* (1985) presented evidence that salmon MCH inhibits the *in vitro* release of ACTH from CRF-stimulated rat pituitary by the rat pituitary gland, although much higher concentrations of MCH were required than were effective on fish pituitaries. Moreover, Kawauchi *et al.* (1986) have demonstrated that the intravenous injection of salmon MCH induces growth hormone secretion in rats.

Several MCH analogs have been synthesized and their structural requirements to act on fish chromatophore have been determined (Hadley *et al.*, 1987; Eberle, 1988). Hopefully some of the MCH-like peptides thus obtained will prove useful in clinical medicine to inhibit ACTH secretion, for example, in Cushing's disease of pituitary origin.

C. Melatonin

Melatonin is synthesized from serotonin by acetylation, followed by methylation catalyzed by a specific enzyme, acetylserotonin methyltransferase (EC 2.1.1.4). In the pineal gland the biosynthesis is controlled by epinephrine, which stimulates adenylate cyclase, which in turn stimulates protein kinase (Weatherhead, 1982). The synthesis is suppressed by light acting via the eyes and nervous system and peaks in the dark (Reiter, 1986). Rats kept in continuous light are in continuous estrus due to the antigonadotropic activity of melatonin. In this connection, Silman and his colleagues (1979) have made the anecdotal observation that a decline in plasma melatonin precedes puberty in boys, but not in girls, and may also be accompanied by darkening of hair and skin.

The complex interactions of endocrine, particularly pineal and pituitary, and neural mechanisms in the regulation of melanophore activity in lower vertebrates has been detailed by Bagnara and Hadley (1973). In all species sensitive to melatonin the response involves the withdrawal of melanophore from the peripheral processes of the melanophores together with their aggregation in the perinuclear region of the cell. Among teleosts, melatonin has a direct paling effect on the embryonic and larval melanophores in some species, for example *Fundulus* (Wyman, 1924), but the adults are quite refractory. Likewise, in *Amphibia* the effects seem to be restricted to embryonic and larval stages in which the potency of melatonin is quite remarkable. It brings about a paling response in larval *Xenopus* when added to their aquarium water at a concentration as low as 100 pg/ml (Quay and Bagnara, 1964). Using cultured melanophores from amphibian embryos, Messenger and Warner (1977) have observed an effect of melatonin on menalosome aggregation at 10 fg/ml. Melatonin is apparently unable to effect pigmentary changes in mammals (Nordlund and Lerner, 1977; Reams *et al.*, 1968; Thody *et al.*, 1984), but it does seem to influence pigmentation in those species which undergo a seasonal pelage color change, for example, snow and varying hares, some lemmings, and some hamsters (Weatherhead, 1988). The hairs of these animals have a winter phase in which they are white and a summer phase in which they are dark, at least over the dorsum. Since the photoperiod is the main determinant of color changes, it is likely that the melanocyte system responds to changes in the level of circulating melatonin. In line with this view, Rust and Meyer (1969) found that subcutaneous implants of melatonin inhibit the pigmentation of the new hair growth following hair plucking in the short-tailed weasel, *Mustela erminea bangsi*. Similar results have been reported by Hoffmann (1978) on the Siberian hamster, *Phodopus leucopus*. However, whether the inhibitory effect of melatonin on melanogenesis was directly exerted peripherally on the melanocyte or centrally on some component of the hypothalamo–hopophyseal system which regulates MSH production was not clear. This issue was addressed by Logan and Weatherhead (1980), who found that in short-term cultures (48 hr) of hair follicles of the Siberian hamster melatonin is a potent inhibitor of melanogenesis. The most interesting feature of this response, however, was that inhibition of pigment formation occurred without a concomitant decrease in tyrosinase activity, consistent with early observations indicating no decrease in the enzyme activity during the autumn molt of hamsters (Logan and Weatherhead, 1978). These results were taken to suggest the existence of additional controlling factors acting after the dopachrome stage, a deduction which is well in line with what we know now about the posttyrosinase regulatory mechanisms of melanogenesis. A possible candidate for inhibition of melanogenesis by melatonin could be DCF or the indole blocking factor (Pawelek *et al.*, 1980), though a decrease in peroxidase activity during the winter phase seems more likely in the author's opinion. Of course, other mechanisms which prevent the enhanced levels of tyrosinase from being expressed as melanogenesis cannot be ruled out. Moreover, the question remains as to how the pineal gland and melatonin exert their effects on the melanocytes. Understanding of this issue ultimately depends upon the characterization of specific receptors and the way in which the receptors are linked to the various intracellular events involved in melanogenesis.

Much of what we know comes from studies on amphibian melanophores which have demonstrated the presence on these cells of melatonin receptors with strict conformational requirements (Heward and Hadley, 1975). The accumulated evidence indicates that the *N*-acetyl group of melatonin is important, but it is the methoxyl in the 5 position that is responsible for the aggregating properties. In line with this view, *N*-acetyl tryptamine was found to have no intrinsic activity, though it can compete with melatonin for binding sites. Receptors with similar characteristics have been found on the melanocytes of cultured hair follicles from the Siberian hamster using a selected range of melatonin analog (Logan and Weatherhead, 1980). As to the location of these receptors, the work by Cohen *et al.* (1978) suggests that they may be found in the cytosol, since they found that specific binding of tritium-labeled melatonin occurs in the supernatant fraction of homogenates from several tissues including skin.

There are various reports that melatonin might affect the pituitary content of melanotropins (Kastin and Schally, 1967), though the daily variation in MSH apparently is not abolished by pinealectomy (Tilders and Smelik, 1975). Rust and Meyer (1969) showed that melatonin inhibits the effects of α-MSH on melanogenesis in the weasel. In amphibian skin, melatonin not only blocks the MSH-induced darkening, but also significantly reduces the MSH-induced accumulation of cAMP (Abe *et al.*, 1969). However, whether it affects the rate of synthesis or breakdown of cAMP is not yet clear, though the former alternative appears more probable since melatonin does not inhibit the skin-darkening response to exogenous cAMP. Caution must be exercised in extrapolating such a mechanism to other melanophores as well as to melanocytes. In this connection, Logan and Weatherhead (1980) reported that in cultured hair follicles both melatonin and cyclic GMP prevent the MSH- or cAMP-induced stimulation of melanogenesis without any apparent change in tyrosinase activity. In any case, interpretation of the intracellular control mechanisms is complicated by interactions with hormones other than MSH. Consideration of this issue is beyond the scope of this discussion and the interested reader is referred to appropriate monographs (Weatherhead, 1988).

D. Sex Hormones

1. Estrogens

The female sex hormones, especially estrogens, have long been known to stimulate skin pigmentation in both animals and humans. Generally the skin, particularly that surrounding the genitalia and the areolae, is very responsive to the effects of sex hormones. In pregnant women, increased pigmentation of the nipples, areolae, and genital region is accompanied, though to a lesser degree, by a darkening of the facial skin, the so-called chloasma, or more properly melasma. There is also a darkening of the middle line of the abdominal wall which becomes known as the ***linea fusca*** (Abdel-Malek, 1988).

Similar effects of steroid hormones on skin pigmentation of other mammals, such as mice and guinea pigs, have been well documented. Snell (1967) found that in the guinea pig ovariectomy results in a marked reduction of the number, size, and melanin content of epidermal melanocytes. There is also a decrease in the visible length of the dendritic processes of the pigment cells. On the other hand, administration of small doses of estrogens to ovariectomized guinea pigs increases the amount of pigment both inside and outside the melanocytes within the skin, particularly in the areolar region. That this is a direct effect of the hormone(s) on the skin rather than an indirect one via the pituitary has been shown by topical application of small doses of estrogens on one nipple. Increased pigmentation of the treated nipple occurs without affecting the other. This selective effect has been explained by either of the following two mechanisms: (1) the melanocytes in the sexual region are more sensitive to estrogen(s) than those situated elsewhere; and (2) the increased metabolic activity of the neighboring keratinocytes that are themselves stimulated by the sex hormones may influence the activity of the melanocytes. This latter explanation seems to be the most likely. Certainly the increased melanogenesis would be facilitated by the increased blood flow that occurs in the sexual areas during pregnancy or as a result of estrogen therapy (Snell, 1972).

The role of estrogens on pigment formation of feathers has also been studied in some detail. Hall (1969a) found that estradiol causes formation of a pinkish-brown pigment, loosely characterized as melanin, in the neck feathers of female brown leghorn chicken. Pigment formation was apparently related to tyrosinase activity and could be inhibited by cycloheximide and puromycin at dosage levels which suppress protein synthesis (Hall, 1969b).

The effects of steroid hormones on melanoma growth in experimental animals have also been described, but the conclusions of these studies are contradictory. Proctor *et al.* (1976), for example, found that in B16 melanoma-bearing mice, estrogen suppresses tumor growth, whereas oophorectomy increases the rate of incidence of lung metastases. Others, however, have presented evidence that estrogen treatment increases the growth of B16 melanoma in male and female mice (Abdel-Malek, 1988).

There is also some disagreement on the reported presence of estrogen receptors in human melanoma cells (Delarue *et al.*, 1987; Walker *et al.*, 1987). It appears that if such receptors are present their concentration is rather low and their structures differ from estrogen receptors of the uterus or of breast carcinomas. In this connection, some of the data in the early literature may be flawed by assays allowing interference by loss of tritium due to tyrosinase activity. As a consequence, even the low level of estrogen receptors reported in some publications may have been overstated. Moreover, in lieu of or in addition to specific protein receptors, it has been suggested that melanin in mammalian tissues may function as a binding agent for estrogens, particularly for those which are more susceptible to oxidation, for example, 2-hydroxyesteradiol (Jacobsohn *et al.*, 1988a). In keeping with this view, incubation of ^{14}C-labeled estradiol or 2-hydroxyestradiol with mouse tyrosinase with dopa as cofactor leads to a melanin with estrogen firmly bound to the pigment (Jacobsohn *et al.*, 1988b). However, whether the steroids are bound as such to melanin or are chemically incorporated into the pigment polymer is not clear. But it is noteworthy that tyrosinase is able to mediate both the hydroxylation of estradiol to 2-hydroxyestradiol and its subsequent oxidation to the corresponding 2,3-orthoquinone; this latter intermediate has been recently isolated and identified by NMR spectroscopy (Jacobsohn *et al.*, 1991). It is hoped that further work in this direction will ultimately give us some insight into the mechanism underlying the effects of estrogens on melanogenesis.

2. Progesterone

Although progesterone also affects skin pigmentation, its effects are much less intense than those caused by estrogens. Interestingly, Snell (1962) reported that at high dosage levels, that is, 5 or 10 mg/day, progesterone has an opposite effect, causing a reduction in the amount of melanin in the areolar region and the anterior abdominal wall. On the other hand, when small doses of estrogen and progesterone were given together melanogenesis was stimulated; the degree of stimulation was greater than that produced by either hormone when given alone in the same dose. This would suggest that estrogen and progesterone may act synergistically to cause variation in skin color as seen during the menstrual cycle.

Resneik (1988) found that 29% of women on contraceptive drugs develop melasma, and of this group 87% develop the pigmentation during a subsequent pregnancy. The melasma is most common in brunettes frequently exposed to sunlight and is prevalent in facial skin, where melanocyte density is maximal. Use of contraceptives has also been considered among the risk factors for the increased incidence of cutaneous melanomas, but not all of the evidence points in the same direction (Evans *et al.*, 1988). A retrospective study based on recall of hormone use and reproductive events in Danish women found no association between risk of melanoma and their reproductive state, pregnancy, and oral contraceptive use (Osterlind *et al.*, 1988).

3. Androgens

The effects of testosterone on melanin pigmentation have been widely studied in both animals and man. Hamilton (1948) found that estrogen or testosterone injection of menopausal women or women who had undergone ovariectomy increases skin pigmentation. Moreover, he reported that castration of human males results in decreased skin pigmentation, whereas injection of androgens causes skin darkening. Similar changes were observed in other mammals, including mice, Long-Evans rats, and the golden hamster (Abdel-

Malek, 1988). In most cases, however, the results are based solely on macroscopic observations of skin color and it is not clear whether the change in pigmentation is due to alteration of the melanocyte activity or alteration of other factors responsible for skin color. According to Hamilton (1948), the relation of testicular secretions to skin pigmentation may well be due to a significant extent to a vascular pheonomenon. In line with this view, Wheeler *et al.* (1953) reported that topical application of testosterone to the nipples and areolae of immature castrated male guinea pigs had no significant effect on melanin pigmentation. Snell (1967) found that, in the intact and orchidectomized male guinea pig, testosterone was unable to produce uniform changes in the melanocytes and melanin content except in the interior abdominal wall of orchidectomized animals, where a definite reduction in melanogenic activity took place. From the foregoing discussion, it is apparent that further work on the relationship between testosterone and melanin pigmentation is needed. How and where testosterone acts are also challenging questions for which there are currently no definite answers.

E. Miscellaneous

Space considerations do not allow a discussion of other known hormones or conditions that have been shown to affect melanogenesis other than to say that they number in the dozen and include thyroid hormones (Snell, 1967), inflammatory mediators such as arachidonic acid and prostaglandins (Nordlund *et al.*, 1986), growth factors (Halaban *et al.*, 1988b; Gordon *et al.*, 1989), and a variety of pharmacological agents that raise the levels of cAMP (Bennett, 1989). β-Adrenergic agonists have also been shown to stimulate tyrosinase activity in hair follicular melanocytes (Burchill and Thody, 1986) and cultured melanoma cell lines (Karg *et al.*, 1991a; McEwan and Parsons, 1987), suggesting that catecholamines may play a regulatory role in melanin pigmentation. Indeed, one might infer that virtually every major pathway of intracellular signaling eventually impinges either directly or indirectly on the functional state of melanocytes and hence on melanogenesis.

CHAPTER 10

PHOTOBIOLOGY AND PHOTOCHEMISTRY OF MELANOGENESIS

I. INTRODUCTION

Exposure of human skin to solar radiation or UV light from artificial sources results in a profound alteration of metabolism, structure, and function of epidermal melanocytes. The visible outcome is a marked increase in skin pigmentation, commonly known as tanning (Pathak *et al.*, 1981). When viewed from the standpoint of photobiology, skin tanning involves two distinct phenomena (Fig. 10.1): (1) immediate tanning (IT), sometimes referred to as immediate pigment darkening (IPD); and (2) delayed tanning (DT). This latter is but one of a complex cascade of events beginning with cell injury and leading to inflammation (erythema) and enhanced keratinocyte proliferation (Parrish, 1981). The end result is an increased protection of skin against the harmful effects of UV radiation. Multiple exposures further increase this protection, but excessive doses of UV radiation result in the long term in a number of pathological changes in skin, such as atrophy, aging, actinic keratoses, and development of many cutaneous cancers, for example, squamous cell carcinoma and basal cell carcinoma. A number of authors have also expressed concern that UV exposure might be responsible for the increased incidence of malignant melanomas (Mackie *et al.*, 1987; Elder, 1989; Koh *et al.*, 1990). However, this view has not reached general consensus (Cascinelli and Marchesini, 1989; Dore and Anders, 1990).

Our incomplete knowledge of skin photobiology makes it difficult to attempt to explain individual susceptibility to the harmful effects of UV radiation. What seems clear is that skin devoid of melanogenic activity, be it constitutive or UV-induced (facultative), is by far less susceptible to actinic damage than white unpigmented skin. Thus, for example, albinotic and vitiliginous skins, which lack melanin, sunburn easily and exhibit little or no ability to tan. Redheads also show a pronounced sensitivity to sunlight, but whether

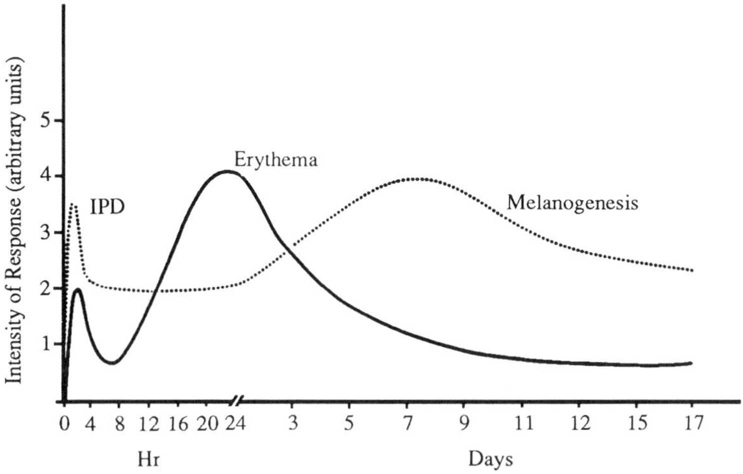

FIG 10.1 Time course of erythema and tanning response of an individual (skin type III) after a single exposure to solar radiation (90 min or approximately 4 × MED). After Pathak *et al.* (1981).

this results from the presence of pheomelanins in the skin is controversial (see Section VI,B).

Kaidbey *et al.* (1979), using spectroscopic techniques, found that the transmission and protection factors of the *stratum corneum* of black and white skin toward UVB rays (280–320 nm) are quite similar, but the dark pigmented epidermis transmits much less and protects much more than the white epidermis. Thus, since the *stratum corneum* and epidermis are of essentially equal thickness in all races, it follows that melanin is the most important skin photoprotective agent. However, as we shall see, the picture is not so simple, since other products of melanocyte activity may be implicated to account for the apparent relationship between skin color and photoprotection.

II. SKIN PHOTOTYPES

The concept of sun-reactive "skin typing" was introduced by Fitzpatrick (1988) for the specific purpose of selecting the correct initial doses of UV light for photochemotherapy. Based on the individual's tendency to sunburn and tan, six main phototypes were recognized. In the proposed classification, persons of skin types I and II have less constitutive pigmentation and less capacity to tan than individuals with skin types III, IV, or V, in that order, while type VI is represented by African and American blacks. Another classification, proposed by Cesarini (1988), takes also into consideration some genetic markers, such as freckles and hair color, which allow a more precise definition of skin phototypes.

Though these classifications serve the purpose for Caucasian subjects, they do not work well when applied to darker populations. For example, in Japan, it is necessary to define a local skin typing system which allows for a burning reaction to UV in the minimal range and a tanning reaction in the moderate to intense range (Kawava, 1986). Currently, when a more objective determination of sensitivity of skin to UV radiation is desired, the usual method is measurement of the minimal erythemal dose (MED), as evaluated either visually or more accurately by chromameters (Shono *et al.*, 1985; Westerhof *et al.*, 1990; Andreassi *et al.*, 1990). In terms of MED, it has been estimated that the protection offered by melanin pigmentation reaches a maximum value of 10–15 for very black individuals, whereas for Hispanics or Kuwaitis or dark mediterraneans it reaches a value of 2.5 (Kollias *et al.*, 1991). Yet, within each skin there may be a wide range of variability in the photobiological response to UVB, suggesting the involvement of another variable which has not been identified so far.

III. IMMEDIATE TANNING

Immediate tanning (IT) is a transient greyish-brown tan of the skin that develops immediately after exposure to UV radiation or sunlight and fades gradually within a few hours (Pathak, 1967). It is best seen in subjects with a darker skin color and can be elicited by both long-wave UV radiation (320–400 nm, UVA) and visible light. With the use of high-intensity UVA radiation sources for photo(chemo)therapy of skin diseases, the IT reaction has attracted renewed interest for the mechanisms of skin photoprotection (Agin *et al.*, 1985). However, results from independent studies have shown that IT has no photoprotective effect against UVB and UVB-induced DNA lesions (Beitner, 1988; Gange, 1988). Its physiological function, if any, is at present unknown.

Early ultrastructural studies of IT on human skin suggested that the reaction involved changes in the distribution pattern of melanosomes mediated by a redistribution of the cytoskeletal elements within the melanocytes (Jimbow *et al.*, 1973, 1975) with an active participation of keratinocytes (Lavker and Kaidbey, 1982). Such a mechanism, which is analogous to that of physiological color changes in amphibians, has been recently challenged by the demonstration that IT does not depend on the structural and functional integrity of melanocytes (Honigsmann *et al.*, 1986). This is consistent with the pioneering observations of Meirowsky (1909) and Lignac (1923) that IT can develop in isolated cadaver skin.

At the chemical level, IT has all the marks of a UVA-inducible photooxidation reaction, which is dependent on oxygenation of the involved tissue at the time of light exposure. In support of this, Rorsman and Tegner (1988) found that an external pressure stopping the blood flow to the skin during UVA irradiation effectively prevents IT reaction, the inhibition being reversed by an extracutaneous supply of oxygen, for example, with a cream containing hydrogen peroxide. In a study of the action spectrum of free-radical formation in freshly excised samples of human skin, Pathak and Stratton (1968) found that irradiation with wavelengths longer than 320 nm augmented the intrinsic melanin free radicals in pigmented skin. Under similar conditions, white skin from a fair Caucasian subject with poor IT response produced no detectable free radicals. Likewise, irradiation with ruby laser (λ_{max} 694 nm) and incoherent light (600–700 nm) resulted in a significant increase of the EPR signal in pigmented skin only, indicating that it was associated with preformed melanin. In both cases, the UV-induced free radicals were stable for 24 hr at 77K but decayed rapidly with a half-life of a few seconds at room temperature, whereas the stable intrinsic melanin signal present in pigmented skin was not destroyed by warming. These results led Pathak and Stratton (1968) to suggest that IT could result from the one-electron oxidation of reduced 5,6-dihydroxyindole units to the fully oxidized state, as indicated in Fig. 10.2.

An alternative mechanism accounting for IT reaction could involve photooxidation of melanogenic precursors, especially DHI or DHICA. When irradiated with pyrex-filtered UV light, these indoles readily undergo photolysis and/or polymerization to give eumelanic pigments (Koch and Chedekel, 1987). The rank order of photolability was found to be DHI > DHICA > 6H5MICA at 300 nm. In the case of DHI, d'Ischia and Prota (1987) showed that in aqueous buffers at neutral pH autooxidation significantly competes with the primary photochemical reactions. To overcome this problem, irradiation was then performed in methanol, using the relatively more stable 1-methyl deriv-

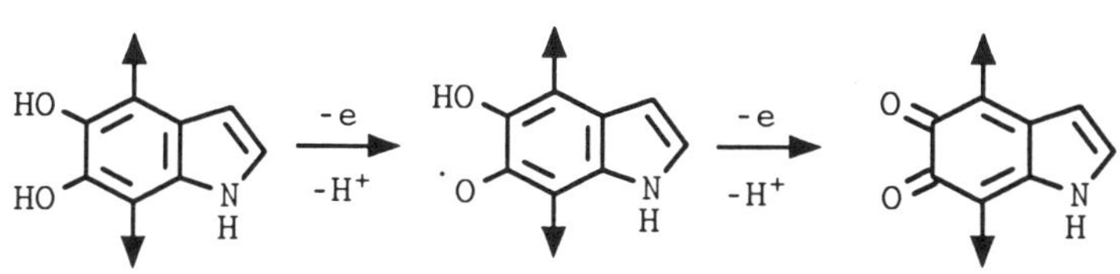

FIG 10.2 One-electron oxidation of 5,6-dihydroxyindole units in melanin during immediate pigment darkening (Pathak, 1967).

ative. Analysis of the reaction mixture, after reduction and acetylation, showed a very complex pattern of products, all in exceedingly small amounts. Five of these could be eventually isolated and identified by NMR and mass spectrometric analysis. They include two isomeric triacetoxyindoles, 5,6-diacetoxy-1-methyloxindole, and the hydroxylated dimers 10.**1** and 10.**2** (Fig. 10.3).

It appears therefore that photooxidation of 5,6-dihydroxyindoles follows a pathway which is basically different from that observed in the enzymic or metal-catalyzed oxidation (Chapter 5, Section III). If these model studies have some bearing on the *in vivo* situation, one could then envisage the possibility that IT (and perhaps DT too) is associated with the formation of melanic pigments that are chemically different from eumelanins enzymically generated in melanocytes.

Recently, Kollias *et al.* (1991) have reported that the spectral characteristics of newly formed melanin in the IT reaction are quite different from those of melanin-constitutive pigmentation. While this latter shows a monotonic increase in the absorbance with decreasing wavelength and a maximum at about 335 nm, IT exhibits a minimum at 365 nm. Kollias *et al.* (1988) also found that UVB-induced tanning is considerably more absorbing in the UVB region than the constitutive melanin pigmentation. Although this may be due in part to the deposition of melanin "dust" in the *stratum corneum*, the possibility that light-inducible pigmentation involves formation of structurally different eumelanins is most attractive.

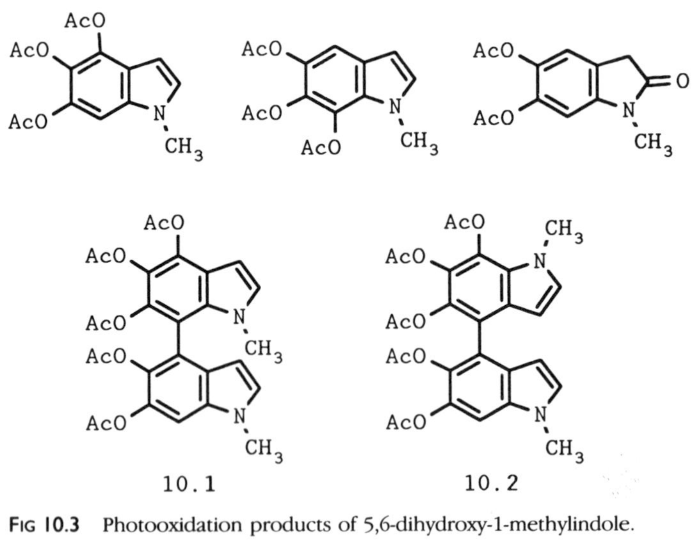

FIG 10.3 Photooxidation products of 5,6-dihydroxy-1-methylindole.

IV. DELAYED TANNING

Unlike immediate tanning, delayed tanning (DT) is not oxygen dependent and produces a long-lasting tanning due to increased synthesis, transfer, and distribution of melanosomes. DT appears 3–5 days after UV radiation exposure and reaches a maximum after about 2 or 3 weeks. From the basal cells of the epidermis the pigment is eventually swept up toward the surface and carried out along with the epidermal cells, which are finally sloughed off as a result of the normal process of desquamation. DT is maximally stimulated by the sunburn spectrum (290–320 nm) and to a lesser extent by exposure to UVA. Tanning induced by UVA shows less photoprotective capacity with respect to UVB-induced pigmentation, although protection against DNA damage does not differ (Gange *et al.*, 1985). In this connection, it may be noted that the newly formed melanin granules in UVA-induced tanning are mainly in the basal layer, while in UVB tanning they are distributed throughout the epidermis, including the dead corneocytes of the *stratum corneum* (Rosen *et al.*, 1990).

At the biological level, an increase in the number of functioning melanocytes is well documented in animals following UVA and UVB exposure and in man following multiple exposures to UVA and UVB (Pathak, 1985). It is still debatable whether mitosis and/or differentiation of stem cells into melanocytes or activation of "dormant" melanocytes or a combination of these events are responsible for the observed increase in active melanocytes (Parrish, 1981; Rosen *et al.*, 1987). However, Quevedo *et al.* (1969), investigating the cellular source and distribution of DT in human skin, were not able to find any evidence in the basal epidermal layer of increased melanocyte mitosis. Rather, their data showed intensification of melanogenesis in the dopa-positive melanocytes. On the other hand, Rosdahl (1979) reported that repeated UVB irradiation induces an increase in the melanocyte population in both irradiated and protected skin in mice and also in humans (Steiner *et al.*, 1989). This would suggest the involvement of a UVB-induced release of a mitogenic factor that reaches the nonexposed melanocytes via circulation. Such a mechanism would be of importance in relation to the appearance of melanoma in covered skin in UV-exposed subjects. However, no further work has appeared supporting the existence of a systemic factor in skin tanning.

Other effects of UVB on the melanocyte population of involved skin include increased branching of the dendritic processes, hypertrophy of the cell perikaryon, and an increase in size and number of melanosomes in both melanocytes and keratinocytes. Tyrosinase activity of melanocytes is also markedly enhanced. But how these changes are brought about is still largely obscure. A major question is whether they are induced by a direct interaction of UV radiation with a specific receptor in melanocytes or whether they result from some indirect mechanism, such as damage of blood vessels and/or keratinocytes, with subsequent release of inflammatory mediators. It is also possible that some of the skin-surface lipids, for example, squalene and cholesterol, which are susceptible to photooxidation, may play an active role in the initiating of events of the IT reaction. In support of this, Picardo *et al.* (1991) found that, when applied to guinea pig skin, UV-generated squalene peroxides induce morphological and biochemical changes mimicking those due to direct UV irradiation. Similar effects were observed with peroxidated cholesterol and other skin lipid components.

To avoid the complex interactions in the whole epidermis, one approach to the mechanism of DT has been to study the UVB-induced changes in cultured human melanocytes. Using this model system, Friedmann and Gilchrest (1987) found that physiological doses of UVB are indeed able to stimulate melanocytes and melanogenesis, presumably via a cAMP-independent mechanism. In another study on Cloudman melanoma cells (Bolognia *et al.*, 1989), the possibility that the UV-induced melanogenesis is mediated through an increase in the availability of MSH on melanocytes was suggested. This, however, has been questioned by Seechurn and Thody (1990), who found that α-MSH has no direct role in mediating UVR-induced melanogenesis *in vivo* in humans or in experimental animals. Prostaglandin E_2 and arachidonic acid, which are apparently able to mediate the UV-induced melanogenesis in epidermal melanocytes of mouse (Nurdlund *et al.*, 1986), also seem to have no effect on human melanocytes. These discrepancies are evidently due to differences between mouse and human or between malignant and normal melanocytes.

Another controversial issue concerns the role of vitamin D in the DT reaction. In 1974 Oikawa and Nakayasu reported that ergocalciferol and cholecalciferol (vitamin D_3) increased the activity of tyrosinase in cultured cells of B16 mouse melanoma. Since these calciferols are formed photochemically from ergosterol and 7-dehydrocholesterol (pro-vitamin D_3) by UVB irradiation, they suggested the appealing hypothesis that vitamin D_3 or its photoproducts might play a physiological role in photo-induced melanogenesis in human skin. This view was later substantiated by Tomita *et al.* (1986), who found that vitamin D_3 is able to stimulate tyrosinase activity and melanogenesis in human cultured melanocytes in a similar fashion as that observed after UVB irradiation. Similar effects on epidermal melanocytes *in vivo* were also observed following topical application of vitamin D_3 on the dorsal pinna of mice (Abdel-Malek *et al.*, 1988).

In a reexamination of these studies, however, Mansur *et al.* (1988) failed to observe any effect of vitamin D_3 and related metabolites on cell growth and melanogenesis in irradiated and nonirradiated melanocytes from human neonatal foreskin and Cloudman S91 melanoma. They therefore concluded that vitamin D is unlikely to play a role in the mechanism of suntanning.

The question thus remains as to what other mechanisms may account for the UVB-induced melanogenesis. An interesting view follows the observation that thioredoxin reductase is relatively susceptible to inactivation by both UVA and UVB (Schallreuter and Wood, 1989). In human subjects, enzyme inactivation was higher with UVA (54%) than with UVB (36%), the degree of inhibition depending significantly upon the skin phototype. When considered in light of the property of reduced thioredoxin to inhibit tyrosinase activity, these results were taken to mean that a change in the thioredoxin reductase/thioredoxin system could account for both the loss of free radical defense and hyperpigmentation in sunburned human skin (Sundarum *et al.*, 1990).

However, as discussed in Chapter 8, a most critical regulatory factor involved in constitutive pigmentation is the redox state of the GSH system, which determines the amount of dopaquinone available for eumelanin synthesis. Thus, a depletion in the cutaneous level of GSH, caused by UVB, either directly or indirectly, for example, by oxygen radicals, would conceivably explain the DT reaction. This would be in keeping with early studies by Halprin and Ohkawara (1967) showing that UVB-induced hyperpigmentation of human skin is preceded by a fall in the cutaneous levels of glutathione reductase activity combined with a fall in GSH and rise in oxidized GSH. These changes start about the third to the fifth day after irradiation and continue for approximately 2 weeks, concomitant with the onset of DT. Similar fluctuations in the epidermal GSH level have been reported by Connor and Wheeler (1987) in irradiated skin of hairless mice. We will address this point again in the following section.

V. PUVA TANNING

The ability of psoralens and related molecules (Fig. 10.4) to enhance the pigmentogenic response of human skin to sunlight has been known since the oldest times and has found widespread application in photochemotherapy of vitiligo and other skin diseases, especially psoriasis (Honigsmann *et al.*, 1987). In more recent years, psoralens have been incorporated with a UVB-absorbing sunscreen to enhance suntanning and photoprotection against vascular and DNA damage (Forlot, 1989; Young *et al.*, 1991). A historical profile of these applications is given by Fitzpatrick (1989).

Psoralens are typical skin photosensitizing compounds and are completely inactive as pigmenting agents in the absence of UVR. The most effective wavelength range is that corresponding to UVA; solar radiation can also be used, but in this case only the UVA fraction is active. Several studies have shown that the effects of PUVA on melanocytes and melanogenesis are qualitatively similar to those seen in UVB-induced DT (Pathak *et al.*, 1976). The

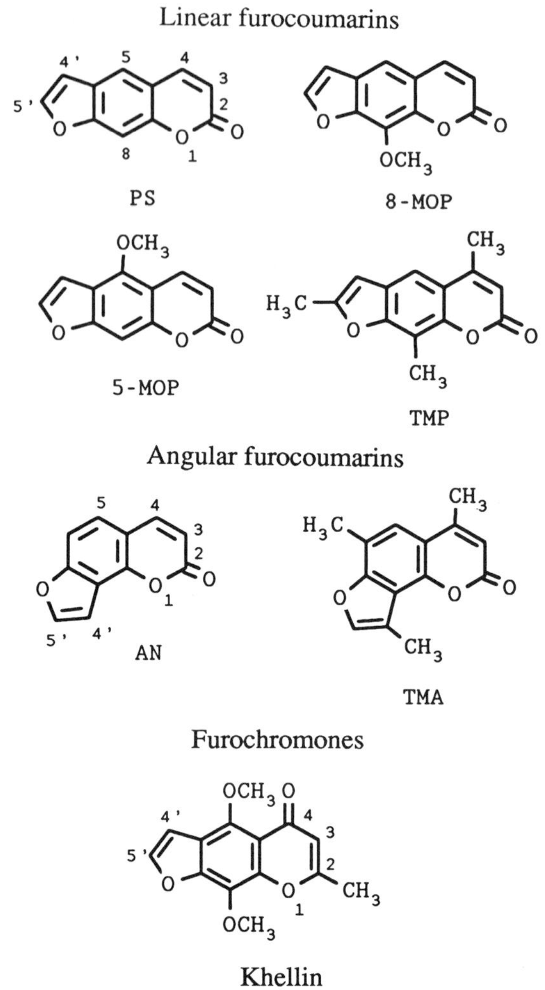

FIG 10.4 Structures of furocoumarins and related products. PS, psoralen; 8-MOP, 8-methoxypsoralen; 5-MOP: 5-methoxypsoralen; TMP, 4,5′,8-trimethylpsoralen; AN, angelicin; TMA, 6,4,4′-trimethylangelicin.

observed changes include (1) an increase in the number of functionally active melanocytes, (2) an enhanced rate of formation of melanosomes, (3) an alteration in the distribution pattern and rate of transfer of melanosomes to the keratinocytes, and (4) the activation of tyrosinase. The hyperpigmenting effect of PUVA is often associated with erythema, and it has been surmised that increased pigmentation is a secondary effect induced by erythema and not a primary effect directly elicited by psoralens; this view, however, has been discounted by the discovery of new psoralen analogs and derivatives which exhibit the typical photosensitizing properties of psoralens, including hyperpigmentation, without inducing erythema (Rodighiero, 1985; Pathak and Dalle Carbonare, 1989).

At the chemical level, the stimulatory effect of PUVA on melanocytes is apparent from an increase in the urine and serum level of colorless metabolites, as represented by 5-cysdopa and 6H5MICA. In studies at Lund (Hansson *et al.*, 1985), it was found that the urinary level of these two metabolites begins to increase contemporarily after 2–3 days of PUVA treatment before any detectable hyperpigmentation is observed, indicating that stimulation of tyrosinase activity has already occurred at that time. A similar, albeit less pronounced, increase in the urinary excretion of 5-cysdopa is also caused by exposure of skin to sunlight, UVA or UVB (Rorsman and Tegner, 1988).

The extensive literature dealing with the various aspects of the chemistry and mode of action of psoralens has been covered in several comprehensive reviews (*e.g.*, Song and Tapley, 1979; Dall'Acqua, 1986) and in symposia proceedings. Classical concepts center largely on the ability of psoralens to form mono- and bifunctional photoadducts with pyrimidine bases of DNA, and the possibility that the delayed division of photodamaged melanocytes may prolong the availability of surface MSH receptors in the G2 phase of the cell cycle. After repair by excision of DNA adducts, the melanocytes would undergo mitosis, giving rise to new functional melanocytes with enhanced tyrosinase activity and further melanization of melanosomes (Straub *et al.*, 1981; Pathak *et al.*, 1986). Such a mechanism, however, does not explain why some psoralen analogs, for example, the angelicins, which are structurally unable to form DNA photoadducts, are nevertheless capable of enhancing the UVA-inducible pigmentation quite efficiently.

According to another view, increased pigment production would result from a psoralen-sensitized type II photooxidation of tyrosine or dopa, mediated by singlet oxygen and superoxide ions as the actual oxidizing agent (De Mol and Beijersbergen van Henegouwen, 1981; Joshi *et al.*, 1987). However, when performed *in vitro* this photochemical reaction is very poor, especially in the case of tyrosine, seeming therefore unlikely to provide more than a minor contribution to the intense pigmentogenic effect of PUVA. In light of what was said previously about the mechanism of constitutive and UVB-induced hyperpigmentation, one would expect that a perturbation of the redox state of the GSH system is also involved in PUVA-induced hyperpigmentation. In support of this, Connor and Wheeler (1987) found that irradiation of hairless mouse skin with UVA combined with oral administration of pigmentogenic psoralens results in a depletion of the cutaneous levels of GSH, the extent of which is dependent on the nature and dose of the psoralen used. In a relevant study (d'Ischia *et al.*, 1989), it was shown that the same psoralens causing GSH depletion *in vivo* are also able to accelerate the kinetics of the UVA-induced photoooxidation of GSH *in vitro* (Fig. 10.5). The relative potency of the psoralens tested was found to follow the order PS > TMP > 8-MOP > 5-MOP, which roughly reflects their pigmenting ability.

Apparently, the accelerating effect is only minimally mediated by active oxygen species and results from the direct interaction of the photoexcited psoralens with GSH. To what extent the proposed mechanism is responsible for the depletion of GSH in cutaneous tissues is not yet clear. It is noteworthy, however, that PUVA-induced hyperpigmentation and suntanning, just like pigmentary variations in constitutive pigmentation, all revolve around the critical role of GSH in melanogenesis.

VI. PHOTOCHEMISTRY OF MELANINS

A. EUMELANINS

UV irradiation of aqueous suspensions of melanin leads to consumption of oxygen and, eventually, to bleaching of the pigment, a phenomenon which is also observed in human hair (Wolfram, 1981). Mechanistic studies carried out

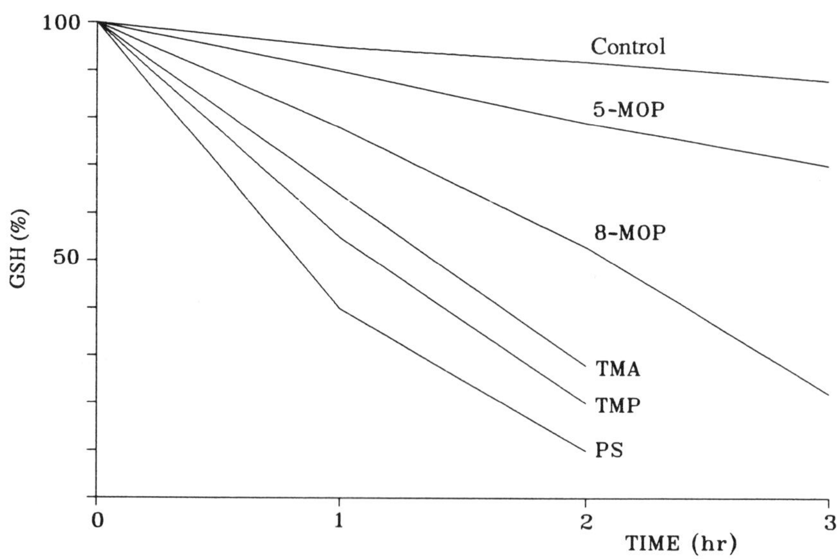

FIG 10.5 Effect of furocoumarins (0.35 m*M*) on the aerial photooxidation of GSH (1 m*M*) in ethanol/0.02 *M* phosphate buffer, pH 6.8 (6:4, v/v).

on both natural and synthetic eumelanins suggest that two distinct processes are involved: reversible oxidation of 5,6-dihydroxyindole units and irreversible degradation of the pigment polymer mediated by hydrogen peroxide, which is the actual bleaching agent (Wolfram and Albrecht, 1987; Korytowski and Sarna, 1990). The rate of production of hydrogen peroxide depends on the fluence rate of the irradiating light. Not unexpectedly, redox metal ions, for example, Cu^{2+}, markedly accelerate the photolysis of melanin, as in the case of chemically induced bleaching.

Several lines of evidence support the active involvement of hydroxyl radicals in melanin photolysis in aerated media, whereas arguments for the possible implication of singlet oxygen are rather weak (Sarna and Swartz, 1991). In fact, the rate of oxygen consumption induced by monochromatic light from either the visible or the UV region seems to be independent of the presence of azide and not sensitive to isotope exchange of the water. On the other hand, Korytowski *et al.* (1987) found that the oxygen consumption induced by light is reduced by 50% when catalase is present in irradiated samples of dopamelanin. Moreover, addition of SOD increased the rate of oxygen consumption by a factor of 2, consistent with the involvement of superoxide ion and hydrogen peroxide in melanin photolysis.

Under anaerobic conditions, irradiation of eumelanins involves mainly reversible changes including, notably, a significant increase in the free radical content in the pigment polymer (Mason *et al.*, 1960; Cope *et al.*, 1963). These transient radicals consist of two pools, one characterized by a low yield (ca. 1–2%) and a decay time of a few seconds and the other with a decay time of a few milliseconds, accounting for about 50% of the signal intensity (Sealy *et al.*, 1980). Under continuous irradiation, the yield of this latter component is 50–100 times that of the former, and the photoexcitation process shows the characteristics of a singlet–triplet mechanism that can be formulated as follows:

$$Q + QH_2 \xrightarrow{h\nu} \text{singlet} \rightarrow \text{triplet} \rightarrow 2\ QH^{\cdot} \rightarrow Q + QH_2$$

where Q, QH_2, and $QH^{\cdot}$ represent quinone, quinol, and semiquinone units in the pigment polymer.

Action spectra measured for natural melanins, for example, bovine eye melanin, are very similar to those found for synthetic dopamelanin and are unaffected by the removal of protein (Felix *et al.*, 1978a). Measurements of the quantum yields indicate that the efficiency of radical production from natural pigments is about three times greater than that from the synthetic material, although the chromophore involved seems to be the same in both systems (Sarna and Sealy, 1984). More recently, Bibang and his associates (1989a,b) have reported on an improved technique for studying the photo-induced radicals on eumelanin samples dispersed in a solid KBr matrix. Under such conditions five different types of radicals, named R1, R1′, R2, R2′, and R3, were detected, the first two of which were assigned to the deprotonated and neutral form of the intrinsic phenoxyl radicals in the eumelanin polymer (Fig. 10.6).

Of the remaining three, R2 and R2′ appeared to be the cation radical and the indolyl radical derived from an excited indolic site in the triplet state, whereas R3 was associated with the formation of an electron-hole pair through a band-to-band excitation of the eumelanin polymer, regarded as a semi-organized solid.

The free-radical properties of eumelanin and its ability to reduce molecular oxygen to superoxide ion and hydrogen peroxide have stimulated speculations about a possible phototoxic role of the pigment in biological systems, for example, the RPE–choroid layer. This is characterized by one of the highest intracellular oxygen concentrations in the entire body, and an excess of incident light could promote the oxidation of the residing melanin with generation of potentially cytotoxic species (Sarna, 1991). It has also been reported that in certain pigmented melanoma cells eumelanin can act as a radiosensitizer of DNA damage (Hill and Hill, 1987), although the observed effect may be due to the existence of photosensitive diffusible substances elaborated by the tumor cells themselves (Hill *et al.*, 1991).

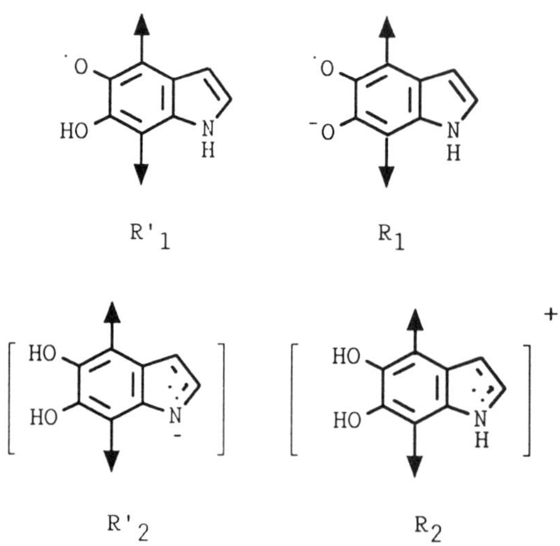

FIG 10.6 Suggested structure of the radicals formed on light exposure of eumelanins from human hair (Bibang *et al.*, 1989b).

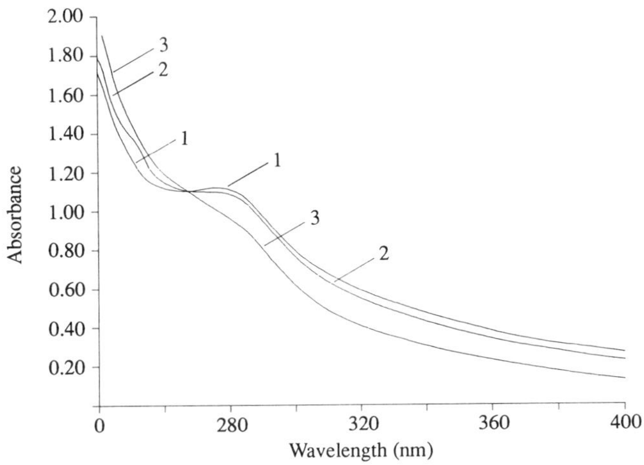

FIG 10.7 Photolysis of pheomelanin (0.43 mg/ml in phosphate buffer, pH 8.0; air-equilibrated in Pyrex test tubes). Spectra 1–3 correspond to photolysis times of 0, 158, and 1113 min, respectively (Chedekel *et al.*, 1978).

B. PHEOMELANINS

Much less is known about the photochemistry of pheomelanins. Chedekel *et al.* (1977) reported that, unlike eumelanins, pheomelanins are "exceptionally" photolabile when irradiated in aerated solution at pH 13 with glass-filtered UV light. The rate of decomposition, as monitored spectrophotometrically at 400 nm, was found to be dependent on pH and the initial concentration of both pigment and oxygen (Chedekel *et al.*, 1978). Notably, at pH 8 in phosphate buffer photolysis of pheomelanin proceeded at a very low rate, more than 2 hr being required before a detectable change in the UV spectrum is observed (Fig. 10.7). Under such conditions a certain amount of superoxide was detected using the nitroblue tetrazolium assay.

These and other studies, including the mutagenic properties of pheomelanin photodegradation product(s) in the Ames test, have led Chedekel (1982) to suggest several mechanisms by which UV light may be deleterious to epidermal pheomelanin-containing tissues: (1) loss of pigment purportedly responsible for photoprotection, (2) formation of dermatitic or carcinogenic photoproducts, and (3) generation of reactive photoproducts inhibiting the enzymes responsible for repair of UV-induced cell damage. These hypotheses have prompted several studies aimed at comparing the photobiological effects of pheomelanins and eumelanins on cellular and enzymatic systems. In one of these studies (Menon *et al.*, 1985) it was found that irradiation of Ehlrich ascites carcinoma cells in the presence of pheomelanin extracted from red human hair lysed the cells to a greater extent than irradiation of the cells in the presence of eumelanin from black hair. The photodynamic effects of pheomelanins were apparently not due to superoxide, but rather to the pigment-free radicals and hydrogen peroxide formed during irradiation. The same authors also reported that irradiation of arachidonic acid and linolenic acid in the presence of pheomelanin produced more peroxides than in the presence of eumelanins. However, the relevance of these observations to the high susceptibility of redheads to the deleterious effects of sunlight is open to question. It is possible, for example, that the observed photobiological differences between pheomelanins and eumelanins may be partly due to the different solubility properties of the two types of pigments. Indeed, Wolfram and Albrecht (1987) found that under comparable conditions native pheomelanin in red hair is less photolabile than eumelanin in black hair, a finding supported by parallel *in vitro* experiments with the isolated pigments (Sarna *et al.*, 1984; Wolfram *et al.*, 1986).

VII. MELANIN, MELANOGENESIS, AND SKIN PHOTOPROTECTION

The role of melanin in photoprotection is either obvious or wholly obscure. Certainly, pigmented skin is by far less susceptible than white skin to sunburn and phototoxic reactions. However, whether this is solely related to the epidermal content of melanin or other products of melanocyte activity is not clear. Confusion also exists as to the actual mechanism by which melanin would provide photoprotection.

A widely accepted view, which was first proposed in 1820 by Sir Everald Home, is that melanin absorbs UVR and therefore acts as a passive screening filter to protect the underlying tissue against actinic damage. Advocates of this view often refer, as supporting evidence, to the umbrellalike distribution of the melanin granules over the keratinocytes and the increased skin photoprotection following suntanning. However, if we analyze the epidermal response to UVR without prejudices, we find that it has all the features of a nonspecific hyperplasia in response to damage and death of cells, in a similar fashion as hyperpigmentation of skin occurs in response to a variety of other traumas such as ionizing radiation, heat, abrasion, and inflammation. In any case, suntanning comes about too long after the UV radiation has done its worst damage. In passing, it may be worth noting that black people tend to avoid sunning because they are not exempt from sunburn, in spite of the fact that their skin is heavily pigmented (Illis, 1988). Kaidbey and Kligman (1978) exposed eight volunteers to UVA twice a week for 8 weeks and compared the photoprotection 1 week after the last irradiation against UVA and sunlight.

Although much pigment was generated, the minimal protection against a MED was only 2–3 times higher with respect to control unpigmented skin.

These observations and others in both human and animal models (Hill, 1989; Vermer *et al.*, 1991) suggest that melanin alone may not be such a good photoprotective agent. In fact, when compared with conventional sunscreens, natural or synthetically produced eumelanins appear to have little effectiveness in protecting skin from UVR exposure (L. J. Wolfram, unpublished). It could be argued, however, that a melanin-containing cream or lotion applied over the skin is far from a reproduction of the potential photoprotection afforded by the strategic distribution and aggregation of melanosomes in the epidermis. In any case, a role of melanin as a neutral sunscreen is unlikely and would not be in keeping with the insoluble character of the pigment granules, giving rise to scattering rather than absorption of light.

An alternative interpretation of the photoprotective role of melanins comes from recent studies, discussed earlier, showing that melanin is by no means an inert passive material, as previously believed, but a rather unstable chemical entity exhibiting an impressive range of physicochemical properties. Among these, perhaps the most relevant from the physiological viewpoint is the ability to scavenge oxygen-derived radicals, such as superoxide anion and hydrogen peroxide, which are normally formed during biochemical or photochemical processes. Both species are known to produce several biological effects, most of which are deleterious to cell homeostasis. While host defenses are adequate to remove these toxic species under physiological conditions, on exposure to UVR or other inflammatory stimuli their concentration increases, and thus the presence of an *in situ* residing quencher such as melanin becomes important.

In this context, the possible contribution of other products of melanocyte activity should not be overlooked. In a preliminary study at Naples (G. Prota *et al.*, unpublished), it was found that DHI is capable of inhibiting lipooxygenase-induced oxidation of arachidonic acid, the major unsaturated fatty acid present in mammalian phospholipids. DHI and DHICA are also endowed with excellent antioxidant properties and are capable of capturing oxygen radicals, themselves being eventually converted to melanin. Viewed in this way, melanin pigmentation could be regarded as a visible indicator of the ability of DHI and related colorless metabolites to counteract noxious effects induced by internal or external stimuli, including UV light.

A further contribution to this picture follows from several studies showing that 5,6-dihydroxyindoles and other melanogenic precursors, for example, cysdopas, undergo photodestruction when irradiated with biologically relevant UV light (*i.e.*, with wavelengths greater than 300 nm). As a result of their photochemical reactivity, a number of free-radical species can be generated. These have been identified by EPR spin trapping, laser flash photolysis, and pulse radiolysis techniques and include hydrated electrons, hydroxyl radicals, semiquinones, aryl, thiyl, and alanine carbon-centered radicals (Land *et al.*, 1986; Chedekel and Zeise, 1988). Whether such photodegradation products of melanin precursors are formed *in vivo* is not yet documented; however, *in vitro* studies suggest that they may play a role in the acute and chronic response of human skin to sunlight.

References

Abdel-Malek, Z. A. (1988). Endocrine factors as effectors of integumental pigmentation. *Dermatologic Clinics* **6,** 175–183.

Abdel-Malek, Z. A., Ross, R., Pike, J., and Nordlund, J. J. (1988). Hormonal effects of vitamin D3 on epidermal melanocytes. *J. Cell. Physiol.* **136,** 273–280.

Abe, K., Butcher, W., Nicholson, W. E., Bairde, C. E., Liddle, R. A., and Liddle, G. W. (1969). Adenosine 3,5-monophosphate (cyclic AMP) as the mediator of melanocyte stimulating hormone (MSH) and norepinephrine on the frog skin. *Endocrinology* **84,** 362–368.

Acquaron, R. R., Rouge, F., and Aubert, C. (1981). Pheomelanin in albino negroes: Urinary excretion in Cameroonian subjects. *In* "Pigment Cell 1981: Phenotypic Expression in Pigment Cells" (M. Seiji, ed.), pp. 97–103. University of Tokyo Press, Tokyo.

Agin, P. P., Desorochers, D. L., and Sayre, R. M. (1985). The relationship of immediate pigment darkening to minimal erythemal dose, skin and eye color. *Photodermatol.* **2,** 288–294.

Agius, C. (1985). The melano-macrophage centres of fish: A review. *In* "Fish Immunology" (M. J. Manning and M. F. Tatner, eds.), pp. 85–105. Academic Press, New York.

Agrup, G., Falck, B., Kennedy, B.-M., Rorsman, H., Rosengren, A.-M., and Rosengren, E. (1975). Formation of cysteinyldopa from glutathionedopa in melanoma. *Acta Derm. Venereol. (Stockh.)* **55,** 1–3.

Agrup, G., Hansson, C., Rorsman, H., Rosengren, A.-M., and Rosengren, E. (1978a). 5-*S*-Cysteinyldopa and trichochromes in red feathers. *Acta Derm. Venereol. (Stockh.)* **58,** 269–282.

Agrup, G., Lindbladh, C., Prota, G., Rorsman, H., Rosengren, A.-M., and Rosengren, E. (1978b). Trichochromes in the urine of melanoma patients. *J. Invest. Dermatol.* **70,** 90–91.

Agrup, G., Hansson, C., Rorsman, H., Rosengren, E., and Tegner, E. (1983a). Methemoglobin catalyzed formation of dopa and 6-OH-dopa from tyrosine. *Acta Derm. Venereol. (Stockh.)* **63,** 152–155.

Agrup, G., Edholm, L.-E., Rorsman, H., and Rosengren, E. (1983b). Diastereoisomers of 5-*S*-cysteinyldopa. *Acta Derm. Venereol. (Stockh.)* **63,** 59–61.

Aime, S., and Crippa, P. R. (1988). ^{13}C solid-state cross polarization magic-angle-spinning nuclear magnetic resonance spectra of natural and synthetic melanins. *Pigment Cell Res.* **1,** 355–357.

Aime, S., Fasano, M., and Croombridge, C. (1990). Solid-state carbon-13 NMR characterization of melanin free acids from biosynthetic and natural melanins. *Gazz. Chim. Ital.* **120,** 663–664.

Alena, F., Jimbow, K., and Ito, S. (1990). Melanocytotoxicity and antimelanoma effects of phenolic amine compounds in mice *in vivo*. *Cancer Res.* **50,** 3743–3747.

Allegri, G., Biasolo, M., Frison, G., Pelli, B., and Traldi, P. (1987). Collisional spectroscopy in structural characterization of melanins. 1. *Pigment Cell Res.* **1,** 87–93.

Allegri, G., Biasolo, M., Frison, G., Pelli, B., and Traldi, P. (1988). Collisional spectroscopy in structural characterization of melanins. 2. Laser desorption experiments on bio- and synthetic tryptophan melanins. *Biomed. Environ. Mass. Spectrom.* **15,** 353–355.

Allegri, G., Arban, R., Costa, C., Biasolo, M., Curcuruto, O., Pozzan, A., and Traldi, P. (1990). Fast atom bombardment mass spectrometry of dopamine melanogenesis intermediates. *Pigment Cell Res.* **3,** 181–186.

Allport, D. C., and Bu'Lock, J. D. (1960). Biosynthetic pathways in *Daldinia concentrica*. *J. Chem. Soc.* 654–662.

Altschule, M. D., and Hegedus, Z. L. (1977). The importance of studying visceral melanins. *Clin. Pharmacol. Ther.* **19,** 124–134.

Ambani, L. M., Van Woert, M. H., and Murphy, S. (1975). Brain peroxidase and catalase in Parkinson's disease. *Arch. Neurol.* **32,** 114–118.

Anders, F. (1991). Contributions of the Gordon-Kosswig melanoma system to the present concept of neoplasia. *Pigment Cell Res.* **3,** 7–29.

Andreassi, L., Casini, S., Simoni, P., Bartalini, P., and Fimiani, M. (1990). Measurement of cutaneous colour and assessment of skin type. *Photodermatol. Photoimmunol. Photomed.* **7,** 20–24.

Aroca, P., Garcia-Borron, J. C., Solano, F., and Lozano, J. A. (1990). Regulation of distal mammalian melanogenesis. I. Partial purification and characterization of a dopachrome converting factor: dopachrome tautomerase. *Biochim. Biophys. Acta* **1035,** 266–275.

Ashida, M. (1971). Purification and characterization of pre-phenoloxidase from hemolymph of the silk worm *Bombyx mori*. *Arch. Biochem. Biophys.* **144,** 749–762.

Aso, Y., Imamura, Y., and Yamasaki, N. (1989). Further studies on dopa quinone imine conversion factor from cuticles of *Manduca sexta L. Insect Biochem.* **19,** 401–407.

Aubert, C., Rosengren, E., Rorsman, H., and Rougé, F. (1975). Differentiation of melanocytes in cultures of primary malignant melanoma indicated by 5-*S*-cysteinyldopa formation. *J. Natl. Cancer Inst.* **55,** 1327–1328.

Aubert, C., Voulot, C., Rougé, F., and Galindo, J. R. (1988). Melanogenesis and malignancy in experimental B16 melanoma variants. *In* "Advances in Pigment Cell Research" (J. T. Bagnara, ed.), pp. 155–167. Alan Liss, New York.

Axelrod, J., and Lerner, A. B. (1963). *O*-Methylation in the conversion of tyrosine to melanin. *Biochim. Biophys. Acta* **71,** 650–655.

Badinand, A., Pasquier, J., Vallon, J. J., Guilluy, R., and Milon, H. (1969). Identification chromatographique en couche mince de deux mélanogènes indolique urinaires. *Clin. Chim. Acta* **25,** 357–364.

Bagnara, J. T. (1983). Developmental aspects of vertebrate chromatophores. *J. Am. Zool.* **23,** 465–478.

Bagnara, J. T., and Hadley, M. E. (1973). "Chromatophores and Color Change." Prentice-Hall, Englewood Cliffs, New Jersey.

Bagnara, J. T., Matsumoto, J., Ferris, W., Frost, S. K., Turner, W. A., Tchen, T. T., and Taylor, J. D. (1979). Common origin of pigment cells. *Science* **203,** 410–415.

Bagnara, J. T., Fukuzawa, T., Jonson, W. C., Law, J. H., Samaraweera, P., and Zuasti, A. (1992). Integumental pigmentary factors of lower vertebrates—Models for human hypo- and hypermelanoses. *In* "The Pigment Cell: From the Molecular to the Clinical Level." (Y. Mishima, ed.). Munksgaord, Copenhagen (In press).

Baker, B. I. (1991). Melanin-concentrating hormone: A general vertebrate neuropeptide. *Int. Rev. Cytol.* **126,** 1–47.

Baker, B. I., Bird, D. J., and Buckingham, J. C. (1985). Salmonid melanin concentrating hormone inhibits corticotropin release. *J. Endocrinol.* **106,** 5–8.

Banda, P. W., Sherry, A. E., and Blois, M. S. (1977). Column cation-exchange separation of melanin-related metabolites in urine. *Clin. Chem.* **23,** 1397–1401.

Baradani, L., Bridelli, M. G., Carbucicchio, M., and Crippa, P. R. (1982). Comparative Mössbauer and infrared analysis of iron-containing melanins. *Biochim. Biophys. Acta* **716,** 8–15.

Barbeau, A., Roy, M., Coultier, T., Plasse, L., and Paris, S. (1986). Environmental and genetic factors in the etiology of Parkinson's disease. *Adv. Neurol.* **45,** 299–306.

Barber, J. I., Townsend, D., Olds, D. P., and King, R. A. (1984). Dopachrome oxidoreductase: A new enzyme in the pigment pathway. *J. Invest. Dermatol.* **83,** 145–149.

Barber, J. I., Townsend, D., Olds, D. P., and King, R. A. (1985). Decreased dopachrome oxidoreductase activity in yellow mice. *J. Hered.* **76,** 59–60.

Barbetta, M., Casnati, G., and Ricca, A. (1966). Findings on aspergillinin. *Rend. Accad. Naz. Lincei* **40,** 450–456.

Barden, H. (1969). The histochemical relationship of neuromelanin and lipofuscin. *J. Neuropathol. Exp. Neurol.* **28,** 419–441.

Barden, H. (1981). The biology and chemistry of neuromelanin. *In* "Age Pigments" (R. S. Sohal, ed.), pp. 155–180. Elsevier-North Holland Press, Amsterdam.

Barden, H., and Brizzee, K. R. (1987). The histochemistry of lipofuscin and neuromelanin. *Adv. Biosci.* **64,** 339–392.

Barden, H., and Martin, E. (1972). Electron probe microanalyses of neuromelanin and lipofuscin. *In* "Pigmentation: Its Genesis and Biologic Control" (V. Riley, ed.), pp. 631–638. Appleton Century Crofts, New York.

Barisas, B. G., and McGuire, J. S. (1974). A proteolytically activated tyrosinase from frog epidermis. *J. Biol. Chem.* **249,** 3151–3156.

Barnicot, N. A. (1956). The pigment, trichosiderin, from human red hair. *Nature* **77,** 528–529.

Barr, F. E. (1983). Melanin: The organizing molecule. *Med. Hypotheses* **11,** 1–140.

Barth, R. F., Soloway, A. S., and Fairchild, R. G. (1990). Boron neutron capture therapy of cancer. *Cancer Res.* **50,** 1061–1070.

Becker, S. W. (1959). Historical background of research on pigmentary diseases of the skin. *J. Invest. Derm.* **32,** 185–196.

Beer, R. J. S., Clarke, K., Khorana, H. G., and Robertson, A. (1948). The chemistry of the melanins. Part I. The synthesis of 5,6-dihydroxyindole and related compounds. *J. Chem. Soc.* 2223–2226.

Beer, R. J. S., McGrath, L., Robertson, A., and Woodier, A. B. (1949). The chemistry of the melanins. Part II. The synthesis of 5,6-dihydroxyindole-2-carboxylic acid and related compounds. *J. Chem. Soc.* 2061–2066.

Beer, R. J. S., Broadhurst, T., and Robertson, A. (1954). The chemistry of the melanins. Part V. The autoxidation of 5,6-dihydroxyindoles. *J. Chem. Soc.* 1947–1953.

Beermann, F., Ruppert, S., Hummler, E., Bosch, F. X., Muller, G., Ruther, U., and Schutz, G. (1990). Rescue of the albino phenotype by introduction of a functional tyrosinase gene into mice. *EMBO J.* **9,** 2819–2826.

Beitner, H. (1988). Immediate pigment-darkening reaction. *Photodermatology* **5,** 96–100.

Benathan, M., and Wyler, H. (1980). Contribution a l'analyse quantitative des melanines. *Yale J. Biol. Med.* **53,** 389. [(Abs.). For details see: Benathan, M. (1980). Ph.D. Thesis, University of Lausanne, Switzerland.]

Benedetto, J. P., Ortonne, J. P., Voulot, C., Khatchadourian, C., Prota, G., and Thivolet, J. J. (1981). Role of thiol compounds in mammalian melanin pigmentation. Part 1. Reduced and oxidized glutathione. *J. Invest. Dermatol.* **77,** 402–405.

Benedetto, J. P., Ortonne, J. P., Voulot, C., Khatchadourian, C., Prota, G., and Thivolet, J. J. (1982). Role of thiol compounds in mammalian melanin pigmentation. Part 2. Glutathione and related enzymatic activities. *J. Invest. Dermatol.* **79,** 422–424.

Bennett, D. C. (1989). Mechanism of differentiation in melanoma cells and melanocytes. *Environ. Health Perspect.* **80,** 49–59.

Bennett, D. C. (1991). Colour genes, oncogenes and melanocyte differentiation. *J. Cell Sci.* **98,** 135–139.

Bennett, D. C., Cooper, P. J., and Hart, I. R. (1987). A line of non-tumorigenic mouse melanocyte, syngenic with the B16 melanoma and requiring a tumor promoter for growth. *Int. J. Cancer* **39,** 414–418.

Bernam, V., Filpula, D., Herber, W., Ribb, M., and Katz, E. (1985). The nucleotide sequence of the tyrosinase gene from *Streptomyces antibioticus* and characterization of the gene product. *Gene* **37,** 101–110.

Bertrand, G. (1896). Sur une nouvelle oxydase, ou ferment soluble oxydant, d'origine vegetale. *Comp. Rend. Acad. Sci. (Paris)* **122,** 1215–1217.

Berzelius, J. J. (1840). *Lehrbuch der Chemie.* **9,** 522.

Bibang, R., Arnaud, R., Lemaire, J., Deflandre, A., and Lang, G. (1989a). Electron spin resonance study of photoinduced radicals in various eumelanin in solid matrix. *Pigment Cell Res.* **2,** 395–400.

Bibang, R., Arnaud, R., Lemaire, J., Deflandre, A., and Lang, G. (1989b). Electron spin resonance of eumelanin from hair: Photoinduced radicals in solid matrix. *Pigment Cell Res.* **2,** 387–394.

Billingham, R. E., and Silvers, W. K. (1960). The melanocytes of mammals. *Q. Rev. Biol.* **35,** 1–40.

Binns, F., Chapman, R. F., Robson, N. C., Swan, G. A., and Waggott, A. (1970a). Studies related to the chemistry of melanins. Part VIII. The pyrrolecarboxylic acids formed by oxidation or hydrolysis of melanins derived from 3,4-dihydroxyphenylethylamine or DOPA. *J. Chem. Soc.* **(C),** 1128–1134.

Binns, F., King, J. A. G., Mishra, S. N., Percival, A., Robson, N. C., Swan, G. A., and Waggott, A. (1970b). Studies related to the chemistry of melanins. Part XIII. Studies on the structure of dopamine-melanin. *J. Chem. Soc.* 2062–2070.

Bjorklund, A., Falck, B., Jacobsson, S., Rorsman, H., Rosengren, A.-M., and Rosengren, E. (1972). Cysteinyldopa in human malignant melanoma. *Acta Derm. Venereol. (Stockh.)* **52,** 357–360.

Bloch, B. (1916). Chemische Untersuchungen über das specifische pigmentbildende Ferment der Haut, die Dopaoxydase. *Z. Physiol. Chem.* **98,** 227–254.

Bloch, B., and Schaaf, F. (1925). Pigment Studien. *Biochem. Z.* **162,** 181–206.

Blois, M. S. (1978). The melanins: Their synthesis and structure. *Photochem. Photobiol. Rev.* **3,** 115–134.

Blois, M. S., Zahlan, A. B., and Maling, J. E. (1964). Electron spin resonance studies on melanin. *Biophys. J.* **4,** 471–490.

Boekelheide, K., Graham, D. G., Mize, P. D., Anderson, C. W., and Jeffs, P. W. (1979). Synthesis of γ L-glutaminyl-[3,5-^{3}H]4-hydroxybenzene and the study of reactions catalyzed by the tyrosinase of *Agaricus bisporus*. *J. Biol. Chem.* **254,** 12185–12191.

Bogacz, A., Buszman, E., and Wilczok, T. (1989). Competition between metal ions for dopa-melanin. *Stud. Biophys.* **132,** 189–195.

Bogerts, B. (1981). A brain stem atlas of catecholaminergic neurons in man, using melanin as a natural marker. *J. Comp. Neurol.* **197,** 63–80.

Boissy, R. E. (1988). The melanocyte: Its structure, function, and subpopulations in skin, eyes and hair. *Dermatologic Clinics* **6,** 161–173.

Boldt, P. (1964). Zur Kenntnis des Trichosiderins eines Pigments aus roten Haaren. *Naturwissenschaften* **51,** 265.

Boldt, P., and Hermestedt, E. (1967). Pyrrotrichole, eine Gruppe farbiger Vergindungen aus rotem Menschenhaar. *Z. Naturforsch.* **22B,** 718–722.

Bolognia, J., Murray, M., and Pawelek, J. (1989). UVB-induced melanogenesis may be mediated through the MSH-receptor system. *J. Invest. Dermatol.* **92,** 651–656.

Bolt, A. G. (1967). Interaction between human melanoprotein and chloropromazine derivatives. I. Isolation and purification of human melanoprotein from hair and melanoma tissue. *Life Sci.* **6,** 1277–1283.

Bonner, T. G., and Duncan, A. (1962). Infrared spectra of some melanins. *Nature* **194,** 1078–1079.

Bouchard, B., Fuller, B. B., Vijiyasaradhi, S., and Houghton, A. N. (1989). Induction of pigmentation in mouse fibroblasts by expression of human tyrosinase cDNA. *J. Exp. Med.* **169,** 2029–2042.

Bouchilloux, S., and Kodja, A. (1960). Combinaison des thiols avec le quinones se formant au cours de la melanogenese. *Bull. Soc. Chim. Biol.* **42,** 1045–1064.

Bouheroum, M., Bruce, J. M., and Land, E. J. (1989). The mechanism of the oxidative decarboxylation of 3,4-dihydroxymandelic acid: Identification of 3,4-mandeloquinone as a key intermediate. *Biochim. Biophys. Acta* **998,** 57–62.

Bourquelot, E., and Bertrand, A. (1895). Le bleuissement et le noircissement des champignons. *Comp. Rend. Soc. Biol. (Paris)* **2,** 582–584.

Bowness, J. M., Morton, R. A., Shakir, M. H., and Stubbs, A. L. (1952). Distribution of copper and zinc in mammalian eyes. Occurrence of metals in melanin fractions from eye tissues. *Biochem. J.* **51,** 521–530.

Boyer, R. F., Mascotti, P. D., and Schori, B. E. (1986). Ferroxidase activity of mushroom tyrosinase. *Phytochemistry* **25,** 1281–1283.

Breathnach, A. S. (1988). Extra-cutaneous melanin. *Pigment Cell Res.* **1,** 234–237.

Breathnach, A. S. (1989). Azelaic acid—Biological activities and therapeutic applications. *Drugs Today* **25,** 463–472.

Bridelli, M. G., De Mattei, M., and Levi, A. C. (1982). Preliminary spectroscopic studies on the neuromelanic pigment of human *substantia nigra* nucleus. *It. J. Biochem.* **31,** 315–321.

Bridelli, M. G., Deriu, A., and Ito, A. S. (1986). X-ray and Mössbauer investigation of iron-containing melanosomes. *Hyperf. Inter.* **29,** 1395–1398.

Bridelli, M. G., Crippa, P. R., and Uguzzoli, F. (1991). X-ray diffraction studies on melanins in lyophilized melanosomes. *Pigment Cell Res.* **3,** 187–191.

Brooks, D. W., and Dawson, C. R. (1966). Aspects of tyrosinase chemistry. *In* "The Biochemistry of Copper" (J. Peisach, P. Aisen, and W. E. Blumberg, eds.), pp. 343–357. Academic Press, New York.

Brown, C., and Davidson, R. M. (1985). 1,4-Benzothiazines, dihydro-1,4-benzothiazines and related compounds. *In* "Advances in Heterocyclic Chemistry" (A. R. Katritsky, ed.), Vol. 38, pp. 135–176. Academic Press, New York.

Brumbaugh, J. A., and Moore, J. W. (1969). The effects of the E alleles upon melanocyte differentiation in the fowl. *Can. J. Gen. Physiol.* **11,** 118–124.

Bu'Lock, J. D. (1960). Intermediates in melanin formation. *Arch. Biochem. Biophys.* **91,** 189–193.

Bu'Lock, J. D. (1961). The formation of melanin from adrenochrome. *J. Chem. Soc.* 52–58.

Bu'Lock, J. D., and Harley-Mason, J. (1951a). Melanin and its precursors. Part II. Model experiments on the reactions between quinones and indole, and consideration of a possible structure for the melanin polymer. *J. Chem. Soc.* 703–716.

Bu'Lock, J. D., and Harley-Mason, J. (1951b). Melanin and its precursors. Part III. New synthesis of 5,6-dihydroxyindole and its derivatives. *J. Chem. Soc.* 2248–2252.

Burchill, S. A., and Thody, A. J. (1986). Melanocyte-stimulating hormone and the regulation of tyrosinase activity in hair follicular melanocytes of the mouse. *J. Endocrinol.* **111,** 225–232.

Burchill, S. A., Thody, A. J., and Ito, S. (1986). Melanocyte-stimulating hormone, tyrosinase activity and the regulation of eumelanogenesis and pheomelanogenesis in the hair follicular melanocytes of the mouse. *J. Endocrinol.* **109,** 15–21.

Burchill, S. A., Virden, R., and Thody, A. J. (1989). Regulation of tyrosinase synthesis and its processing in the hair follicular melanocytes of the mouse during eumelanogenesis and pheomelanogenesis. *J. Invest. Dermatol.* **93,** 235–240.

Burchill, S. A., Marks, J. M., and Thody, A. J. (1990). Tyrosinase synthesis in different skin types and the effects of α-melanocyte stimulating hormone and cyclic AMP. *J. Invest. Dermatol.* **95,** 558–561.

Burchill, S. A., Ito, S., and Thody, A. J. (1991). Tyrosinase expression and its relationship to eumelanin and pheomelanin synthesis in human hair follicles. *J. Dermatol. Sci.* **2,** 281–286.

Burnett, J. B. (1971). The tyrosinase of mouse melanoma. Isolation and molecular properties. *J. Biol. Chem.* **246,** 3079–3091.

Cabanes, J., Garcia-Canovas, F., Lozano, J. A., and Garcia-Carmona, F. (1987). A kinetic study of the melanization pathway between L-tyrosine and dopachrome. *Biochim. Biophys. Acta* **923,**187–195.

Cable, J., and Steel, K. P. (1991). Identification of two types of melanocyte within the *stria vascularis* of the mouse inner ear. *Pigment Cell Res.* **4,** 87–101.

Carraro, C., and Pathak, M. A. (1988). Characterization of superoxide dismutase from mammalian skin epidermis. *J. Invest. Dermatol.* **90,** 31–36.

Carstam, R., Hansson, C., Krook, G., Rorsman, H., Rosengren, E., and Wirestrand, L. E. (1985). Oxidation of dopa in human albinism. *Acta Derm. Venereol. (Stockh.)* **65,** 435–437.

Carstam, R., Hansson, C., Lindbladh, C., Rorsman, H., and Rosengren, E. (1987). Dopaquinone addition products in cultured human melanoma cells. *Acta Derm. Venereol. (Stockh.)* **67,** 100–105.

Carstam, R., Brinck, C., Fornstedt, B., Rorsman, H., and Rosengren, E. (1990). 5-*S*-Cysteinyldopac in human urine. *Acta Derm. Venereol. (Stockh.)* **70,** 373–377.

Carstam, R., Brinck, C., Hindemith-Augustsson, A., Rorsman, H., and Rosengren, E. (1991). The neuromelanin of the human *substantia nigra*. *Biochim. Biophys. Acta* **1097,** 152–160.

Cascinelli, N., and Marchesini, R. (1989). Increasing incidence of cutaneous melanoma, ultraviolet radiation and the clinician. *Photochem. Photobiol.* **50,** 497–505.

Castagnoli, N., Jr., Chiba, K., and Trevor, A. J. (1985). Potential bioactivation pathways for the neurotoxin 1-methyl-4-phenyl-1,2,3,6-tetrahydropyridine (MPTP). *Life Sci.* **36,** 225–230.

Castrucci, A. M. L., Hadley, M. E., Wilkes, B. C., Zechel, C., and Hruby, V. J. (1987). Melanin concentrating hormone exhibits both MSH and MCH activities on individual melanophores. *Life Sci.* **40,** 1845–1851.

Cesarini, J. P. (1988). Photo-induced events in the human melanocytic system: Photoaggression and photoprotection. *Pigment Cell Res.* **1,** 223–233.

Chacon, J. N., Chedekel, M. R., Land, E. J., and Truscott, T. G. (1987). Chemically induced Parkinson's disease: Intermediates in the oxidation of 1-methyl-4-phenyl-1,2,3,6-tetrahydropyridine to the 1-methyl-4-phenylpyridinium ion. *Biochem. Biophys. Res. Commun.* **144,** 957–964.

Chakraborty, A. K., Mishima, Y., Inazu, M., Hatta, S., and Ichihashi, M. (1989). Melanogenic regulatory factors in coated vesicles from melanoma cells. *J. Invest. Dermatol.* **93,** 616–620.

Chakraborty, A. K., Ichihashi, M., Mishima, Y. (1991). Effect of dopa-loading on glutathione metabolizing enzymes and tyrosinase in relation to 5-*S*-cysteinyldopa genesis in cultured B-16 melanoma cells. *J. Dermatol. Sci.* **2,** 325–335.

Chedekel, M. R., (1982). Photochemistry and photobiology of epidermal melanins. *Photochem. Photobiol.* **35,** 881–885.

Chedekel, M. R., and Zeise, L. (1988). Sunlight, melanogenesis and radicals in the skin. *Lipids* **23,** 587–591.

Chedekel, M. R., Post, P. W., Diebel, R. M., and Klaus, M. (1977). Photodestruction of pheomelanin. *Photochem. Photobiol.* **26,** 651–653.

Chedekel, M. R., Smith, S. K., Post, P. W., Pokora, A., and Vessel, D. L. (1978). Photodestruction of pheomelanin: Role of oxygen. *Proc. Natl. Acad. Sci. U.S.A.* **75,** 5395–5399.

Chedekel, M. R., Land, E. J., Thompson, A., and Truscott, T. G. (1984). Early steps in the free radical polymerization of 3,4-dihydroxyphenylalanine (dopa) into melanin. *J. Chem. Soc. Chem. Commun.* 1170–1172.

Chedekel, M. R., Bahn, P., Patil, D., Wolfram, L., and Schultz, T. (1986). Melanin free acid. A chemical standard for eumelanin research. *J. Invest. Dermatol.* **87,** 397.

Chedekel, M. R., Subbarao, K. V., Bahan, P., and Schultz, T. M. (1987). Biosynthetic and structural studies on pheomelanin. *Biochim. Biophys. Acta* **912,** 239–243.

Chedekel, M. R., Patil, D. G., Rao, K. V., Murphy, B. P., Clark, M., Gardella, J., and Schultz, T. M. (1988). Solid phase carbon-13 NMR of ^{13}C-enriched eumelanins: The fate of the pyrrolic ring. *Pigment Cell Res.* **1,** 282–293.

Chen, Y. M., and Chavin, W. (1965). Radiometric assay of tyrosinase and theoretical considerations of melanin formation. *Anal. Biochem.* **13,** 234–258.

Chen, Y. M., and Chavin, W. (1972). *In* "Pigmentation: Its Genesis and Biologic Control" (V. Riley, ed.), pp. 593–606. Appleton Century Crofts, New York.

Chen, Y. M., and Chavin, W. (1975). Melanogenesis in human melanomas. *Cancer Res.* **35,**606–612.

Cheng, A. C., Shulgin, A. T., and Castognoli, N., Jr. (1982). Studies on the chemical reactivity on the quinone methide derived from the oxidative cyclization of α-methyl-3,4-dihydroxyphenylalanine ethyl ester. *J. Org. Chem.* **47,** 5258–5262.

Chio, S. S. (1977). Ph.D. Thesis, University of Houston.

Chioccara, F., and Novellino, E. (1986). A convenient one-step synthesis of 5-cystein-*S*-yldopa using ceric ammonium nitrate. *Synth. Commun.* **16,** 967–971.

Chioccara, F., Novellino, E., Prota, G., and Sodano, G. (1977). A trimeric aldolization product of 1,4-benzothiazine. *J. Chem. Soc., Chem. Commun.* 50–51.

Cicero, R., Sciuto, S., Chillemi, R., and Sichel, G. (1982). Melanosynthesis in the Kupffer cells of *Amphibia. Comp. Biochem. Physiol.* **73A,** 477–479.

Clark, M. B. J., Gardella, J. A. J., Schultz, T. M., Patil, D. G., and Salvati, L. J. (1990). Solid-state analysis of eumelanin biopolymers by electron spectroscopy for chemical analysis. *Anal. Chem.* **62,** 949–956.

Cleffman, G. (1963). Agouti pigment cells *in situ* and *in vitro. Ann. N.Y. Acad. Sci.* **100,** 749–761.

Cleffman, G. (1964). Function-specific changes in the metabolism of agouti pigment cells. *Exp. Cell Res.* **35,** 590–600.

Clemo, G. R., and Weiss, J. (1945). Melanin formation. *J. Chem. Soc.* 702–705.

Cohen, G. N. (1945). Sur le premier produit de transformation de l'adrenochrome au cours de la melanisation de l'adrenaline. *Comp. Rend. Acad. Sci.* **220,** 796–797.

Cohen, G. (1983). The pathobiology of Parkinson's disease: Biochemical aspects of dopamine neuron senescence. *J. Neural Transm.* **19S,** 89–103.

Cohen, M., Roselle, D., Chabner, B., Schmidt, T. J., and Lippman, M. (1978). Evidence for a cytoplasmic melatonin receptor. *Nature* **274,** 894–895.

Cohen, T., Muller, R. M., Tomita, Y., and Shibahara, S. (1990). Nucleotide sequence of the cDNA encoding human tyrosinase-related protein. *Nucleic Acids Res.* **18,** 2807–2808.

Commoner, B., Townsend, J., and Pake, J. E. (1954). Free radicals in biological material. *Nature* **174,** 689–691.

Connor, M. J., and Wheeler, L. A. (1987). Depletion of cutaneous glutathione by ultraviolet radiation. *Photochem. Photobiol.* **46,** 239–245.

Cope, F. W., Sever, R. J., and Polis, B. D. (1963). Reversible free radical generation in the melanin granules of the eye by visible light. *Arch. Biochem. Biophys.* **100,** 171–177.

Corradini, M. G., Napolitano, A., and Prota, G. (1986). A biosynthetic approach to the structure of eumelanins. The isolation of oligomers from 5,6-dihydroxy-1-methylindole. *Tetrahedron* **42,** 2083–2088.

Corradini, M. G., Crescenzi, O., and Prota, G. (1988). A reinvestigation of the anaerobic conversion of adrenochrome to "adrenaline black." *Tetrahedron* **44,** 1803–1808.

Cory, J. G., and Frieden, E. (1967a). Faster oxidation of tyrosine-26 of oxidized B-chain of insulin by tyrosinase. *Biochemistry* **6,** 116–120.

Cory, J. G., and Frieden, E. (1967b). Differential reactivities of tyrosine residues of protein to tyrosinase. *Biochemistry* **6,** 121–126.

Costantini, C., Crescenzi, O., Prota, G., and Palumbo, A. (1990). New intermediates of phaeomelanogenesis *in vitro* beyond the 1,4-benzothiazine stage. *Tetrahedron* **46,** 6831–6838.

Costantini, C., Crescenzi, O., and Prota, G. (1991). Mechanism of the rearrangement of dopachrome to 5,6-dihydroxyindole. *Tetrahedron Lett.* **31,** 3849–3850.

Cowen, D. (1986). The melanoneurons of the human cerebellum (*nucleus pigmentosus cerebellaris*) and homologues in the monkey. *J. Neuropathol. Exp. Neurol.* **45,** 205–221.

Crescenzi, O., Costantini, C., and Prota, G. (1990). Evidence for the intermediacy of quinone-methides in the rearrangement of aminochromes to 5,6-dihydroxyindoles. *Tetrahedron Lett.* **42,** 6095–6096.

Crippa, P. R., Horak, V., Prota, G., Svoronos, P., and Wolfram, L. (1989). Chemistry of melanins. *In* "The Alkaloids" (A. Brossi, ed.), Vol. 36, pp. 253–323. Academic Press, New York.

Cromartie, R. I. T., and Harley-Mason, J. (1953). Melanin and its precursors. Part VII. Synthesis of methylated 5,6-dihydroxyindoles, β-(4,5-dihydroxy-2-methylphenyl) alanine, and related amines. *J. Chem. Soc.* 3525–3528.

Cromartie, R. I. T., and Harley-Mason, J. (1957). Melanin and its precursors. 8. The oxidation of methylated 5,6-dihydroxyindoles. *Biochem. J.* **66,** 713–720.

Culp, C. H., Eckels, D. E., and Sidles, P. H. (1975). Threshold switching in melanin. *J. Appl. Phys.* **46,** 3658–3659.

Dall'Acqua, F. (1986). Furocumarin photochemistry and its main biological implications. *In* "Current Problems in Dermatology: Therapeutic Photomedicine" (H. Honigsman and C. Stingl, eds.), Vol. 15, pp. 137–163. Karger, Basel.

D'Amato, R. J., Lipman, Z. P., and Snyder, S. H. (1986). Selectivity of the parkinsonian neurotoxin MPTP: Toxic metabolite MPP^+ binds to neuromelanin. *Science* **231,** 987–989.

Daniels, F., Jr., Post, P. W., and Johnson, B. E. (1972). Theories of the role of pigment in the evolution of human races. *In* "Pigmentation: Its Genesis and Biologic Control" (V. Riley, ed.), pp. 13–22. Appleton Century Crofts, New York.

Das, K. C., Abramson, M. B., and Katzman, R. (1976). A new chemical method for quantifying melanin. *J. Neurochem.* **26,** 695–699.

Das, K. C., Abramson, M. B., and Katzman, R. (1978). Neuronal pigments: Spectroscopic characterization of human brain melanin. *J. Neurochem.* **30,** 601–606.

Dawson, C. R., and Tarpley, W. B. (1951). Copper oxidase. *In* "The Enzymes" (B. Summer and K. Myrback, eds.), Vol. II, Part I, pp. 454–498. Academic Press, New York.

D'Azur, V. (1786). Traitè d'anatomie et de physiologie. Cited by C. D. Marsden (1969). Brain melanin. *In* "Pigments in Pathology" (M. Wolman, ed.). Academic Press, New York.

Deibel, R. B. (1983). Biosynthetic and structural studies on pheomelanin. Ph.D. Thesis. Johns Hopkins University, Baltimore, Maryland.

Deibel, R. B., and Chedekel, M. R. (1982). Biosynthetic and structural studies on pheomelanin. *J. Am. Chem. Soc.* **104,** 7306–7309.

Deibel, R. B., and Chedekel, M. R. (1984). Biosynthetic and structural studies on pheomelanin. 2. *J. Am. Chem. Soc.* **106,** 5884–5888.

Delarue, J. C., Avril, M. F., Charpentier, P., and Prade, M. (1987). Estrogen receptors in malignant melanoma. *Arch. Dermatol.* **123,** 159–160.

De Mol, N. J., and Beijersbergen van Henegouwen, G. M. J. (1981). Relation between some photobiological properties of furocoumarins and their extent of singlet oxygen production. *Photochem. Photobiol.* **33,** 815–819.

Dencker, L., Larsson, B., Olander, K., Ullberg, S., and Yokota, M. (1979). False precursors of melanin as selective melanoma seekers. *Br. J. Cancer* **39,** 449–452.

Devi, C. C., Tripathi, R. K., and Ramaiah, A. (1987). pH-dependent interconvertible allosteric forms of murine melanoma tyrosinase: Physiological implications. *Eur. J. Biochem.* **166,** 705–711.

Devi, C. C., Tripathi, R. K., and Ramaiah, A. (1989). pH-dependent interconvertible forms of mushroom tyrosinase with different kinetic properties. *Pigment Cell Res.* **2,** 8–13.

Dietler, C., and Lerch, K. (1982). Reaction inactivation of tyrosinase. *In* "Oxidases and Related Redox Systems" (T. E. Kind, M. Morrison, and H. S. Mason, eds.), pp. 305–317. Pergamon Press, New York.

d'Ischia, M., and Prota, G. (1987). Photooxidation of 5,6-dihydroxy-1-methyl-indole. *Tetrahedron* **43,** 431–434.

d'Ischia, M., Palumbo, A., and Prota, G. (1985). 5,6-dihydroxyindole-2-carboxylic acid by treatment of sepiomelanin with sodium borohydride. *Tetrahedron Lett.* **26,** 2801–2804.

d'Ischia, M., Napolitano, A., and Prota, G. (1987a). Sulfhydryl compounds in melanogenesis. Part I. Reaction of cysteine and glutathione with 5,6-dihydroxyindoles. *Tetrahedron* **43,** 5351–5356.

d'Ischia, M., Napolitano, A., and Prota, G. (1987b). Sulfhydryl compounds in melanogenesis. Part II. Reactions of cysteine and glutathione with dopachrome. *Tetrahedron* **43,** 5357–5362.

d'Ischia, M., Palumbo, A., and Prota, G. (1988). Adrenalin oxidation revisited. New products beyond the adrenochrome stage. *Tetrahedron* **44,** 6441–6446.

d'Ischia, M., Napolitano, A., and Prota, G. (1989). Psoralens sensitize glutathione photooxidation *in vitro*. *Biochim. Biophys. Acta* **993,** 143–147.

d'Ischia, M., Napolitano, A., Tsiakas, K., and Prota, G. (1990). New intermediates in the oxidative polymerisation of 5,6-dihydroxyindole to melanin promoted by the peroxidase/H_2O_2 system. *Tetrahedron* **46,** 5789–5796.

d'Ischia, M., Napolitano, A., and Prota, G. (1991). Peroxidase as an alternative to tyrosinase in the oxidative polymerization of 5,6-dihydroxyindoles to melanin(s). *Biochim. Biophys. Acta* **1073,** 423–430.

Dohke, K. (1973). Studies on prephenoloxidase-activating enzyme from cuticle of the silkworm *Bombyx mori*. I. Activation reaction by the enzyme. *Archiv. Biochem. Biophys.* **157,** 203–209.

Donaldsson, J., McGregor, D., and La Bella, F. (1982). Manganese neurotoxicity: A model for free radical mediated neurodegeneration? *Can. J. Physiol. Pharmacol.* **60,** 1398–1405.

Dore, J. F., and Anders, F. (1990). Contributions of experimental models to understanding of human melanoma. *In* "Cutaneous Melanoma: Biology and Management" (N. Cascinelli, M. Santinami, and U. Veronesi, eds.), pp. 63–71. Masson, Milan.

Dorea, J. G., and Pereira, S. E. (1983). The influence of hair color on the concentration of zinc and copper in boys' hair. *J. Nutrition* **113,** 2375–2381.

Dorey, C. K., Torres, X., and Swart, T. (1990). Evidence of melanogenesis in porcine retinal pigment epithelial cells *in vitro*. *Exp. Eye Res.* **50,** 1–10.

Dorris, F. (1939). The production of pigment by chick neural crest in grafts to the 3-day limb bud. *J. Exp. Zool.* **80,** 315–345.

Drager, U. C., and Balkema, G. W. (1987). Does melanin do more than protect from light? *Neurosci. Res.* **6S,** 75–86.

Dryja, T. P., O'Neil-Dryja, M., Pawelek, J. M., and Albert, D. M. (1978). Demonstration of tyrosinase in the adult bovine uveal tract and retinal pigment epithelium. *Invest. Ophthalmol. Visual Sci.* **17,** 511–514.

Dryja, T. P., O'Neil-Dryja, M., and Albert, D. M. (1979). Elemental analysis of melanins from bovine hair, iris, choroid, and retinal pigment epithelium. *Invest. Ophthalmol. Vis. Sci.* **18,** 231–236.

Duchon, J. (1987). Urinary melanogens as a mirror of melanogenesis *in vivo*. *In* "Cutaneous Melanoma" (U. Veronesi, N. Cascinelli, and M. Santinami, eds.), pp. 225–231. Academic Press, New York.

Duchon, J., and Matous, B. (1967). Identification of two new metabolites in melanoma urine: 5-Hydroxy-6-methoxyindole-2-carboxylic and 5-methoxy-6-hydroxyindole-2-carboxylic acids. *Clin. Chim. Acta* **16,** 397–402.

Duchon, J., and Pechan, Z. (1963). The biochemical and clinical significance of melanogenuria. *Ann. N.Y. Acad. Sci.* **100,** 1048–1068.

Duchon, J., Matous, B., and Pechan, Z. (1966). On the chemical nature of urinary melanogens. *In* "Structure and Control of the Melanocyte" (G. Della Porta and O. Muhlbock, eds.), pp. 175–184. Springer Verlag, Berlin.

Duchon, J., Matous, B., and Pavel, S. (1981). Melanogenuria. *In* "Pigment Cell 1981: Phenotypic Expression in Pigment Cells" (M. Seiji, ed.), pp. 609–615. University of Tokyo Press, Tokyo.

Duckworth, H. W., and Coleman, J. E. (1970). Physicochemical and kinetic properties of mushroom tyrosinase. *J. Biol. Chem.* **245,** 1613–1625.

Duff, G. A., Roberts, J. E., and Foster, N. (1988). Analysis of the structure of synthetic and natural melanins by solid-phase NMR. *Biochemistry* **27,** 7112–7116.

Duncker, H. R. (1985). The neural crest. *In* "Pigment Cell 1985: Biological, Molecular and Clinical Aspects" (J. T. Bagnara, S. Klaus, E. Paul, and M. Schartl, eds.), Vol. 7, pp. 255–269. University of Tokyo Press, Tokyo.

DuShane, G. P. (1935). An experimental study of the origin of pigment cells in Amphibia. *J. Exp. Zool.* **72,** 1–32.

Dworzanski, J. P., and Debowski, M. (1985). Pyrolysis-gas chromatography of pheomelanins. *J. Anal. Appl. Pyrolysis* **8,** 463–472.

Dworzanski, J. P., Debowski, M. T., and Waclawek, R. (1984). Analysis of melanosomes from cattle eyes by pyrolysis-gas chromatography. *J. Anal. Appl. Pyrolysis* **6,** 391–400.

Eberle, A. N. (1988). The Melanotropins. Chemistry, Physiology and Mechanism of Action. Karger, Basel.

Eberle, A. N., Atherton, E., Dryland, A., and Sheppard, R. C. (1986). Peptide synthesis. Part 9. Solid-phase synthesis of melanin-concentrating hormone using a continuous-flow polyamide method. *J. Chem. Soc. Perkin Trans.* **I,** 361–367.

Eiberg, H., and Mohr, J. (1987). Major locus for red hair color linked to MNS blood groups on chromosome 4. *Clin. Genet.* **32,** 125–128.

Eisinger, M., and Marko, O. (1982). Selective proliferation of normal human melanocytes *in vitro* in the presence of phorbol ester and cholera toxin. *Proc. Natl. Acad. Sci. U.S.A.* **79,** 2018–2022.

Elder, D. E. (1989). Human melanocytic neoplasms and their etiologic relationship with sunlight. *J. Invest. Dermatol.* **92S,** 297–303.

Enami, M. (1955). Melanophore-concentrating hormone (MCH) of possible hypothalamic origin in the catfish *Parasilurus. Science* **121,** 36–37.

Evans, W. C., and Raper, H. S. (1937). The accumulation of L-3,4-dihydroxyphenylalanine in the tyrosinase–tyrosine reaction. *Biochem. J.* **31,** 2162–2170.

Evans, R. D., Kopf, A. W., Lew, R. A., Rigel, D. S., Bart, R. S., Friedman, R. J., and Rivers, J. K. (1988). Risk factors for the development of malignant melanoma. I. Review of case control studies. *J. Dermatol. Surg. Oncol.* **14,** 393–408.

Faray, B. A., Lawson, D. H., and Nixon, D. W. (1981). Melanoma detection by enzyme radioimmunoassay of L-dopa, dopamine, and 3-*O*-methyldopamine in urine. *Clin. Chem.* **27,**108–112.

Fattorusso, E., and Cimino, G. (1971). Sul processo di polimerizzazione dei 5,6-diossindoli. *Rend. Accad. Sci. Fis. Mat. (Napoli)* **38,** 173–182.

Fattorusso, E., Piattelli, M., and Nicolaus, R. A. (1965). Some natural black pigments. *Rend. Accad. Sci. Fis. Mat.* **32,** 57–58.

Fattorusso, E., Minale, L., De Stefano, S., Cimino, G., and Nicolaus, R. A. (1968). Struttura e biogenesi delle feomelanine. Nota V. Sulla struttura della gallofeomelanina-1. *Gazz. Chim. Ital.* **98,** 1443–1463.

Fattorusso, E., Minale, L., De Stefano, S., Cimino, G., and Nicolaus, R. A. (1969a). Struttura e biogenesi delle feomelanine. Nota IX. Feomelanine biosintetiche. *Gazz. Chim. Ital.* **99,** 969–992.

Fattorusso, E., Minale, L., Cimoni, G., De Stefano, S., and Nicolaus, R. A. (1969b). Struttura e biogenesi delle feomelanine. Nota VI. Sulla struttura della gallofeomelanina. *Gazz. Chim. Ital.* **99,** 29–45.

Fattorusso, E., Minale, L., De Stefano, S., and Nicolaus, R. A. (1970a). Struttura e biogenesi delle feomelanine. Nota XIII. Sulla struttura della gallofeomelanina. *Gazz. Chim. Ital.* **100,** 880–887.

Fattorusso, E., Minale, L., and Sodano, G. (1970b). Feomelanine e eumelanine da nuove fonti naturali. *Gazz. Chim. Ital.* **100,** 452–460.

Feeney-Burns, L. (1980). The pigment of the retinal pigment epithelium. *In* "Current Topics in Eye Research" (J. A. Zadunaisky and H. Davson, eds.), Vol. 2, pp. 119–178. Academic Press, New York.

Fehling, C., Hansson, C., Poulsen, J., Rorsman, H., and Rosengren, E. (1981). Formation of glutathionedopa in albino rats after DOPA injection. *Acta Derm. Venereol. (Stockh.)* **61,** 339–342.

Feldman, J. F., and Thayer, J. P. (1974). Cyclic AMP-induced tyrosinase synthesis in *Neurospora crassa. Biochem. Biophys. Res. Commun.* **61,** 977–982.

Felix, C. C., Hyde, J. S., Sarna, T., and Sealy, R. C. (1978a). Melanin photoreaction in aerated media: Electron spin resonance evidence for production of superoxide and hydrogen peroxide. *Biochem. Biophys. Res. Commun.* **84,** 335–341.

Felix, C. C., Hyde, J. S., Sarna, T., and Sealy, R. C. (1978b). Interactions of melanin with metal ions: Electron spin resonance evidence for chelate complexes of metal ions with free radicals. *J. Am. Chem. Soc.* **100,** 3922–3926.

Ferrini, U., Mileo, A. M., and Hearing, V. J. (1987). Microheterogeneity of melanosome bound tyrosinase from the Harding-Passey murine melanoma. *Int. J. Biochem.* **19,** 227–234.

Figge, F. H. (1939). Melanin: A natural reversible oxidation–reduction system and indicator. *Proc. Soc. Exp. Biol. Med.* **41,** 127–129.

Filatovs, J., McGuiness, J., and Corby, P. (1976). Thermal and electronic contributions to switching in melanins. *Biopoymers* **15,** 2309–2312.

Fitzpatrick, T. B. (1988). The validity and practicality of sun-reactive skin types I through VI. *Arch. Dermatol.* **124,** 869–871.

Fitzpatrick, T. B. (1989). The psoralen story: Photochemotherapy and photochemoprotection. *In* "Psoralens, Past, Present and Future of Photochemoprotection and Other Biological Activities" (T. B. Fitzpatrick, P. Forlot, M. A. Pathak, and F. Urbach, eds.), pp. 5–10. J. Libbey Eurotext, Montrouge, France.

Fitzpatrick, T. B., and Breathnach, A. S. (1963). Das epidermale Melanin-Einheit System. *Derm. Wschr.* **147,** 481–489.

Fitzpatrick, T. B., Brunet, P. C. J., and Kukita, A. (1958). The nature of hair pigment. *In* "The Biology of Hair Growth" (W. Montagna, ed.), pp. 255–303. Academic Press, New York.

Flesch, P. (1968). Epidermal iron pigments in red species. *Nature* **217,** 1056–1057.

Flesch, P. (1970). The red pigmentary system and its relation to black melanin genesis. *J. Soc. Cosmet. Chem.* **21,** 77–83.

Flesch, P., and Rothman, S. (1945). Isolation of an iron pigment from human red hair. *J. Invest. Dermatol.* **6,** 257–270.

Flesch, P., and Rothman, S. (1948). Role of sulphydryl compounds in pigmentation. *Science* **108,**505–506.

Flesch, P., Esoda, E. J., and Katz, S. A. (1967). The color of red hair. *J. Cosmet. Chem.* **18,** 777–784.

Foley, J. M., and Baxter, D. (1958). On the nature of pigment granules in the cells of the locus coerulus and substantia nigra. *J. Neuropathol. Exp. Neurol.* **17,** 586–598.

Fontaine, A. R. (1962). Pigments in *Ophiocomina nigra.* II. Presence of melanin and fluorescent pigments. *J. Mar. Biol. Assoc. U.K.* **42,** 9–31.

Forlot, P. (1989). Psoralen induced pigmentation in human skin: An overview of clinical evidence, mechanisms of induction and future research. *In* "Psoralens, Past, Present and Future of Photochemoprotection and Other Biological Activities" (T. B. Fitzpatrick, P. Forlot, M. A. Pathak, and F. Urbach, eds.), pp. 63–71. J. Libbey Eurotext, Montrouge, France.

Forrest, F. M. (1974). Evolutionary origin of extrapyramidal disorders in drug treated mental patients, its significance, and the role of neuromelanin. *Adv. Biochem. Psychopharmacol.* **9,** 255–268.

Fox, D. L. (1974). Animal biochromes. *In* "Biochemical and Biophysical Perspectives in Marine Biology" (D. C. Malins and J. R. Sargent, eds.), Vol. 1, pp. 170–209. Academic Press, New York.

Fox, D. L. (1976). "Animal Biochromes and Structural Colours." University of California Press, Berkeley.

Fox, D. L., and Millot, T. (1954). The pigmentation of the jelly-fish, *Pelagia noctiluca. Proc. R. Soc. London, B* **142,** 392–408.

Fox, D. L., and Pantin, C. F. A. (1944). Pigments in the *Coelenterata. Biol. Rev. Camb. Philos. Soc.* **19,** 121–134.

Fox, D. L., and Uppdegraff, D. M. (1943). Adenochrome, a glandular pigment in branchial hearts of the *Octopus. Arch. Biochem.* **1,** 339–356.

Fox, H. M., and Vevers, H. G. (1960). "The Nature of Animal Colours." Sidgwick and Jackson, London.

Frenk, E., and Ott, F. (1971). Evaluation of the toxicity of the monoethyl ether of hydroquinone for mammalian melanocytes and melanoma cells. *J. Invest. Dermatol.* **56,**287–293.

Friedmann, P. S., and Gilchrest, B. A. (1987). Ultraviolet radiation directly induces pigment production by cultured human melanocytes. *J. Cell. Physiol.* **133,** 88–94.

Friedmann, P. S., and Thody, A. J. (1986). Disorders of pigmentation. *In* "Scientific Basis of Dermatology. A Physiological Approach" (A. G. Thody and P. S. Friedmann, eds.), pp. 244–261. Churchill Livingstone, Edinburgh.

Fritsch, P., and Varga, J. M. (1976). Melanocyte-stimulating receptors on cultured guinea-pig melanocytes. *J. Invest. Dermatol.* **76,** 538.

Froncisz, W., Sarna, T., and Hyde, J. S. (1980). Cu^{2+} probe of metal-ion binding sites in melanin using electron paramagnetic resonance spectroscopy. *Arch. Biochem. Biophys.* **202,** 289–303.

Fujita, K., Ito, S., Inoue, S., Yamamoto, Y., Takeuchi, J., Shamoto, M., and Nagatsu, T. (1980). Selective toxicity of 5-*S*-cysteinyldopa, a melanin precursor, to tumor cells *in vitro* and *in vivo. Cancer Res.* **40,** 2543–2546.

Fukuzawa, T., and Bagnara, J. T. (1989). Control of melanoblast differentiation in amphibia by a melanocyte stimulating hormone, a serum melanization factor, and a melanization inhibiting factor. *Pigment Cell Res.* **2,** 171–181.

Fukuzawa, T., and Ide, H. (1988). A ventrally localized inhibitor of melanization in *Xenopus laevis* skin. *Dev. Biol.* **129,** 25–36.

Fuller, B. B., and Viskochil, D. H. (1979). The role of RNA and protein synthesis in mediating the action of MSH on mouse melanoma cells. *Life Sci.* **24,** 2405–2416.

Fuller, B. B., Lunsford, J. B., and Iman, D. S. (1987). Alpha-melanocyte stimulating hormone regulation of tyrosinase in Cloudman S91 mouse melanoma cell cultures. *J. Biol. Chem.* **262,** 4024–4033.

Galvao, D. S., and Caldas, M. J. (1988). Polymerization of 5,6-indolequinone: A view into the band structure of melanins. *J. Chem. Phys.* **88,** 4088–4091.

Galvao, D. S., and Caldas, M. J. (1990a). Theoretical investigation of model polymers for eumelanins. II. Isolated defects. *J. Chem. Phys.* **93,** 2848–2853.

Galvao, D. S., and Caldas, M. J. (1990b). Theoretical investigation of model polymers for eumelanins. I. Finite and infinite polymers. *J. Chem. Phys.* **92,** 2630–2636.

Gan, E. V., Haberman, H. F., and Menon, I. A. (1974). Oxidation of NADH by melanin and melanoproteins. *Biochim. Biophys. Acta* **370,** 62–69.

Gan, E. V., Haberman, H. F., and Menon, I. A. (1976). Electron transfer properties of melanin. *Arch. Biochem. Biophys.* **173,** 666–672.

Gan, E. V., Lam, K. M., Haberman, H. F., and Menon, I. A. (1977). Electron transfer properties of melanins. *Br. J. Dermatol.* **96,** 25–28.

Gange, R. W. (1988). Comparison of pigment responses in human skin to UVB and UVA radiation. *In* "Advances in Pigment Cell Research" (J. T. Bagnara, ed.), pp. 475–485. Alan Liss, New York.

Gange, R. W., Blackett, A. D., Matzinger, E. A., Sutherland, B. M., and Kochevar, I. E. (1985). Comparative protection efficiency of UVA and UVB-induced tans against erythema and formation of endonuclease-sensitive sites in DNA by UVB in human skin. *J. Invest. Dermatol.* **85,** 362–364.

Garcia-Carmona, F., Carcia-Canovas, F., and Lozano, J. A. (1980). Epidermis tyrosinase is a hysteretic enzyme. *Int. J. Biochem.* **11,** 325–327.

Geremia, E., Corsaro, C., Baratta, D., Santoro, C., Scalia, M., and Sichel, G. (1989). Antioxidant enzymatic systems in pigment tissues of *Amphibia. Pigment Cell Res.* **2,** 208–212.

Gerst, J. E., Sole, J., Hazum, E., and Salomon, Y. (1988). Identification and characterization of melanotropin binding proteins from M2R melanoma cells by covalent photoaffinity labeling. *J. Endocrinol.* **123,** 1792–1797.

Geschwind, I. I., Huseby, R. A., and Nishioka, R. (1972). The effect of melanocyte-stimulating hormone on coat color in the mouse. *Recent Prog. Horm. Res.* **28,** 91–130.

Gessard, C. (1903). Sur la formation du pigment melanique dans le tumeurs du cheval. *Compt. Rend. Acad. Sci. (Paris)* **136,** 1086–1088.

Ghanem, G. E., Comunale, G., Libert, A., Vercammen-Grandjcan, A., and Lejeune, F. J. (1988). Evidence for alpha-melanocyte stimulating hormone receptors on human malignant melanoma cells. *Int. J. Cancers* **41,** 248–255.

Giebel, L. B., Strunk, K. M., King, R. A., Hanifin, J. M., and Spritz, R. A. (1990). A frequent tyrosinase gene mutation in classic, tyrosinase-negative (type IA) oculocutaneous albinism. *Proc. Natl. Acad. Sci. U.S.A.* **87,** 3255–3258.

Gilchrest, B. A. (1985). Ageing of differentiated cells in skin. *In* "Ageing and the Skin" (A. K. Balin and A. M. Kligman, eds.), pp. 77–83. Raven Press, New York.

Giordano, F., Mazzarella, L., Prota, G., Santacroce, C., Sica, D. (1971). Synthesis and stereochemistry of $\Delta^{2,2'}$-bi(2*H*-1,4-benzothiazine) derivatives. Crystal structures of *cis*- and *trans*-$\Delta^{2,2'}$-bi(3-*p*-bromophenyl)-2*H*-1,4-benzothiazines. *J. Chem. Soc.* **(C),** 2610–2616.

Giroud, P. (1882). Recherches sur la poche du noir des Cephalopodes des cotes de France. *Arch. Zool. Exp. Gen.* **10,** 1–100.

Goetz, N., Arnaud, J. C., and Boré, P. (1981). Human hair pigment: A mild isolation procedure and method for characterization. *In* "Hair Research" (C. Orfanos and W. Montagna, eds.), pp. 140–145. Springer-Verlag, Berlin.

Goldgeier, M. H., Klein, L. E., Klein-Angerer, S., Moellmann, G., and Nordlund, J. J. (1984). The distribution of melanocytes in the leptomeninges of the human brain. *J. Invest. Dermatol.* **82,** 235–238.

Goodall, M. C., and Alton, H. (1972). Metabolism of 3,4-dihydroxyphenylalanine (L-dopa) in human subjects. *Biochem. Pharmacol.* **21,** 2401–2408.

Goodwin, T. W. (1972). Pigments of *Mollusca*. *In* "Chemical Zoology" (M. Florkin and B. T. Scheer, eds.), Vol. VII, pp. 187–198. Academic Press, New York.

Gordon, J. (1979). The expression of the dominant spotting (*W*) gene in chimeric mice. *In* "Pigment Cell: Biologic Basis of Pigmentation" (S. N. Klaus, ed.), Vol. 4, pp. 159–163. Karger, Basel.

Gordon, P. R., Mansur, C. P., and Gilchrest, B. A. (1989). Regulation of human melanocyte growth dendricity, and melanization by keratinocyte derived factors. *J. Invest. Dermatol.* **92,** 565–572.

Görnitz, K. (1923). Versuch einer Klassifikation der haufigsten Federfarbuger. *Jahrb. f. Ornithol.* **71,** 127–131.

Görtner, R. A. (1910). Studies on melanin. I. Methods of isolation. The effect of alkali on melanin. *J. Biol. Chem.* **8,** 341–361.

Görtner, R. A. (1911). On two different types of melanin. *Proc. Soc. Exp. Biol. Med.* **9,** 120–121.

Graham, D. G. (1978). Oxidative pathways for catecholamines in the genesis of neuromelanin and cytotoxic quinones. *Mol. Pharmacol.* **14,** 633–643.

Graham, D. G. (1979). On the origin and significance of neuromelanin. *Arch. Phatol. Lab. Med.* **103,** 359–362.

Graham, D. G. (1984). Catecholamine toxicity: A proposal for the molecular pathogenesis of manganese neurotoxicity and Parkinson's disease. *Neurotoxicology* **5,** 83–96.

Graham, D. G., Tiffany, S. M., and Vogel, S. F. (1978). The toxicity of melanin precursors. *J. Invest. Dermatol.* **70,** 113–116.

Granholm, N. H., Opbroek, A. J., Harvison, G. A., and Kappenman, K. E. (1990). Tyrosinase activity and melanin production in regenerating hairbulb melanocytes of lethal yellow, black, agouti and albino mice. *Pigment Cell Res.* **3,** 233–242.

Greenstein, G. P. (1948). Chemistry of melanomas. *Spec. Pubs. N.Y. Acad. Sci.* **4,**433–436.

Gutteridge, S., and Robb, D. (1975). The catecholase activity of *Neurospora* tyrosinase. *Eur. J. Biochem.* **54,** 107–116.
Hack, M. H., and Helmy, F. M. (1964). A comparative chemical study of the liver of various vertebrates. *Acta Histochem.* **19,** 316–328.
Hack, M. H., and Helmy, F. M. (1983). The melanin and lipofuscin. *Comp. Biochem. Physiol.* **3,** 399–407.
Hadley, M. E. (1988). "The Melanotropic Peptides." Vols. I–III. CRC Press, Boca Raton, Florida.
Hadley, M. E., and Dawson, B. V. (1988). Biomedical applications of synthetic melanotropins. *Pigment Cell Res.* **1S,** 69–78.
Hadley, M. E., and Quevedo, W. C., Jr. (1966). Vertebrate epidermal melanin unit. *Nature* **209,** 1334–1335.
Hadley, M. E., Zechel, C., Wilkes, B. C., Castrucci, A. M. L., Visconti, M. A., Pozo-Alonso, M., and Hruby, V. J. (1987). Differential structural requirements for the MSH and MCH activities of melanin concentrating hormone. *Life Sci.* **40,** 1139–1145.
Halaban, R., and Moellmann, G. (1990). Murine and human *b* locus pigmentation genes encode a glycoprotein (gp75) with catalase activity. *Proc. Natl. Acad. Sci. U.S.A.* **87,** 4809–4813.
Halaban, R., Pomerantz, S. H., Marshall, S., Lambert, D. T., and Lerner, A. B. (1983). Regulation of tyrosinase in human melanocytes grown in culture. *J. Cell Biol.* **97,** 480–488.
Halaban, R., Moellmann, G., Tamura, A., Kwon, B. S., Kuklinska, E., Pomerantz, S. H., and Lerner, A. B. (1988a). Tyrosinases of murine melanocytes with mutations at the albino locus. *Proc. Natl. Acad. Sci. U.S.A.* **85,** 7241–7245.
Halaban, R., Langdon, R., Birchall, N., Cuono, C., Baird, A., Scott, G., Moellmann, G., and McGuire, J. (1988b). Basic fibroblast growth factor from human keratinocytes is a natural mitogen for melanocytes. *J. Cell Biol.* **107,** 1611–1619.
Hall, P. F. (1969a). The influence of hormones on melanogenesis. *Aust. J. Dermatol.* **10,** 125–139.
Hall, P. F. (1969b). Hormonal control of melanin synthesis in birds. *Gen. Comp. Endocrinol. Suppl.* **2,** 452–458.
Hall, K., and Wolfram, L. J. (1975). Isolation and identification of the protein component of hair melanin. *J. Soc. Cosmet. Chem.* **26,** 247–254.
Halprin, K. M., and Ohkawara, A. (1967). Human pigmentation: The role of glutathione. *In* "Advances in Biology of Skin" (W. Montagna and F. Hu, eds.), Vol. 8, pp. 241–251. Pergamon Press, Oxford.
Hamilton, J. B. (1948). Influence of the endocrine status upon pigmentation in man and in mammals. *Spec. Pub. N.Y. Acad. Sci.* **4,** 341–357.
Hansson, C. (1981). Electrochemical synthesis of cysteinyldopas. *Experientia* **37,** 1253–1254.
Hansson, C. (1988). Some indolic compounds as markers of the melanocyte activity. *Acta Derm. Venereol. Suppl. (Stockh.)* **138,** 1–60.
Hansson, C., Agrup, G., Rorsman, H., Rosengren, A.-M., and Rosengren, E. (1979a). Electron spin resonance studies on pheomelanins. *Acta Derm. Venereol. (Stockh.)* **59,** 453–456.
Hansson, C., Agrup, G., Rorsman, H., Rosengren, A.-M., and Rosengren, E. (1979b).

Analysis of cysteinyldopas, dopa, dopamine, noradrenaline, and adrenaline in serum and urine using high-performance liquid chromatography and electrochemical detection. *J. Chromatogr.* **162,** 7–22.

Hansson, C., Rorsman, H., and Rosengren, E. (1980). 5-hydroxydopa. A new compound in the Raper–Mason scheme of melanogenesis. *Acta Derm. Venereol. (Stockh.)* **60,** 281–286.

Hansson, C., Wirestrand, L. E., Aronsson, A., Rorsman, H., and Rosengren, E. (1985). Urinary excretion of 6-hydroxy-5-methoxyindole-2-carboxylic acid and 5-*S*-cysteinyldopa during PUVA treatment. *Photodermatology* **2,** 52–57.

Harley-Mason, J. (1950). The chemistry of adrenochrome and its derivatives. *J. Chem. Soc.* 1276–1282.

Harley-Mason, J. (1953). Melanin and its precursors. Part VI. Further syntheses of 5,6-dihydroxyindole and its derivatives. *J. Chem. Soc.* 200–203.

Hashiguchi, H., and Takahashi, H. (1977). Inhibition of two copper-containing enzymes, tyrosinase and dopamine β-hydroxylase, by L-mimosine. *Mol. Pharmacol.* **13,** 362.

Hashimoto, A., Ichihashi, M., and Mishima, Y. (1984). The mechanism of depigmentation by hydroquinone, a study on suppression and recovery process of tyrosinase activity in pigment cells *in vivo* and *in vitro*. *Jpn. J. Dermatol.* **94,** 797–804. (In Japanese.)

Hashimoto, A., Hasegawa, K., Asai, T., Masamoto, Y., Ichihashi, M., and Mishima, Y. (1988). Depigmenting potency of hydroquinone fatty acid esters on the skin. *J. Dermatol.* **15,** 37–43.

Hatta, S., Mishima, Y., Ichihashi, M., and Ito, S. (1988). Melanin monomers within coated vesicles and premelanosomes in melanin synthesizing cells. *J. Invest. Dermatol.* **91,** 181–184.

Havens, M. B., and Tramposch, K. M. (1986). *J. Invest. Dermatol.* **87,** 409.

Hawley, M. D., Tatawawadi, S. V., Pieraaki, S., and Adams, R. N. (1967). Electrochemical studies of the oxidation pathways of catecholamines. *J. Am. Chem. Soc.* **89,** 447–450.

Heacock, R. A. (1965). The aminochromes. *Adv. Heterocycl. Chem.* **5,** 205–290.

Hearing, V. J. (1973). Tyrosinase activity in subcellular fractions of black and albino mice. *Nature New Biol.* **245,** 81–83.

Hearing, V. J., and Ekel, T. M. (1975). Involvement of tyrosinase in melanin formation in murine melanoma. *J. Invest. Dermatol.* **64,** 80–83.

Hearing, V. J., and Jimenez, M. (1987). Mammalian tyrosinase: The critical regulatory control point in melanocyte pigmentation. *Int. J. Biochem.* **19,** 1141–1147.

Hearing, V. J., and Jimenez, M. (1989). Analysis of mammalian pigmentation at the molecular level. *Pigment Cell Res.* **2,** 75–85.

Hearing, V. J., and Tsukamoto, K. (1991). Enzymatic control of pigmentation in mammals. *FASB J.* **5,** 2902–2909.

Hearing, V. J., Ekel, T. M., Montague, P. M., and Nicholson, J. M. (1980). Mammalian tyrosinase. Stoichiometry and measurement of reaction products. *Biochim. Biophys. Acta* **611,** 251–268.

Hearing, J. V., Ekel, T. M., and Montague, P. M. (1981). Mammalian tyrosinase: Isozymic forms of the enzyme. *J. Biochem.* **13,** 99–103.

Hearing, J. V., Korner, A. M., and Pawelek, J. M. (1982). New regulators of melanogenesis are associated with purified tyrosinase isoenzymes. *J. Invest. Dermatol.* **79,** 16–18.

Hegedus, Z. L. (1977). Studies on rheomelanins. IV. The apparent occurrence *in vivo* of rheomelanins in human blood. *Arch. Int. Physiol. Biochem.* **85,** 539–555.

Heikkila, R. E., Hess, A., and Duvoisin, R. C. (1984). Dopaminergic neurotoxicity of 1-methyl-4-phenyl-1,2,5,6-tetrahydropyridine in mice. *Science* **224,** 1451–1453.

Hempel, K. (1966). Investigation on the structure of melanin in malignant melanoma with ^{3}H- and ^{14}C-dopa labeled at different positions. *In* "Structure and Control of the Melanocyte" (G. Della Porta and O. Muhlbock, eds.), pp. 162–175. Springer-Verlag, Berlin.

Hempel, K. (1967). Über Biosynthese und Struktur des tierischen Melanins. *Z. Naturforsch.* **22B,** 173–180.

Heward, C. B., and Hadley, M. E. (1975). Structure–activity relationship of melatonin and related indoleamines. *Life Sci.* **17,** 1167–1178.

Hill, H. Z. (1989). The relationship of the photobiology of skin cancer and melanins to the radiation biology of melanomas: A selective review. *Comments Mol. Cell. Biophys.* **6,** 141–174.

Hill, H. Z., and Hill, G. J. (1987). Eumelanin causes DNA strand breaks and kills cells. *Pigment Cell Res.* **1,** 163–170.

Hill, H. Z., Cathcart, K. N., Bargellini, Z. T., Hill, G. J., Schallreuter, K. U., and Wood, J. M. (1991). Does melanin affect the low LET radiation response of Cloudman S91 mouse melanoma cell lines? *Pigment Cell Res.* **4,** 80–86.

Himmelwright, R. S., Eickman, N. C., Lubien, C. D., Lerch, K., and Solomon, E. I. (1980). Chemical and spectroscopic studies of the binuclear copper active site of *Neurospora* tyrosinase: Comparison to hemocyanins. *J. Am. Chem. Soc.* **102,** 7339–7344.

Hinterberger, H., Freedman, A., and Bartholomew, R. J. (1973). Precursors of melanin in the urine and 3,4-dihydroxyphenylalanine in the blood of patients with malignant melanoma. *In* "Pigment Cell: Mechanisms in Pigmentation" (V. G. McGovern and P. Russell, eds.), Vol. 1, pp. 312–316. Karger, Basel.

Hirsch, M. M. (1954). Environmental factors influencing the differentiation of protoperithecia and their relation to tyrosinase and melanin formation in *Neurospora crassa. Physiol. Plantarum* **7,** 72–97.

Hirsch, E., Graybiel, A. M., and Agid, Y. A. (1988). Melanized dopaminergic neurons are differentially susceptible to degeneration in Parkinson's disease. *Nature* **334,** 345–348.

Hochstein, P., and Cohen, G. (1963). The cytotoxicity of melanin precursor. *Ann. N.Y. Acad. Sci.* **100,** 876–886.

Hoffman, K. (1978). Photoperiodic mechanism in hamsters: The participation of the pineal gland. *In* "Environmental Endocrinology" (I. Assenmacher and D. S. Farner, eds.), pp. 94–110. Springer-Verlag, Berlin.

Hogben, L. T., and Slome, D. (1931). The pigmentary effector system VI. The dual character of endocrine coordination in amphibian color change. *Proc. R. Soc. (London), B* **108,** 10–53.

Holbrook, K. A., Vogel, A. M., Underwood, R. A., and Foster, C. A. (1988). Melanocytes in human embryonic and fetal skin: A review and new finding. *Pigment Cell Res.* **1S** 6–17.

Honigsmann, H., Schuler, G., and Aberer, W. (1986). Immediate pigment darkening phenomenon: A reevaluation of its mechanism. *J. Invest. Dermatol.* **87,** 648–652.

Honigsmann, H., Wolff, K., Fitzpatrick, T. B., and Pathak, M. A. (1987). Oral photochemotherapy with psoralens and UVA (PUVA). Principles and practice. *In* "Dermatology in General Medicine" (T. B. Fitzpatrick, A. Z. Eisen, and K. Wolff, eds.), pp. 1533–1588. McGraw-Hill, New York.

Horak, V., and Gillette, J. R. (1971). A study of oxidation–reduction state of synthetic 3,4-dihydroxy-DL-phenylalanine melanin. *Mol. Pharmacol.* **7,** 429–433.

Horcicko, J., Borovansky, J., Duchon, J., and Prochazkova, B. (1973). Distribution of zinc and copper in pigmented tissues. *Hoppe-Seyler's Z. Physiol. Chem.* **354,** 203–204.

Horikoshi, T., Balin, A. K., and Carter, D. M. (1991). Effects of oxygen tension on the growth and pigmentation of normal human melanocytes. *J. Invest. Dermatol.* **96,** 841–844.

Horowitz, N. H., Fling, M., and Horn, G. (1970). Tyrosinase (*Neurospora crassa*). *Meth. Enzymol.* **17A,** 615–620.

Hruby, V. J., Wilkes, B. C., Cody, W. L., Sawyer, T. K., and Hadley, M. E. (1984). Melanotropins: Structural, conformational and biological considerations in the development of superpotent and superprolonged analogs. *In* "Peptide Protein Revue" (M. T. W. Hearn, eds.), Vol. 3, pp. 1–64. Marcel Dekker, New York.

Hruby, V. J., Wilkes, B. C., Hadley, M. E., Al-Obeidi, F., Sawyer, T. K., Staples, D. J., deVaux, A. E., Dym, O., Castrucci, A. M. L., Hintz, M. E., Riehm, J. P., and Rango Rao, K. (1988). Alpha-melanotropin; the minimal active sequence. I. Frog skin bioassay. *J. Med. Chem.* **30,** 2126–2130.

Hu, F., and Buxman, M. M. (1981). Gamma-glutamyl transpeptidase in the pigment cells of *Rhesus* eyes. *J. Invest. Dermatol.* **76,** 371–375.

Hu, F., Staricco, R. J., Pinkus, H., and Fosnauch, R. P. (1957). Human melanocytes in tissue culture. *J. Invest. Dermatol.* **28,** 15–32.

Huang, J. S., Sung, J., Eisner, M., Moss, S. C., and Gallas, J. (1990). The fractal structure and the dynamics of aggregation of synthetic melanin in low pH aqueous solution. *J. Chem. Phys.* **90,** 25–29.

Huber, M., Hintermann, G., and Lerch, K. (1985). Primary structure of tyrosinase from *Streptomyces glaucescens. Biochemistry* **24,** 6038–6044.

Hurrell, R. F., and Finot, P. A. (1984). Nutritional consequences of the reactions between proteins and oxidized polyphenolic acid. *Adv. Exp. Med. Biol.* **177,** 423–435.

Illis, I. (1988). Photosensitivity reactions in black skin. *Dermatologic Clinics* **6,** 369–375.

Imokawa, G., and Mishima, Y. (1981). Isolation and biochemical characterization of tyrosinase rich GERL and coated vesicle in the melanin synthesizing cells. *Br. J. Dermatol.* **104,** 160–178.

Imokawa, G., and Mishima, Y. (1984). Functional analysis of tyrosinase isozymes of cultured malignant melanoma cells during the recovery period following interrupted melanogenesis induced by glycosylation inhibitors. *J. Invest. Dermatol.* **83,** 196–201.

Ingraham, L. L. (1959). Polyphenoloxidase action at low pH values. *In* "Pigment Cell Biology" (M. Gordon, ed.), pp. 609–617. Academic Press, New York.

Ings, R. M. J. (1984). The melanin binding of drugs and its implications. *Drug Metabol. Rev.* **15,** 1183–1212.

Inoue, S., Ito, S., Wakamatsu, K., Kimbow, K., and Fujita, K. (1990). Mechanism of growth inhibition of melanoma cells by 4-*S*-cysteaminylphenol and its analogs. *Biochem. Pharmacol.* **39,** 1077–1083.

Ito, S. (1986). Reexamination of the structure of eumelanin. *Biochim. Biophys. Acta* **883,** 155–161.

Ito, S. (1989). Optimization of conditions for preparing synthetic pheomelanin. *Pigment Cell Res.* **2,** 53–56.

Ito, S., and Fujita, K. (1981). New possible routes for the biosynthesis of cysteinyldopas and related metabolites. *In* "Pigment Cell 1981: phenotypic Expression in Pigment Cells" (M. Seiji, ed.), pp. 85–91. University of Tokyo Press, Tokyo.

Ito, S., and Fujita, K. (1985). Microanalysis of eumelanin and pheomelanin in hair and melanomas by chemical degradation and liquid chromatography. *Anal. Biochem.* **144,** 527–536.

Ito, S., and Nicol, J. A. C. (1974). Isolation of oligomers of 5,6-dihydroxyindole-2-carboxylic acid from the eye of the catfish. *Biochem. J.* **143,** 207–217.

Ito, S., and Nicol, J. A. C. (1977). A new amino acid, 3-(2,5-*S*,*S*-dicysteinyl-3,4-dihydroxyphenyl)alanine, from the *tapetum lucidum* of the gar (*Lepisosteidae*) and its enzymic synthesis. *Biochem. J.* **161,** 499–507.

Ito, S., and Prota, G. (1977a). A facile one-step synthesis of cysteinyldopas using mushroom tyrosinase. *Experientia* **33,** 1118–1119.

Ito, S., and Prota, G. (1977b). A novel reaction of cysteine with phenolic amino acids in hydrobromic acid: Reversible formation of 3-*S*-cysteinyltyrosine and *S*-cysteinyldopas. *J. Chem. Soc., Chem. Commun.* 251–252.

Ito, S., Novellino, E., Chioccara, F., Misuraca, G., and Prota, G. (1980). Co-polymerization of dopa and cysteinyldopa in melanogenesis *in vitro. Experientia* **36,** 822–823.

Ito, S., Inoue, S., Yamamoto, Y., and Fujita, K. (1981). Synthesis and antitumor activity of cysteinyl-3,4-dihydroxyphenylalanines and related compounds. *J. Med. Chem.* **24,** 673–677.

Ito, S., Inoue, S., and Fujita, K. (1983). The mechanism of toxicity of 5-*S*-cysteinyldopa to tumour cells. Hydrogen peroxide as a mediator of cytotoxicity. *Biochem. Pharmacol.* **32,** 2079–2081.

Ito, S., Fujita, K., Takahashi, K., and Jimbow, K. (1984a). Characterization of melanogenesis in mouse and guinea pig hair by chemical analysis of melanin and of free and bound dopa and 5-*S*-cysteinyldopa. *J. Invest. Dermatol.* **83,** 12–14.

Ito, S., Kato, T., Shinpo, K., and Fujita, K. (1984b). Oxidation of tyrosine residues in proteins by tyrosinase. *Biochem. J.* **222,** 407–411.

Ito, S., Palumbo, A., and Prota, G. (1985a). Tyrosinase-catalysed conjugation of dopa with glutathione. *Experientia* **41,** 960–961.

Ito, S., Kato, T., Maruta, K., Jimbow, K., and Fujita, K. (1985b). Total acidic metabolites of catecholamines in urine as determined by hydrolysis with hydroiodic acid and liquid chromatography: Application to patients with neuroblastoma and melanoma. *Clin. Chem.* **31,** 1185–1188.

Ito, S., Imai, Y., Jimbow, K., and Fujita, K. (1988). Incorporation of sulfhydryl compounds into melanins *in vitro. Biochim. Biophys. Acta* **964,** 1–7.

Jackson, I. J. (1988). A cDNA encoding tyrosinase-related protein maps to the *brown* locus in mouse. *Proc. Natl. Acad. Sci. U.S.A.* **85,** 4392–4396.

Jackson, I. J., and Bennett, D. C. (1990). Identification of the albino mutation of mouse tyrosinase by analysis of an *in vitro* revertant. *Proc. Natl. Acad. Sci. U.S.A.* **87,** 7010–7014.

Jackson, I. J., Chambers, D. M., Tsukamoto, K., Copeland, N. G., Gilbert, D. J., Jenkins, N. A., and Hearing, V. (1992). A second tyrosinase-related protein, TPR-2, maps to and is mutated at the mouse *slaty* locus. *EMBO J.* **11,** 527–535.

Jacobsohn, G. M., Chiartas, P. L., Hearing, V. J., and Jacobsohn, M. K. (1988a). Role of estradiol and 2-hydroxyestradiol in melanin formation *in vitro. Biochim. Biophys. Acta* **966,** 222–230.

Jacobsohn, M. K., Dobre, V. C., Branan, C., and Jacobsohn, G. M. (1988b). Oxidation of 2-hydroxyestradiol and its incorporation into melanin by mushroom tyrosinase. *J. Steroid Biochem.* **31,** 377–385.

Jacobsohn, M. K., Byler, D. M., and Jacobsohn, G. M. (1991). Isolation of estradiol-2,3-quinone and its intermediacy in melanin formation. *Biochim. Biophys. Acta* **1073,** 1–10.

Jacobson, F. W., and Millot, N. (1953). Phenolases and melanogenesis in the coelomic fluid of the echinoid *Diadema antillarum philippi. Proc. R. Soc.* **141 B,** 231–247.

Jacques, S. L., and McAuliffe, D. J. (1991). The melanosome: Threshold temperature for explosive vaporization and internal absorption coefficient during pulsed laser irradiation. *Photochem. Photobiol.* **53,** 769–1141.

Jara, J. R., Solano, F., and Lozano, J. A. (1988a). Assays for mammalian tyrosinase: A comparative study. *Pigment Cell Res.* **1,** 332–339.

Jara, J. R., Aroca, P., Solano, F., Martinez, J. H., and Lozano, J. A. (1988b). The role of sulfydryl compounds in mammalian melanogenesis: The effect of cysteine and glutathione upon tyrosinase and the intermediates of the pathway. *Biochim. Biophys. Acta* **967,** 296–303.

Jara, J. R., Solano, F., Garcia-Borron, J. G., Aroca, P., and Lozano, J. A. (1990). Regulation of mammalian melanogenesis II: The role of metal cations. *Biochim. Biophys. Acta* **1035,** 276–285.

Jerbil, B., Lindbladh, C., Rorsman, H., and Rosengren, E. (1983). Tyrosinase activity in the medium of human melanoma cell cultures. *Acta Derm. Venereol. (Stockh.)* **63,** 205–208.

Jimbow, K., and Takeuchi, T. (1979). Ultrastructural comparison of pheo- and eumelanogenesis in animals. *In* "Pigment Cell 1979: Biologic Basis of Pigmentation" (S. N. Klaus, ed.), Vol. 4, pp. 308–317. Karger, Basel.

Jimbow, K., Pathak, M. A., and Fitzpatrick, T. B. (1973). Effect of ultraviolet light on the distribution pattern of microfilament and microtubules and on the nucleus in human melanocytes. *Yale J. Biol. Med.* **46,** 411–426.

Jimbow, K., Obata, H., Pathak, M. A., and Fitzpatrick, T. B. (1974). Mechanism of depigmentation by hydroquinone. *J. Invest. Dermatol.* **62,** 436–449.

Jimbow, K., Roth, S. I., Fitzpatrick, T. B., and Szabo, G. (1975). Mitotic activity in non-neoplastic melanocytes *in vivo* as determined by histochemical, autoradiographic, and electron microscopic studies. *J. Cell Biol.* **66,** 633–671.

Jimbow, K., Quevedo, W. C., Jr., Fitzpatrick, T. B., and Szabo, G. (1976). Some aspects of melanin biology: 1950–1975. *J. Invest. Dermatol.* **67,** 72–89.

Jimbow, M., Kanoh, H., and Jimbow, K. (1982). Characterization of biochemical properties of melanosomes for structural and functional differentiation: Analysis of the compositions of lipids and proteins in melanosomes and their subfractions. *J. Invest. Dermatol.* **79,** 97–102.

Jimbow, K., Miura, T., Ito, S., and Ishikawa, K. (1989). Phenolic melanin precursors provide a rational approach to the design of antitumor agents for melanoma. *Pigment Cell Res.* **2,** 34–39.

Jimbow, K., Fitzpatrick, T. B., and Wick, M. M. (1991). Biochemistry and physiology of melanin pigmentation. *In* "Biochemistry and Physiology of Skin" (L. A. Goldsmith, ed.), Vol. 2, pp. 873–909. Oxford University Press.

Jimenez, M., Garcia-Carmona, F., Garcia-Canovas, F., Iborra, J. L., Lozano, J. A., and Martinez, F. (1984). Chemical intermediates in dopamine oxidation by tyrosinase and kinetic studies of the process. *Arch. Biochem. Biophys.* **235,** 438–448.

Jimenez, M., Maloy, W. L., and Hearing, V. J. (1989). Specific identification of an authentic tyrosinase clone. *J. Biol. Chem.* **264,** 3397–3403.

Jimenez, M., Tsukamoto, K., and Hearing, V. J. (1991). Tyrosinases from two different loci are expressed by normal and by transformed melanocytes. *J. Biol. Chem.* **226,** 1147–1156.

Jolley, R. L., Robb, D. A., and Mason, H. S. (1969). The multiple forms of mushroom tyrosinase—association dissociation phenomena. *J. Biol. Chem.* **244,** 1593–1599.

Jolley, R. L., Evans, L. H., and Mason, H. S. (1972). Reversible oxygenation of tyrosinase. *Biochem. Biophys. Res. Commun.* **46,** 878–884.

Joshi, P. C., Carraro, C., and Pathak, M. A. (1987). Involvement of reactive oxygen species in the oxidation of tyrosine and dopa to melanin and in skin tanning. *Biochem. Biophys. Res. Commun.* **142,** 265–274.

Kable, E. P. W., and Parsons, P. G. (1988). Potency, selectivity, and cell cycle dependence of catechols in human tumour cells *in vitro. Biochem. Pharmacol.* **9,** 1711–1715.

Kable, E. P. W., and Parsons, P. G. (1989). Melanin synthesis and the action of L-dopa and 3,4-dihydroxybenzylamine in human melanoma cells. *Cancer Chemother. Pharmacol.* **23,** 1–7.

Kagedal, B., and Petterson, A. (1985). Determination of urinary 5-*S*-cysteinyldopa by high-performance liquid chromatography. *J. Chromatogr.* **272,** 287–297.

Kagedal, B., Kallberg, M., and Arstrand, K. (1989). Automated high-performance liquid chromatographic determination of 5-*S*-cysteinyl-3,4-dihydroxyphenylanine in urine. *J. Chromatogr.* **473,** 359–370.

Kahn, V., and Andrawis, A. (1985). Inhibition of mushroom tyrosinase by tropolone. *Phytochemistry* **24,** 905–908.

Kaidbey, K. H., and Kligman, A. M. (1978). Sunburn protection by long wave ultraviolet light upon human skin. *Arch. Dermatol.* **114,** 46–48.

Kaidbey, K. H., Agin, P. P., Sayre, R. M., and Kligman, A. M. (1979). Photoprotection by melanin—a comparison of Black and Caucasian skin. *J. Am. Acad. Dermatol.* **1,**249–260.

Kalyanaraman, B., Felix, C. C., and Sealy, R. C. (1985). Semiquinone anion radicals of catechol(amine)s, catechol estrogens, and their metal ion complexes. *Environ. Health Perspect.* **64,** 185–198.

Kalyanaraman, B., Koryotowski, W., Pilas, B., Sarna, T., Land, E. J., and Truscott, T. G. (1988). Reaction between *ortho*-semiquinones and oxygen: Pulse radiolysis, electron spin resonance, and oxygen uptake studies. *Arch. Biochem. Biophys.* **266,** 277–284.

Kamayama, K., Jimenez, M., Muller, J., Ishida, Y., and Hearing, V. J. (1989). Regulation of mammalian melanogenesis by tyrosinase inhibition. *Differentiation* **42,** 28–36.

Karg, E., Odh, G., Rosengren, E., Wittbjer, A., and Rorsman, H. (1991a). Melanin-related biochemistry of IGR 1 human melanoma cells. *Melanoma Res.* **1,** 5–13.

Karg, E., Rosengren, E., and Rorsman, H. (1991b). Hydrogen peroxide as a mediator of dopac-induced effects on melanoma cells. *J. Invest. Dermatol.* **96,** 224–227.

Kastin, A. J., and Schally, A. V. (1967). Autoregulation of release of MSH from the rat pituitary. *Nature* **213,** 1238–1240.

Kastin, A. J., and Schally, A. V. (1972). MSH release in mammals. *In* "Pigmentation: Its Genesis and Biologic Control" (V. Riley, ed.), pp. 207–223. Appleton Century Crofts, New York.

Kastin, A. J., Kuzemchak, B., Schally, A. V., and Eggert, M. W., Jr. (1979). Rat brain melanin at different ages and after various treatments. *Brain Res. Bull.* **4,** 793–797.

Kaul, B. L. (1974). Studies on heterocyclic colouring matters. Part II. $\Delta^{2,2'}$-Bi(2*H*-1,4-benzothiazines). *Helv. Chim. Acta* **57,** 2664–2678.

Kawauchi, H. (1989). Structure and biosynthesis of melanin-concentrating hormone. *Life Sci.* **45,** 1133–1140.

Kawauchi, H., and Kawazoe, I. (1988). Structure–activity studies on melanin-concentrating hormone. *In* "Advances in Pigment Cell Research" (J. T. Bagnara, ed.), pp. 517–530. Alan Liss, New York.

Kawauchi, H., Kawazoe, I., Tsubokawa, M., Kishida, M., and Baker, B. I. (1983). Characterization of melanin-concentrating hormone from chum salmon pituitaries. *Nature* **305,** 321–323.

Kawauchi, H., Kawazoe, I., Hayashi, Y., Minamitake, Y., and Tanaka, S. (1986). Melanin-concentrating hormone: Chemistry and biology. *In* "Natural Products and Biological Activities" (H. Imura, T. Goto, T. Murachi, and R. Nakajima, eds.), pp. 303–313. University of Tokyo Press, Tokyo.

Kawava, A. (1986). UVB-induced erythema, induced tanning, and UVA-induced immediate tanning in Japanese skin. *Photodermatol.* **3,** 372–333.

Kayser, H. (1985). Pigments. *In* "Comprensive Insect Physiology, Biochemistry, and Pharmacology" (G. A. Kerkut, and L. I. Gilbert, eds.), Vol. 10, pp. 367–415. Pergamon Press, Oxford.

Kean, E. A. (1964). A procedure which demonstrates substrate inhibition of tyrosinase. *Biochim. Biophys. Acta* **92,** 602–204.

Keilin, D., and Mann, T. (1938). Polyphenol oxydase. Purification, nature and properties. *Proc. R. Soc. London, Ser. B.* **125,** 187–204.

Kemali, M., and Gioffré, D. (1985). Anatomical localisation of neuromelanin in the brains of the frog and tadpole. Ultrastructural comparison of neuromelanin with other melanins. *J. Anat.* **142,** 73–83.

Kennedy, G. Y. (1969). Pigments of *Anellida, Echiuroidea, Sipunculoidea, Priapuloidea,* and *Phoronidea. In* "Chemical Zoology" (M. Florkin and B. T. Scheer, eds.), Vol. IV, pp. 311–376. Academic Press, New York.

Kertesz, D., and Zito, R. (1965). Mushroom polyphenol oxidase. I. Purification and general properties. *Biochim. Biophys. Acta.* **96,** 447–462.

Kidson, S. H., and Fabian, B. C. (1981). The effect of temperature on tyrosinase activity in Himalayan mouse skin. *J. Exp. Zool.* **215,** 91–97.

King, J. A. G., Percival, A., Robson, N. C., and Swan, G. A. (1970). Studies related to the chemistry of melanins. XI. The distribution of polymeric linkages in dopamelanin. *J. Chem. Soc.* **(C),** 1418.

King, R. A., Olds, D. P., and Witkop, C. J. (1978). Characterization of human hair bulb tyrosinase: Properties of normal and albino enzyme. *J. Invest. Dermatol.* **71,** 136–139.

King, R. A., Mentink, M. M., and Oetting, W. S. (1991a). Non-random distribution of missense mutations within the human tyrosinase gene in type I (tyrosinase-related oculocutaneous albinism. *Mol. Biol. Med.* **8,** 19–29.

King, R. A., Townsend, D., Oetting, W., Summers, C. G., Olds, D. P., White, J. G., and Spritz, R. A. (1991b). Temperature-sensitive tyrosinase associated with peripheral pigmentation in oculocutaneous albinism. *J. Clin. Invest.* **87,** 1046–1053.

Kinnear, P. E., Jay, B., and Witkop, C. J., Jr. (1985). Albinism. *Surv. Ophthalmol.* **30,** 75–101.

Kirby, G. W., and Ogunkoya, L. (1965). Structure of melanin derived from (±)-3,4-dihydroxy-1(^{14}C,^{3}H)-phenylalanine by oxidation with tyrosinase. *J. Chem. Soc., Chem. Comm.* **21,** 546–547.

Kittredge, J. S., Takahashi, F. T., Linsey, J., and Lasker, R. (1974). Chemical signals in the sea: Marine allelochemics and evolution. *Fish. Bull.* **72,** 1–11.

Koch, W. H., and Chedekel, M. R. (1987). Photochemistry and photobiology of melanogenic metabolites: Formation of free radicals. *Photochem. Photobiol.* **46,** 229–238.

Koh, H. K., Klinger, B. E., and Lew, R. A. (1990). Sunlight and cutaneous malignant melanoma: Evidence for and against causation. *Photochem. Photobiol.* **51,** 765–779.

Kollias, N., and Baqer, A. (1985). Spectroscopic characteristics of human melanin *in vivo. J. Invest. Dermatol.* **85,** 38–42.

Kollias, N., Baqer, A., and Chedekel, M. R. (1988). *In vivo* spectroscopic investigation of UVB induced melanogenesis. *Photochem. Photobiol.* **47,** 95.

Kollias, N., Sayre, R. M., Zeise, L., and Chedekel, M. R. (1991). Photoprotection by melanin. *J. Photochem. Photobiol. B.: Biol.* **9,** 135–160.

Kono, R., Yamaoka, T., Yoshizaki, H., and McGinness, J. E. (1979). Acoustic and dispersion of sound waves in diethylamine melanin. *J. Appl. Phys.* **50,** 1236–1244.

Kopin, I. J., Burns, S. R., Chiueh, C. C., and Markey, S. P. (1986). MPTP induced Parkinsonian syndrome in human and animals. *In* "Alzheimer and Parkinson's Disease. Strategies for Research and Development" (A. Fisher, I. Hanin, and C. H. Lachman, eds.), pp. 519–530. Plenum Press, New York.

Korner, A. M., and Gettins, P. (1985). Synthesis *in vitro* of 5,6-dihydroxyindole-2-carboxylic acid by dopachrome conversion factor from Cloudman S91 melanoma cells. *J. Invest. Dermatol.* **85,** 229–231.

Korner, A., and Pawelek, J. (1980). Dopachrome conversion: A possible control point in melanin biosynthesis. *J. Invest. Dermatol.* **75,** 192–195.

Korner, A. M., and Pawelek, J. M. (1982). Mammalian tyrosinase catalyzes three reactions in the biosynthesis of melanin. *Science* **217,** 1163–1165.

Korytowski, W., and Sarna, T. (1990). Bleaching of melanin pigments. Role of copper ions and hydrogen peroxide in autooxidation and photooxidation of synthetic dopa-melanin. *J. Biol. Chem.* **265,** 12410–12416.

Korytowski, W., Hintz, P., Sealy, R. C., and Kalyanaraman, B. (1985). Mechanism of dismutation of superoxide produced during autooxidation of melanin pigments. *Biochem. Biophys. Res. Commun.* **131,** 659–665.

Korytowski, W., Pilas, B., Sarna, T., and Kalyanaraman, B. (1987). Photoinduced generation of superoxide anion and hydroxyl radical in melanins. *Photochem. Photobiol.* **45,** 185–190.

Kreutzfeld, K. T., Fukuzawa, T., and Bagnara, J. T. (1989). Effect of ventrally localized inhibitor of melanization on cultured S91 and B16 mouse melanoma. *Pigment Cell Res.* **2,** 123–125.

Krüger, R. C. (1955). The inhibition of tyrosinase. *Arch. Biochem. Biophys.* **57,** 52–60.

Kubowitz, F. (1938). Spaltung und Resynthese der Polyphenoloxydase und des Hämocyanins. *Biochem. Z.* **299,** 32–57.

Kupper, U., Niedermann, D. M., Travaglini, G., and Lerch, K. (1989). Isolation and characterization of the tyrosinase gene from *Neurospora crassa. J. Biol. Chem.* **264,** 17250–17258.

Kurtz, S. K., Albrecht, L., Schultz, T., and Wolfram, L. J. (1987). The physical origin of color in melanin pigment dispersions. *Pigment Cell Res.* **1,** 261–262.

Kurtz, S., Wolfram, L., and Albrecht, L. (1989). A contribution to understanding the nature of hair color. Scientific Seminar of the Society of Cosmetic Chemists, Orlando, Florida.

Kwon, B. S., Haq, A. K., Pomerantz, S. H., and Halaban, R. (1987). Isolation and sequence of a cDNA clone for human tyrosinase that maps at the mouse *c*-albino locus. *Proc. Natl. Acad. Sci. U.S.A.* **84,** 7473–7477.

Kwon, B. S., Wakulchik, M., Haq, A. K., Halaban, R., and Kestler, D. (1988). Sequence analysis of mouse tyrosinase cDNA and the effect of melanotropin on its gene expression. *Biochem. Biophys. Res. Commun.* **153,** 1301–1309.

Kwon, B. S., Halaban, R., and Chintamaneni, S. (1989). Molecular basis of mouse Himalayan mutation. *Biochem. Biophys. Res. Commun.* **161,** 252–260.

Lacy, M. E. (1981). Neuromelanin: A hypothetical component of bioelectronic mechanism in brain function. *Physiol. Chem. Phys.* **13,** 319–324.

Lambert, C., Chacon, J. N., Chedekel, M. R., Land, E. J., Riley, P. A., Thompson, A., and Truscott, G. (1989). A pulse radiolysis investigation of the oxidation of indolic melanin precursors: Evidence for indolequinones and subsequent intermediates. *Biochim. Biophys. Acta* **993,** 12–20.

Lamoreaux, M. L., and Mayer, T. C. (1975). Site of gene action in the development of hair pigment in recessive yellow (*e/e*) mice. *Dev. Biol.* **46,** 160–166.

Lamoreaux, M. L., Woolley, C., and Pendergast, P. (1986). Genetic controls over activities of tyrosinase and dopachrome conversion factor in murine melanocytes. *Genetics* **113,** 967–984.

Land, E. J. (1988). Pulse irradiation studies of some reactive intermediates of melanogenesis. *Revs. Chem. Interm.* **10,** 219–240.

Land, E. J., Thompson, A., Truscott, T. G., Subbarao, K. V., and Chedekel, M. R. (1986). Photochemistry of melanin precursors: Dopa, 5-*S*-cysteinyldopa and 2,5-*S,S*-dicysteinyldopa. *Photochem. Photobiol.* **44,** 697–702.

Land, E. J., Cooksey, C. J., and Riley, P. A. (1990). Kinetics of 4-methoxyorthobenzoquinone in relation to its mechanism of cytotoxicity: A pulse radiolysis study. *Biochem. Pharmacol.* **39,** 1133–1135.

Landolt, H. (1899). Über das Melanin der Augenhaute. *Zeitschr. f. Physiol. Chem.* **28,** 192–210.

Langston, J. W., Irwin, I., and Ricaurte, G. A. (1987). Neurotoxins, parkinsonism and Parkinson's disease. *Pharmacol. Ther.* **32,** 19–49.

Larsson, B. S. (1991). Melanin-affinic thioureas as selective melanoma seekers. *Melanoma Res.* **2,** 85–90.

Larsson, B., and Tjälve, H. (1978). Studies on the melanin-affinity of metal ions. *Acta Physiol. Scand.* **104,** 479–484.

Larsson, B., and Tjälve, H. (1979). Studies on the mechanism of drug-binding to melanin. *Biochem. Pharmacol.* **28,** 1181–1187.

Larsson, P., Larsson, B. S., and Tjälve, H. (1988). Binding of aflatoxin B_1 to melanin. *Fd. Chem. Toxic.* **26,** 579–586.

Laskin, J. D., and Piccinini, L. A. (1986). Tyrosinase isozyme heterogeneity in differentiating B16/c3 melanoma. *J. Biol. Chem.* **261,** 16626–16635.

Laskin, J. D., Mufson, R. A., Weinstein, I. B., and Engelhardt, D. L. (1980). Identification of a distinct phase during melanogenesis that is sensitive to extracellular pH and ionic strength. *J. Cell. Physiol.* **103,** 467–474.

Lavker, R., and Kaidbey, K. M. (1982). Redistribution of melanosomal complexes within keratinocytes following UV-A irradiation: A possible mechanism for cutaneous darkening in man. *Arch. Dermatol. Res.* **272,** 215–228.

Le Cat, C. N. (1765). Traité de la Couleur de la Peau Humaine en general, de celle des Negres en particulier, et de la Metamorphose d'une de ces Couleurs en l'autre, soit de Naissance, soit accidentellement. *M. M. Rey. Amsterdam.*

Legros, F., Coel, J., Doyen, A., Hanson, P., Van Tieghein, N., Vercammen-Grandjean, A., Fruhling, J., and Lejeune, F. (1981). Alpha-melanocyte stimulating hormone binding and biological activity in a human melanoma cell line. *Cancer Res.* **41,** 1539–1544.

Lejeune, F. J., Vercammen-Grandjean, A., and Ghanem, G. (1987). Alpha-melanocyte stimulating hormone in human melanoma. *In* "Cutaneous Melanoma" (U. Veronesi, N. Cascinelli, and M. Santinami, eds.), pp. 637–638. Academic Press, New York.

Leonard, L. J., Townsend, D., and King, R. A. (1988). Function of dopachrome oxidoreductase and metal ions in dopachrome conversion in the eumelanin pathway. *Biochemistry* **27,** 6156–6159.

Leonhardi, G. (1953). Zum Mechanismus der Thormählenschen Reaktion. *Naturwissenschaften* **40,** 621–622.

Leonhardi, G. (1954). Die Trennung des Oxyprolins and des Allo-Oxyprolins durch Papierchromatographie. *Naturwissenschaften* **41,** 141.

Leonhardi, G. (1955). Über Struktur and Lichtabsorption der Hahrmelanogene. *Naturwissenschaften* **42,** 17–18.

Lerch, K. (1981). Copper monooxygenases: Tyrosinase and dopamine beta-hydroxylase. *In* "Metal Ions in Biological Systems" (H. Sigel, ed.), Vol. 13, pp. 143–186. Marcel Dekker, New York.

Lerch, K. (1982). Primary structure of tyrosinase from *Neurospora crassa. J. Biol. Chem.* **257,** 6414–6419.

Lerch, K. (1983). *Neurospora* tyrosinase: Structural, spectroscopic and catalytic properties. *Mol. Cell. Biochem.* **52,** 125–138.

Lerch, K. (1988). Protein and active-site structure of tyrosinase. *In* "Advances in Pigment Cell Research" (J. T. Bagnara, ed.), pp. 85–98. Alan Liss, New York.

Lerch, K., and Germann, U. A. (1988). Evolutionary relationships among copper proteins containing coupled binuclear copper sites. *Prog. Clin. Biol. Res.* **274,** 331–348.

Lerner, A. B. (1961). Hormones and skin color. *Sci. Amer.* **205,** 98–108.

Lerner, A. B., and Fitzpatrick, T. B. (1950). Biochemistry of melanin formation. *Physiol. Rev.* **30,**91–126.

Lerner, A. B., and McGuire, J. S. (1961). Effect of alpha and beta melanocyte stimulating hormones on the skin color of man. *Nature (London)* **189,** 176–179.

Lerner, A. B., Shizume, K., and Bunding, I. (1954). The mechanism of endocrine control of melanin pigmentation. *J. Clin. Endocrinol. Metab.* **14,** 1463.

Lerner, A. B., Case, J. D., and Heinzelman, R. V. (1959). Structure of melatonin. *J. Am. Chem. Soc.* **81,** 6084–6085.

Lerner, A. B., Snell, R. S., Chanco-Turner, M. L., and Mcguire, J. S. (1966). The effect of sympathectomy and α-MSH on human epidermal melanocytes. *Arch. Dermatol.* **94,** 269–278.

Lignac, G. O. E. (1923). Über den Chemismus und die Biologie des menschlichen Hautpigments. *Virchow's Arch. Pathol. Anat.* **240,** 386–416.

Lillie, R. D. (1955). The basophilia of melanins. *J. Histochem. Cytochem.* **3,** 453–455.

Lillie, R. D. (1957). Ferrous ion uptake. *Arch. Path.* **64,** 100–104.

Lindquist, N. G. (1973). Accumulation of drug in melanin. *Acta. Radiol. (Stockh.)* **325,** 1–92.

Lindquist, N. G. (1986). Melanin affinity of xenobiotics. *Uppsala J. Med. Sci.* **91,** 238–288.

Lindquist, N. G., Larsson, B. S., and Lyden-Sokolowski, A. (1987). Neuromelanin and its possible protective and destructive properties. *Pigment Cell Res.* **1,** 133–136.

Link, E., and Carpenter, R. N. (1990). At-211-methylene blue for targeted radiotherapy of human melanoma xenografts—Treatment of micrometastases. *Cancer Res.* **50,** 2963–2967.

Link, E. M., Brown, I., Carpenter, R. N., and Mitchell, J. S. (1989). Uptake and therapeutic effectiveness of ^{125}I- and ^{211}At-methylene blue for pigmented melanoma in an animal model system. *Cancer Res.* **49,** 4332–4337.

Logan, A., and Weatherhead, B. (1978). Pelage color cycles and hair follicle tyrosinase activity in the Siberian hamster. *J. Invest. Dermatol.* **71,** 295–298.

Logan, A., and Weatherhead, B. (1980). Post-tyrosinase inhibition of melanogenesis by melatonin in hair follicles *in vitro*. *J. Invest. Dermatol.* **74,** 47–50.

Longuet-Higgins, H. C. (1960). On the origin of the free radical properties of melanins. *Arch. Biochem. Biophys.* **86,** 225.

Lucas, A. M., Thody, A. J., and Shuster, S. (1987). The role of calcium in stimulated melanosome dispersion. *Peptides* **8,** 955–960.

L'vov, Y. M., Krivadin, A. V., Fedorovich, I. B., and Ostrovskii, M. A. (1983). Discovery of periodicity in structure of melanoprotein granules of eye pigment epithelium cell. *Dokl. Biophys.* (English translation) 114–116.

Lyden, A., Larsson, B. S., and Lindquist, N. G. (1982). Studies on the melanin affinity of haloperidol. *Arch. Int. Pharmacodyn. Ther.* **259,** 230–243.

Lyden, A., Larsson, B. S., and Lindquist, N. G. (1984). Melanin affinity of manganese. *Acta Pharmacol. Toxicol.* **55,** 133–138.

Lyttkens, L., Larsson, B., Stahale, J., and Englesson, S. (1979). Accumulation of substances with melanin affinity to the internal ear. *Adv. Otorhinolaryngol.* **25,**17–25.

MacGinitie, G. E., and MacGinitie, N. (1968). "Natural History of Marine Animals." McGraw-Hill, New York.

Mackie, R. M., Elwood, J. M., and Hawk, J. L. M. (1987). Links between exposure to ultraviolet radiation and skin cancer. *J. Royal Coll. Phys. (London)* **21,** 91.

Marsden, C. D. (1983). Neuromelanin and Parkinson's disease. *J. Neural Transmission* **19S,** 121–141.

Martinez, J. H., Solano, F., Arocas, A., Garcia-Borron, J. C., Iborra, J. L., and Lozano, J. A. (1987). The existence of apotyrosinase in the cytosol of Harding-Passey mouse melanoma melanocytes and characteristics of enzyme reconstitution by Cu(II). *Biochim. Biophys. Acta* **923,**413–420.

Mascagna, D., Ghanem, G., Morandini, R., d'Ischia, M., Misuraca, G., Lejeune, F., and Prota, G. (1992). Synthesis and cytotoxic properties of new N-substituted 4-aminophenol derivatives with a potential as antimelanoma agents. *Melanoma Res.* **2,** 25–32.

Mason, H. S. (1948). The chemistry of melanin. III. Mechanism of the oxidation of dihydroxyphenylalanine by tyrosinase. *J. Biol. Chem.* **172,** 83–92.

Mason, H. S. (1955). Comparative biochemistry of the phenolase complex. *In* "Advances in Enzymology" (F. F. Nord, ed.), Vol. 16, pp. 105–184. Interscience, New York.

Mason, H. S. (1959). Structure of melanins. *In* "Pigment Cell Biology" (M. Gordon, ed.), pp. 563–582. Academic Press, New York.

Mason, H. S. (1967). The structure of melanin. *In* "The Pigmentary System: Advances in Biology of Skin" (W. Montagna and F. Hu, eds.), Vol. 8, pp. 293–312. Pergamon Press, Oxford.

Mason, H. S. (1976). Binuclear copper clusters as active sites for oxidases. *In* "Iron and

Copper Proteins" (K. T. Yasunobu, H. F. Mower, and O. Hayaishi, eds.), Plenum, New York.

Mason, H. S., and Peterson, E. W. (1965). Melanoprotein 1. Reactions between enzyme-generated quinones and amino acids. *Biochim. Biophys. Acta* **111,** 134–146.

Mason, H. S., Fowlks, W. L., and Peterson, E. (1955). Oxygen transfer and electron transport by the phenolase complex. *J. Am. Chem. Soc.* **77,** 2914–2915.

Mason, H. S., Ingram, J. E., and Allen, B. (1960). The free radical property of melanins. *Arch. Biochem. Biophys.* **86,** 225–230.

Masson, P. (1914). Le glande endocrine de l'intestine chez l'homme. *Acad. Sci. (Paris)* **158,** 59.

Masson, P. (1948). Pigment cells in man. *In* "The Biology of Melanomas" (R. W. Miner and H. Gordon, eds.), Vol. IV, pp. 15–51. N.Y. Acad. Sci., New York.

Matous, B., Budesinska, A., Budesinsky, M., Duchon, J., and Pavel, S. (1981). The identification of Thormählen positive melanogen "A" from the urine of melanoma patients as 6-methoxy-5-indolylglucosiduronate. *Neoplasma* **28,** 271–274.

Maul, G. G., and Brumbaugh, J. A. (1971). On the possible function of coated vesicles in melanogenesis of the regenerating fowl feather. *J. Cell Biol.* **48,** 41–48.

Mazza, F. P., and Stolfi, G. (1931). Ricerche sul pigmento di *Halla parthenopaea Costa. Arch. Sci. Biol.* **16,** 183–197.

Mazzocco, F., and Pifferi, P. G. (1976). An improvement of the spectrophotometric method for the determination of tyrosinase catecholase activity by Besthorn's hydrazone. *Anal. Biochem.* **72,** 643–647.

McCord, C. P., and Allen, F. P. (1917). Evidences associating pineal gland function with alterations in pigmentation. *J. Exp. Zool.* **23,** 207–224.

McEwan, M., and Parsons, P. G. (1987). Inhibition of melanization in human melanoma cells by a serotonin uptake inhibitor. *J. Invest. Dermatol.* **89,** 82–86.

McGinness, J., and Proctor, P. (1973). The importance of the fact that melanin is black. *J. Theor. Biol.* **39,** 677–678.

McGinness, J. E., Corry, P., and Proctor, P. (1974). Amorphous semiconductor switching in melanins. *Science* **183,** 853–854.

Meirowsky, E. (1909). Über Pigmentbildung in vom Körper losgeloster Haut. *Z. Pathol.* **2,** 439–448.

Meister, A., and Anderson, M. E. (1983). Glutathione. *Ann. Rev. Biochem.* **52,** 711–760.

Menon, A., Gan, E. V., and Haberman, H. F. (1976). Electron transfer properties of melanin and melanoproteins. *In* "Unique Properties of Melanocytes" (V. Riley, ed.), Vol. 3, pp. 69–81. Karger, Basel.

Menon, A., Leu, S. L., and Haberman, H. F. (1977). Electron transfer properties of melanin. Optimum conditions and the effects of various chemical treatments. *Can. J. Biochem.* **55,** 783–787.

Menon, I. A., Persad, S., Ranadive, N. S., and Haberman, H. F. (1985). Photobiological effects of eumelanin and pheomelanin. *In* "Pigment Cell 1985: Biological, Molecular and Clinical Aspects of Pigmentation" (J. T. Bagnara, S. N. Klaus, E. Paul, and M. Schartl, eds.), pp. 77–85. University of Tokyo Press, Tokyo.

Menon, I. A., Trope, G. E., Basu, P. K., Persad, S., Avaria, M., Felix, C. C., and Kalyanaraman, B. (1987). Is there any difference in the photobiological properties of melanins isolated from human blue and brown eyes? *Br. J. Ophthalmol.* **71,** 549–552.

Messenger, E. A., and Warner, A. E. (1977). The action of melatonin on single amphibian pigment cells in tissue culture. *Br. J. Pharmacol.* **61,** 607.

Meyer zum Gottesberge, A. M. (1988). Physiology and pathophysiology of inner ear melanin. *Pigment Cell Res.* **1,** 238–249.

Mikkelsen, R. B., and Triplett, E. L. (1975). Tyrosinase in *Rana pipiens.* Purification and physical properties. *J. Biol. Chem.* **250,** 638–643.

Minale, L., Fattorusso, E., Cimino, G., De Stefano, S., and Nicolaus, R. A. (1967). Struttura e biogenesi delle feomelanine. Nota III. Prodotti di degradazione. *Gazz. Chim. Ital.* **97,** 1636–1663.

Minale, L., Fattorusso, E., Cimino, G., De Stefano, S., and Nicolaus, R. A. (1969). Struttura e biogenesi delle feomelanine. Nota VIII. Sulla struttura della gallofeomelanina-1. *Gazz. Chim. Ital.* **99,** 431–449.

Minale, L., Fattorusso, E., De Stefano, S., and Nicolaus, R. A. (1970). Struttura e biogenesi delle feomelanine. Nota XI. Ulteriori ricerche sulla biogenesi delle feomelanine. *Gazz. Chim. Ital.* **100,** 461–466.

Mintz, B. (1967). Gene control of mammalian pigmentary differentiation. I: Clonal origin of melanocytes. *Proc. Natl. Acad. Sci. U.S.A.* **58,** 344–351.

Miranda, M., Botti, D., Bonfigli, A., and Arcadi, A. (1985). 5,6-Dihydroxyindole oxidation by mammalian, mushroom and amphibian tyrosinase preparations. *Biochim. Biophys. Acta* **841,**159–165.

Mishima, Y. (1973). Neutron capture tretment of malignant melanoma using a ^{10}B-chlorpromazine compound. *In* "Pigment Cell: Mechanisms in Pigmentation" (V. J. McGovern and P. Russel, eds.), Vol. 1, pp. 215–221. Karger, Basel.

Mishima, Y. (1990). A post-melanosoma era: Control of melanogenesis and melanoma growth. XIV Int. Pigment Cell Conference, Kobe, Japan, p. 70, (Abs.).

Mishima, Y., and Imokawa, G. (1985). Role of glycosylation in initial melanogenesis: Post inhibition dynamics. *In* "Pigment Cell 1985: Biological and Molecular Aspects of Pigmentation" (J. Bagnara, S. N. Klaus, E. Paul, and M. Scharl, eds.), pp. 17–30. University of Tokyo Press, Tokyo.

Mishima, Y., Imokawa, G., and Ogura, R. (1979). Functional three-dimensional differentiation of smooth membrane structure in melanogenesis. *In* "Pigment Cell: Biological Basis of Pigmentation" (S. N. Klaus, ed.), Vol. 4, pp. 277–290. Karger, Basel.

Mishima, Y., Hatta, S., Ohyama, Y., and Inazu, M. (1988). Induction of melanogenesis suppression: Cellular pharmacology and mode of differential action. *Pigment Cell Res.* **1,** 367–374.

Mishima, Y., Honda, C., Ichihashi, M., Obara, H., Hiratsuka, J., Fukuda, H., Karashima, H., Kobayashi, T., Kanda, K., and Yoshino, K. (1989a). Treatment of malignant melanoma by single thermal neutron capture therapy with melanoma-seeking ^{10}B-compound. *Lancet* **2,** 388–389.

Mishima, Y., Ichihashi, M., Hatta, S., Honda, C., Yamamura, K., and Nakagawa, T. (1989b). New thermal neutron capture therapy for malignant interaction from *in vitro* to first clinical trial. *Pigment Cell Res.* **2,** 226–234.

Misuraca, G., Nicolaus, R. A., Prota, G., and Ghiara, G. (1969). A cytochemical study of phaeomelanin formation in feather papillae of New Hampshire chick embryos. *Experientia* **25,** 920–922.

Misuraca, G., Prota, G., Lauvergne, J. J. (1974). Pigments melaniques du pelage de quelques races bovines: Le schema possible de certaines actions geniques. *Ann. Genet. Sel. Animal* **6,** 399–404.

Misuraca, G., Prota, G., Bagnara, J. T., and Frost, S. K. (1977). Identification of the leaf-frog melanophore pigment, rhodomelanochrome, as pterorhodin. *Comp. Biochem. Physiol.* **57B,** 41–43.

Miyake, Y., Izumi, Y., Tsutsumi, A., and Jimbow, K. (1986). Chemico-physical properties of melanin. IV. *In* "Structure and Functions of Melanins" (K. Jimbow, ed.), Vol. 2, pp. 3–18. Fuji-Shoin Co. Ltd., Sapporo, Japan.

Miyashita, Y., and Moriya, T. (1990). Calcium ion in chromatic nerve transmission and melanophore movements in teleost. *J. Exp. Zool.* **256,** 121–129.

Mizutani, U., Massalski, T., McGuiness, J. E., and Corry, P. M. (1976). Low temperature specific heat anomalies in melanins and tumour melanosomes. *Nature* **259,** 505–507.

Moellmann, G., McGuire, J., and Lerner, A. B. (1973). Intracellular dynamics and the fine structure of melanocytes with special reference to the effect of MSH and cyclic AMP on microtubules and 10 nm filaments. *Yale J. Biol. Med.* **46,** 337–360.

Moellmann, G., Slominski, A., Kuklinska, E., and Lerner, A. B. (1988). Regulation of melanogenesis in melanocytes. *Pigment Cell Res.* **1S,** 79–87.

Mojamdar, M., Ichihashi, M., and Mishima, Y. (1982). Tyrosinase and γ-glutamyl transpeptidase in 5-*S*-cysteinyldopa genesis within melanotic and amelanotic melanomas. *J. Dermatol.* **9,** 73–77.

Mojamdar, M., Ichihashi, M., and Mishima, Y. (1983). γ-glutamyl transpeptidase, tyrosinase and 5-*S*-cysteinyldopa production in melanoma cells. *J. Invest. Dermatol.* **81,** 119–121.

Molokhia, M., and Portnoy, B. (1973). Trace elements and skin pigmentation. *Br. J. Dermatol.* **89,** 207–209.

Morgan, B. D. G., O'Neill, T., Dewey, D. L., Galpine, A. R., and Riley, P. A. (1981). Treatment of malignant melanoma by intravascular 4-hydroxyanisole. *Clin. Oncol.* **7,** 227–234.

Morishima, T., and Hanawa, S. (1981). Urinary excretion of 5-*S*-cysteinyldopa in healthy Japanese. *Acta Derm. Venereol., (Stockh.)* **61,** 149–150.

Morrison, M. E., and Cohen, G. (1983). Novel substrates for tyrosinase serve as precursors of 6-hydroxydopamine and 6-hydroxydopa. *Biochemistry* **22,** 5465–5467.

Mosher, D. B., Fitzpatrick, T. B., Ortonne, J. P., and Hori, Y. (1987). Abnormalities of pigmentation. *In* "Dermatology in General Medicine" (T. B. Fitzpatrick, A. Z. Eisen, K. Wolff, I. W. Freedberg, and K. F. Austen, eds.), pp. 794–876. McGraw-Hill, New York.

Mottaz, J. M., and Zelickson, A. S. (1967). Melanin transfer: A possible phagocytic process. *J. Invest. Derm.* **49,** 605–610.

Movaghar, M. (1989). Tyrosinase activity in the first coat of agouti and black mice. *Pigment Cell Res.* **2,** 401–407.

Moyer, F. H. (1966). Genetic variation in the fine structure and ontogeny of mouse melanin granules. *Am. Zool.* **6,** 43–66.

Mufson, R. A. (1975). The tyrosinase activity of melanosomes from the Harding-Passey melanoma: The absence of a peroxidase component *in vitro*. *Arch. Biochem. Biophys.* **167,** 738–743.

Müller, G., Ruppert, S., Schmid, E., and Schutz, G. (1988). Functional analysis of alternatively spliced tyrosinase gene transcripts. *EMBO J.* **7,** 2723–2730.

Munn, E. A., and Bufton, S. F. (1973). Purification and properties of a phenol oxidase from the blowfly *Calliphora erythrocephala*. *Eur. J. Biochem.* **35,** 3–10.

Murray, M., Pawelek, J., and Lamoreaux, M. L. (1983). New regulatory factors for melanogenesis: Developmental changes in neonatal mice of various genotypes. *Dev. Biol.* **100,** 120–126.

Nambudiri, A. M. D., Bhat, J. V., and Subba Rao, P. V. (1972). Conversion of *p*-coumarate into caffeate by *Streptomyces nigrifaciens*. *Biochem. J.* **130,** 425–433.

Napolitano, A., Chioccara, F., and Prota, G. (1985a). A re-examination of the zinc-catalysed rearrangement of dopachrome using immobilized tyrosinase. *Gazz. Chim. Ital.* **115,** 357–359.

Napolitano, A., Corradini, M. G., and Prota, G. (1985b). A reinvestigation of the structure of melanochrome. *Tetrahedron Lett.* **26,** 2805–2808.

Nazzaro-Porro, M., and Passi, S. (1978). Identification of tyrosinase inhibitors in cultures of *Pityrosporum*. *J. Invest. Dermatol.* **71,** 205–208.

Nazzaro-Porro, M., Passi, S., Morpurgo, G., and Breathnach, A. S. (1979). Identification of tyrosinase inhibitors in cultures of *Pityrosporum* and their melanocytotoxic effect. *In* "Pigment Cell: Biologic Basis of Pigmentation" (S. N. Klaus, ed.), Vol. 4, pp. 234–243. Karger, Basel.

Nazzaro-Porro, M., Zina, G., Breathnach, A., Passi, S., Bernengo, A., Gallagher, S., Morpurgo, G. (1980). Effect of azelaic acid on human malignant melanoma. *Lancet* 1109–1111.

Needham, A. E. (1974). "The Significance of Zoochromes." Springer-Verlag, Berlin.

Nelson, J. M., and Dawson, C. R. (1944). Tyrosinase. *In* "Advances in Enzymology" (F. F. Nord and C. K. Wercman, eds.), Vol. 4, pp. 99–152. Interscience, New York.

Nicholls, E. M. (1973). Pigment spotting in man and the number of genes determining skin and eye colour. *Human Hered.* **23,** 1–12.

Nicolaus, R. A. (1962). Biogenesi delle melanine. *Rass. Med. Sperim.* **9,** 1–32.

Nicolaus, R. A. (1968). "Melanins." Hermann, Paris.

Nicolaus, R. A. (1978). Melanins. *In* "Metodicum Chimicum" (F. Korte and M. Goto, eds.), Vol. 11, pp. 190–199. Georg Thieme, Stuttgart.

Nicolaus, R. A., and Piattelli, M. (1962). Structure of melanins and melanogenesis. *J. Pol. Sci.* **58,** 1133.

Nicolaus, R. A., Piattelli, M., and Fattorusso, E. (1964). The structure of melanins and melanogenesis. IV. On some natural melanins. *Tetrahedron* **20,** 1163–1172.

Nicolaus, R. A., Prota, G., Santacroce, C., Scherillo, G., and Sica, D. (1969). Struttura e biogenesi delle feomelanine. Nota VII. Sulla struttura delle tricosiderine. *Gazz. Chim. Ital.* **99,** 323–350.

Niebauer, G. (1968). "Dendritic Cells of Human Skin." Monographs on Experimental Biology and Medicine. Vol. 2. Karger, Basel.

Nimmo, J. E., Hunter, J. A. A., Percy-Robb, I. W., Jay, B., Phillips, C. I., and Taylor, W.

O. G. (1985). Plasma 5-*S*-cysteinyldopa concentrations in oculocutaneous albinism. *Acta Derm. Venereol., (Stockh.)* **65,** 169–171.

Nimmo, J. E., Gawkrodger, D. J., O'Docherty, C. S. J., Going, S. M., Percy-Robb, I. W., and Hunter, J. A. A. (1988). Plasma 5-*S*-cysteinyldopa as an index of melanogenesis. *Br. J. Dermatol.* **118,** 487–495.

Nishioka, K. (1978). Particulate tyrosinase of human malignant melanoma. *Eur. J. Biochem.* **85,**137–146.

Nordlund, J. J. (1985). The pigmentary system: New interpretations of old data. *J. Dermatol.* **12,** 105–116.

Nordlund, J. J., and Ackles, A. (1985). Melanocytotoxins and fatty acids that stimulate proliferation of melanocytes in murine pinneal epidermis—A possible role for arachidonic acid metabolism in regulation of melanocyte growth. *In* "Pigment Cell 1985: Biological, Molecular, and Clinical Aspects of Pigmentation" (J. T. Bagnara, S. N. Klaus, E. Paul, and M. Schartl, eds.), pp. 501–513. University of Tokyo Press, Tokyo.

Nordlund, J. J., and Lerner, A. B. (1977). The effects of oral melatonin in skin color and on the release of pituitary hormones. *J. Clin. Endocrinol. Metab.* **45,** 768.

Nordlund, J. J., Collins, C. E., and Rheins, L. A. (1986). Prostaglandin E_2 and D_2 but not MSH stimulate the proliferation of pigment cells in the pinnal epidermis of the DBA/2 mouse. *J. Invest. Dermatol.* **86,** 433–437.

Nordlund, J. J., Amornisiripanitch, S., Rheins, L. A., Abdelk-Malek, Z. A., Boissy, R. E., and Bell, M. (1988). Role of the melanocyte in epidermal inflammatory and immune responses. *Pigment Cell Res.* **1S,** 101–112.

Novales, R. R. (1983). Cellular aspects of hormonally controlled pigment translocations within chromatophores of poikilothermic vertebrates. *Am. Zool.* **23,** 559–568.

Novellino, E., Ortonne, J. P., Voulot, C., Chioccara, F., Misuraca, G., and Prota, G. (1981). Identification of cysteinyldopa derived units in eumelanins from mammalian eyes. *FEBS Lett.* **125,** 101–103.

O'Donohue, T. L., and Dorsa, D. M. (1982). The opiomelanotropinergic neuronal and endocrine systems. *Peptides* **3,** 353–395.

Oikawa, A., and Nakayasu, M. (1974). Stimulation of melanogenesis in cultured melanoma cells by calciferols. *FEBS Lett.* **41,** 32–35.

Oikawa, A., and Nakayasu, M. (1975). Solutions of eumelanin showing no light scattering. *Anal. Biochem.* **63,** 634–637.

Okamoto, K., Yasumara, K., Fujitani, K., Kiso, Y., Kawauchi, H., Kawazoe, I., and Yajima, H. (1984). Synthesis of the heptadecapeptide corresponding to the entire amino acid sequence of salmon melanin-concentrating hormone (MCH). *Chem. Pharm. Bull.* **32,** 2963–2970.

Okazaki, K., Uzuka, M., Morikawa, F., Toda, K., and Seiji, M. (1976). Transfer, mechanism, of melanosomes in epidermal cell culture. *J. Invest. Dermatol.* **57,** 541–547.

Okazaki, M., Kuwata, K., Miki, Y., Shiga, S., and Shiga, T. (1985). Electron spin relaxation of synthetic melanin and melanin-containing human tissues as studied by electron spin echo and electron spin resonance. *Arch. Biochem. Biophys.* **242,** 197–205.

Okun, M., Edelstein, L., Or, N., Hamada, G., and Donnelan, B. (1970). The role of peroxidase versus the role of tyrosinase in enzymatic conversion of tyrosine to melanin in melanocytes, mast cells, and eosinophils. An autoradiographic-histochemical study. *J. Invest. Dermatol.* **55,** 1–12.

Okun, M. R., Edelstein, L. M., Hamada, G., Blumental, G., Donnelan, B., and Burnett, J. (1972). Oxidation of tyrosine and dopa to melanin mediated by mammalian peroxidase: The possible role of peroxidase in melanin synthesis and catecholamine synthesis *in vivo*. *In* "Pigmentation, Its Genesis and Biologic Control" (V. Riley, ed.), pp. 571–592. Appleton Century Crofts, New York.

Okun, M. R., Edelstein, L. M., Patel, R. P., and Donnellan, B. (1973). A revised concept of mammalian melanogenesis: The possible synergistic functions of aerobic dopa oxidase and peroxidase. *Yale J. Biol. Med.* **46,** 535–540.

Ortonne, J. P. (1990). Pigmentary changes of the ageing skin. *Br. J. Dermatol.* **35S,** 21–28.

Ortonne, J. P., Voulot, C., Khatchadourian, C., Palumbo, A., and Prota, G. (1981). A reexamination of melanogenesis in the ink gland of cephalopods. *In* "Pigment Cell 1981: Phenotypic Expression in Pigment Cells" (M. Seiji, ed.), pp. 49–57. University of Tokyo Press, Tokyo.

Osterlind, A., Tucker, M. A., Stone, B. J., and Jensen, O. M. (1988). The Danish case-control study of cutaneous malignant melanona. III. Hormonal and reproductive factors in women. *Int. J. Cancer* **42,** 821–824.

Packer, S. (1984). Tumor detection with radiopharmaceuticals. *Seminar Nucl. Med.* **14,** 21–29.

Palumbo, A., Nardi, G., d'Ischia, M., Misuraca, G., and Prota, G. (1983). Non enzymic oxidation of cysteinyldopa catalyzed by metallic ions. *Gen. Pharmacol.* **14,** 253–257.

Palumbo, A., Misuraca, G., d'Ischia, M., and Prota, G. (1985). Effect of metal ions on the kinetics of tyrosine oxidation catalysed by tyrosinase. *Biochem. J.* **228,** 647–651.

Palumbo, A., d'Ischia, M., Misuraca, G., and Prota, G. (1987a). Effect of metal ions on the rearrangement of dopachrome. *Biochim. Biophys. Acta* **925,** 203–209.

Palumbo, P., d'Ischia, M., Crescenzi, O., and Prota, G. (1987b). Isolation of a new intermediate in the oxidative conversion of 5,6-dihydroxyindole-2-carboxylic acid to melanin. *Tetrahedron Lett.* **28,** 467–470.

Palumbo, P., d'Ischia, M., and Prota, G. (1987c). Tyrosinase promoted oxidation of 5,6-dihydroxyindole-2-carboxylic acid to melanin. Isolation and characterization of oligomer intermediates. *Tetrahedron* **43,** 4203–4206.

Palumbo, A., d'Ischia, M., Misuraca, G., Prota, G., and Schultz, T. (1988). Structural modifications in biosynthetic melanins induced by metal ions. *Biochim. Biophys. Acta* **964,** 193–199.

Palumbo, A., d'Ischia, M., Misuraca, G., and Prota, G. (1989). A new look at the rearrangement of adrenochrome under biomimetic conditions. *Biochim. Biophys. Acta* **990,** 297–302.

Palumbo, A., d'Ischia, M., Misuraca, G., and Prota, G. (1990a). Selective uptake of 2-thiouracil into melanin producing system depends on chemical binding to enzymically generated dopaquinone. *Biochim. Biophys. Acta* **1036,** 221–227.

Palumbo, A., d'Ischia, M., Misuraca, G., Carratu, L., and Prota, G. (1990b). Activation of mammalian tyrosinase by ferrous ions. *Biochim. Biophys. Acta* **1033,** 256–260.

Palumbo, A., d'Ischia, M., Misuraca, G., and Prota, G. (1991a). Mechanism of inhibition of melanogenesis by hydroquinone. *Biochim. Biophys. Acta* **1073,** 85–90.

Palumbo, A., Solano, F., Misuraca, G., Aroca, P., Garcia Borron, J. C., Lozano, J. A., and Prota, G. (1991b). Comparative action of dopachrome tautomerase and metal ions on the rearrangement of dopachrome. *Biochim. Biophys. Acta* **1115,** 1–5.

Parakkal, P. F. (1967). The fine structure of melanocytes of the hair follicles of the guinea-pig. *In* "Advances in Biology of Skin" (W. Montagna and H. Hu, eds.), Vol. 8, pp. 179–188. Pergamon Press, New York.

Parrish, J. (1981). Photobiology of melanin pigmentation. *In* "Psoralens in Cosmetics and Dermatology" (J. Cahn, P. Forlot, C. Grupper, A. Meybeck, and F. Urbach, eds.), pp. 17–30. Pergamon Press, Paris.

Parsons, P. G. (1985). Modification of dopa toxicity in human tumour cells. *Biochem. Pharmacol.* **34,** 1801–1807.

Parsons, P. G., and Morrison, L. E. (1982). DNA damage and selective toxicity of dopa and ascorbate: Copper in human melanoma cells. *Cancer Res.* **42,** 3783–3788.

Passi, S., and Nazzaro-Porro, M. (1981). Molecular basis of substrate and inhibitory specificity of tyrosinase: Phenolic compounds. *Br. J. Dermatol.* **104,** 659–665.

Passi, S., Picardo, M., and Nazzaro-Porro, M. (1987). Comparative cytotoxicity of phenols *in vitro*. *Biochem. J.* **245,** 537–542.

Passi, S., Picardo, M., Mingrone, G., Breathnach, A. S., and Nazzaro-Porro, M. (1989). Azelaic acid: Biochemistry and metabolism. *Acta Derm. Venereol. (Stockh.)* **143S,** 8–13.

Patel, R. P., Okun, M. R., Edelstein, L. M., and Epstein, D. (1971). Biochemical studies on the peroxidase-mediated oxidation of tyrosine to melanin: Demonstration of the hydroxylation of tyrosine by plant and human peroxidases. *J. Biochem.* **124,** 439–441.

Patel, R. P., Okun, M. R., Edelstein, L. M., and Cariglia, N. (1974). Peroxidatic conversion of tyrosine to dopachrome. *J. Invest. Dermatol.* **63,** 374–377.

Pathak, M. A. (1967). Photobiology of melanogenesis: Biophysical aspects. *In* "Advances in Biology of Skin" (W. Montagna and F. Hu, eds.), Vol. 8, pp. 397–420. Pergamon Press, Oxford.

Pathak, M. A. (1985). Activation of the melanocyte system by ultraviolet radiation and cell proliferation. *Ann. N.Y. Acad. Sci.* **453,** 328–329.

Pathak, M. A., and Dalle Carbonare, M. (1989). Melanogenic potential of various furocoumarins in normal and vitiliginous skin. *In* "Psoralens" (T. B. Fitzpatrick, P. Forlot, M. A. Pathak, and F. Urbach, eds.), pp. 87–101. J. Libbey Eurotext, Paris.

Pathak, M. A., and Stratton, K. (1968). Free radicals in human skin before and after exposure to light. *Archiv. Biochem. Biophys.* **123,** 468–476.

Pathak, M. A., Jimbow, K., Parrish, J. A., Kaidbey, K. H., Kligman, A. L., and Fitzpatrick, T. B. (1976). Effect of UV-A, UV-B, and psoralen on *in vivo* human melanin pigmentation. *In* "Pigment Cell: Unique Properties of Melanocytes" (V. Riley, ed.), Vol. 3, pp. 291–298. Karger, Basel.

Pathak, M. A., Jimbow, K., and Fitzpatrick, T. B. (1981). Photobiology of pigment cell. *In* "Pigment Cell 1981: Phenotypic Expression in Pigment Cells" (M. Seiji, ed.), pp. 655–670. University of Tokyo Press, Tokyo.

Pathak, M. A., Zarebska, Z., Mihm, M. C., Jr., Jarzabek-Chorzelska, M., Chorzelski, T., and Jablonska, S. (1986). Detection of DNA–psoralen photoadducts in mammalian skin. *J. Invest. Dermatol.* **86,** 308–315.

Patil, D. G., and Chedekel, M. R. (1984). Synthesis and analysis of pheomelanin degradation products. *J. Org. Chem.* **49,** 997–1000.

Pavel, S. (1989). Analysis of eumelanin-related indolic compounds. *Trends Anal. Chem.* **8,** 177–183.

Pavel, S., Matous, B., Duchon, J., and Bubnova, E. (1978). 5-Hydroxy-6-methoxy-2-indolylcarboxylic and 6-hydroxy-5-methoxy-2-indolylcarboxylic acids in the urine of hamsters with melanotic melanoma. *Neoplasma* **25,** 343–347.

Pavel, S., Muskiet, F. A. J., Nagel, G. T., Schwippelova, Z., and Duchon, J. (1981). Identification of two Thormählen-positive compounds from melanocytic urine by gas chromatography–mass spectrometry. *J. Chromatogr.* **222,** 329–336.

Pavel, S., Elzinga, H., Muskiet, F. A. J., Smit, J. M., Mulder, N. H., and Schraffordt Koops, H. (1986). Eumelanin related indolic compounds in the urine of treated melanoma patients. *J. Clin. Chem. Clin. Biochem.* **24,** 167–173.

Pawelek, J. M. (1976). Factors affecting growth and pigmentation of melanoma cells. *J. Invest. Dermatol.* **66,** 201–209.

Pawelek, M. J. (1985). Studies on the Cloudman melanoma cell line as a model for the action of MSH. *Yale J. Biol. Med.* **58,** 571–578.

Pawelek, J. M. (1990). Dopachrome conversion factor functions as an isomerase. *Biochem. Biophys. Res. Commun.* **166,** 1328–1333.

Pawelek, J. M. (1991). After dopachrome? *Pigment Cell Res.* **4,** 53–62.

Pawelek, J. M., and Korner, A. M. (1982). The biosynthesis of mammalian melanin. *Amer. Sci.* **70,** 136–145.

Pawelek, J., and Lerner, A. B. (1978). 5,6-dihydroxyindole is a melanin precursor showing potent cytotoxicity. *Nature* **276,** 627–628.

Pawelek, J., Korner, A., Bergstrom, A., and Bologna, J. (1980). New regulators of melanin biosynthesis and the autodestruction of melanoma cells. *Nature* **286,** 617–619.

Pawelek, J., Chakraborty, A. K., and Orlow, S. J. (1991). Purification of high molecular weight melanogenic complex containing MSH receptors, dopachrome isomerase, gp75, and tyrosinase. *J. Clin. Invest.* (in press).

Penafiel, R., Galindo, J. D., Pedreno, E., and Lozano, J. A. (1982). The process for the activation of frog epidermis pro-tyrosinase. *Biochem. J.* **205,** 397–404.

Peter, M. G. (1989). Chemical modification of biopolymers by quinones and quinones methides. *Angew. Chemie Int. Ed. Engl.* **28,** 555–570.

Peter, M. G., and Förster, H. (1989). The structure of eumelanins: Identification of composition patterns by solid-state NMR spectroscopy. *Angew. Chemie Int. Ed. Engl.* **28,** 741–743.

Pfiffner, E., and Lerch, K. (1981). Histidine at the active site of *Neurospora* tyrosinase. *Biochemistry* **20,** 6030–6035.

Piattelli, M., Fattorusso, E., Magno, S., and Nicolaus, R. A. (1962). The structure of melanins and melanogenesis. Nota II. Sepiomelanin and synthetic pigments. *Tetrahedron* **18,** 941–949.

Piattelli, M., Fattorusso, E., Magno, S., and Nicolaus, R. A. (1963). The structure of melanins and melanogenesis. Nota III. The structure of sepiomelanin. *Tetrahedron* **19,** 2061–2072.

Piattelli, M., Fattorusso, E., Nicolaus, R. A., and Magno, S. (1965). The structure of melanins and melanogenesis. Nota V. Ustilago melanin. *Tetrahedron* **21,** 3229–3236.

Picardo, M., Passi, S., Nazzaro-Porro, M., Breathnach, A., Zompetta, C., Faggioni, A., and Riley, P. (1987). Mechanism of antitumoral activity of catechols in culture. *Biochem. Pharmacol.* **36,**417–425.

Picardo, M., Zombetta, C., De Luca, C., Cirone, M., Faggioni, A., Nazzaro-Porro, M., Passi, S., and Prota, G. (1991). Role of skin surface lipids in UV-induced epidermal cell changes. *Arch. Dermatol. Res.* **283,** 191–197.

Poirer, J., Donaldson, J., and Barbeau, A. (1985). The specific vulnerability of the *substantia nigra* to MPTP is related to the presence of transition metals. *Biochem. Biophys. Res. Commun.* **128,** 25–33.

Pomerantz, S. H. (1966). The tyrosine hydroxylase activity of mammalian tyrosinase. *J. Biol. Chem.* **241,** 161–168.

Pomerantz, S. H. (1976). A sensitive new assay for the oxidation of 3,4-dihydroxy-L-phenylalanine by tyrosinase. *Anal. Biochem.* **75,** 86–90.

Pomerantz, S. H., and Li, J. P. C. (1973). Purification and properties of tyrosinase isoenzymes from hamster melanoma. *Yale J. Biol. Med.* **46,** 541–552.

Pomerantz, S. H., and Li, J. P. C. (1974). Tyrosinase in the skin of albino hamsters and mice. *Nature* **252,** 241–243.

Pomerantz, S. H., and Murthy, V. V. (1974). Purification and properties of tyrosinase from *Vibrio tyrosinaticus. Arch. Biochem. Biophys.* **160,** 73–82.

Pomerantz, S. H., and Warner, M. C. (1967). 3,4-dihydroxy-L-phenylalanine as the tyrosinase cofactor. *J. Biol. Chem.* **242,** 5308–5314.

Porter, S., and Mintz, B. (1991). Multiple alternative spliced transcripts of the mouse tyrosinase gene. *Gene* **79,** 277–282.

Potts, A. M. (1962). The concentration of phenothiazines in the eyes of experimental animals. *Invest. Ophthalmol.* **1,** 522–524.

Potts, A. M., and Au, P. C. (1976). The affinity of melanin for inorganic ions. *Exp. Eye Res.* **22,** 487–491.

Presse, F., Nahon, J. L., Fischer, W. H., and Vale, W. (1990). *Mol. Endocrinol.* **4,**632–637.

Prezioso, J. A., Fitzgerald, G. B., and Wick, M. M. (1990). Effects of tyrosinase activity on the cytotoxicity of 3,4-dihydroxybenzylamine and buthionine sulfoximine in human melanoma cells. *Pigment Cell Res.* **3,** 49–54.

Proctor, T., Auclair, B. G., and Stokowski, L. (1976). Endocrine factors and the growth and spread of B16 melanoma. *J. Natl. Cancer Inst.* **57,** 1197–1198.

Prota, G. (1972). Structure and biogenesis of pheomelanins. *In* "Pigmentation: Its Genesis and Biologic Control" (V. Riley, ed.), pp. 615–630. Appleton Century Crofts, New York.

Prota, G. (1980a). Recent advances in the chemistry of melanogenesis in mammals. *J. Invest. Derm.* **75,** 122–127.

Prota, G. (1980b). Nitrogenous pigments in marine invertebrates. *In* "Marine Natural Product" (J. P. Scheuer, ed.), Vol. 3, pp. 141–178. Academic Press, New York.

Prota, G. (1988a). Progress in the chemistry of melanins and related metabolites. *Med. Res. Rev.* **8,** 525–556.
Prota, G. (1988b). Some new aspects of eumelanin chemistry. *In* "Advances in Pigment Cell Research" (J. T. Bagnara ed.), pp. 101–124. Alan Liss, New York.
Prota, G. (1989a). Melanin and pigmentation. *In* "Coenzymes and Cofactors" (D. Dolphin, R. Paulson, and O. Abramovic, eds.), Vol. 3, pp. 441–466. Wiley, New York.
Prota, G. (1989b). Mechanisms in melanin pigmentation: An up-to-date view. *In* "Psoralens: Past, Present and Future of Photochemoprotection and other Biological Activities" (T. B. Fitzpatrick, P. Forlot, M. A. Pathak, and F. Urbach, eds.), pp. 13–23. John Libbey Eurotext, Paris.
Prota, G. (1990).
Prota, G., and Nicolaus, R. A. (1967a). On the biogenesis of pheomelanins. *In* "Advances in Biology of Skin" (W. Montagna and F. Hu, eds.), Vol. 8, pp. 323–328. Pergamon Press, New York.
Prota, G., and Nicolaus, R. A. (1967b). Struttura e biogenesi delle feomelanine. Nota I. Isolamento e proprietà dei pigmenti delle piume. *Gazz. Chim. Ital.* **97,** 666–684.
Prota, G., and Searle, A. G. (1978). Biochemical sites of gene action for melanogenesis in mammals. *Ann. Genet. Sel. Anim.* **10,** 1–8.
Prota, G., and Thomson, R. H. (1976). Melanin pigmentation in mammals. *Endeavour* **35,** 31–38.
Prota, G., Piattelli, M., and Nicolaus, R. A. (1966). Preliminary results in the study of phaeomelanins. *Rend. Accad. Sci. Mat. Fis.* **33,** 146–150.
Prota, G., Scherillo, G., Napolano, E., and Nicolaus, R. A. (1967). Struttura e biogenesi delle feomelanine. Nota II. Sulla reazione tra *o*-chinoni e cisteina. *Gazz. Chim. Ital.* **97,** 1451–1478.
Prota, G., Scherillo, G., and Nicolaus, R. A. (1968a). On the structure of "trichosiderins." *Rend. Accad. Sci. Mat. Fis.* **35,** 1–4.
Prota, G., Scherillo, G., and Nicolaus, R. A. (1968b). Struttura e biogenesi delle feomelanine. Nota IV. Sintesi e proprietà della 5-*S*-dicysteinildopa. *Gazz. Chim. Ital.* **98,** 495–510.
Prota, G., Scherillo, G., Petrillo, O., and Nicolaus, R. A. (1969). Struttura e biogenesi delle feomelanine. Nota X. Sulla struttura delle tricosiderine. *Gazz. Chim. Ital.* **93,** 1193–1207.
Prota, G., Petrillo, O., Santacroce, C., and Sica, D. (1970a). 2*H*-1,4-benzothiazines from the reaction of 2-aminoethanethiol with nascent 4-methyl-*o*-benzoquinone. *J. Heterocycl. Chem.* **7,** 555–561.
Prota, G., Crescenzi, S., Misuraca, G., and Nicolaus, R. A. (1970b). New intermediates of pheomelanogenesis *in vitro*. *Experientia* **26,** 1058–1059.
Prota, G., Suarato, A., Nicolaus, R. A. (1971). The isolation and structure of trichosiderin B. *Experientia* **27,** 1381–1383.
Prota, G., D'Agostino, M., and Misuraca, G. (1972). The structure of hallachrome: 7-hydroxy-8-methoxy-6-methyl-1,2-anthraquinone. *J. Chem. Soc.* **1,** 1614–1616.
Prota, G., Ponsiglione, E., and Ruggiero, R. (1974). Synthesis and photochromism of $\Delta^{2,2'}$-bi-(2*H*-1,4-benzothiazine). *Tetrahedron* **30,** 2781–2784.
Prota, G., Rorsman, H., Rosengren, A.-M., and Rosengren, E. (1976a). Occurrence of trichochromes in the urine of a melanoma patient. *Experientia* **32,** 1122–1124.

Prota, G., Rorsman, H., Rosengren, A.-M., and Rosengren, E. (1976b). Phaeomelanin pigments from a human melanoma. *Experientia* **32,** 970–971.

Prota, G., Ito, S., and Nardi, G. (1977a). The chemistry of adenochromes. *In* "Marine Natural Products Chemistry" (D. J. Faulkner and W. H. Fenical, eds.), NATO Conf. Ser. IV Marine Sciences, Vol. 1, pp. 45–59. Plenum Press, New York.

Prota, G., Rorsman, H., Rosengren, A.-M., and Rosengren, E. (1977b). Isolation of 2-*S*-cysteinyldopa and 2,5-*S,S*-dicysteinyldopa from the urine of patients with melanoma. *Experientia* **33,**720–721.

Prota, G., Ortonne, J. P., Voulot, C., Khatchadourian, C., Nardi, G., and Palumbo, A. (1981). Occurrence and properties of tyrosinase in the ejected ink of cephalopods. *Comp. Biochem. Physiol.* **68B,** 415–419.

Prota, G., d'Ischia, M., and Napolitano, A. (1988). The regulatory role of sulfhydryl compounds in melanogenesis. *Pigment Cell Res.* **1S,** 48–53.

Prunieras, M. (1986). Melanocytes, melanogenesis, and inflammation. *Int. J. Dermatol.* **25,** 624–628.

Pullman, A., and Pullman, B. (1961). The band structure of melanins. *Biochim. Biophys. Acta* **54,** 384–385.

Quay, W. B., and Bagnara, J. T. (1964). Relative potencies of indolic and related compounds in the body-lightening reaction of larval *Xenopus*. *Arch. Int. Pharmacodyn. Ther.* **150,** 137–143.

Quevedo, W. C., Jr., Szabo, G., Virks, J., and Sinesi, S. J. (1969). Influence of age and UV in the populations of DOPA-positive melanocytes in human skin. *J. Invest. Dermatol.* **52,** 287–290.

Quevedo, W. C., Jr., Fitzpatrick, T. B., Pathak, M. A., and Jimbow, K. (1974). Light and skin color. *In* "Sunlight and Man" (T. B. Fitzpatrick, M. A. Pathak, L. C. Herber, M. Seiji, and A. Kukita, eds.), pp. 165–194. University of Tokyo Press, Tokyo.

Quevedo, W. C., Jr., Fleischmann, R. D., and Holstein, T. J. (1981). Pheomelanogenesis in the mouse: A review of its genetic and developmental features. *In* "Pigment Cell 1981: Phenotypic Expression in Pigment Cells" (M. Seiji, ed.), pp. 129–137. University of Tokyo Press, Tokyo.

Quevedo, W. C., Jr., Fitzpatrick, T. B., and Jimbow, K. (1985). Human skin color; origin, variation and significance. *J. Human Evolution* **14,** 43–56.

Quevedo, W. C., Jr., Fitzpatrick, T. B., Szabo, G., and Jimbow, K. (1987). Biology of the melanin pigmentary system. *In* "Dermatology in General Medicine" (T. B. Fitzpatrick, H. S. Eisen, K. Wolff, M. Freedberg, and K. F. Austen, eds.), Vol. 1, pp. 224–251. McGraw-Hill, New York.

Quevedo, W. C., Jr., Holstein, T. J., Dyckman, J., and Nordlund, J. J. (1990). Influence of depigmenting agents on hair and skin color in yellow (pheomelanic) and black (eumelanic) mice. *Pigment Cell Res.* **3,** 71–79.

Quilico, A. (1937). "I Pigmenti Neri Animali e Vegetali." Fusi, Pavia.

Rabey, J. M., and Hefti, F. (1990). Neuromelanin synthesis in rat and human *substantia nigra*. *J. Neural. Transm.* **2,** 1–14.

Raghavan, P. R., Zane, P. A., and Tripp, S. L. (1990). Calculation of drug-melanin binding energy using molecular modeling. *Experientia* **46,** 77–80.

Raper, H. S. (1927). The tyrosinase–tyrosine reaction. VI. Production from tyrosine of

5,6-dihydroxyindole and 5-6-dihydroxyindole-2-carboxylic acid the precursor of melanin. *Biochem. J.* **21,** 89–96.

Raper, H. S. (1928). The aerobic oxidases. *Physiol. Rev.* **8,** 245–282.

Raper, H. S. (1938). Some problems of tyrosine metabolism. *J. Chem. Soc.* 125–130.

Rast, D. M., Stussi, H., Hegnauer, H., and Nyhlen, L. E. (1981). Melanins. *In* "The Fungal Spore: Morphogenetic Controls" (G. Turian and H. R. Hohl, eds.), pp. 507–531. Academic Press, New York.

Rawles, M. E. (1948). Origin of melanophores and their role in development of color patterns in vertebrates. *Physiol. Rev.* **28,** 383–408.

Ray, A. C., and Eakin, R. E. (1975). Studies on the biosynthesis of aspergillin by *Aspergillus niger*. *Appl. Microbiol.* **30,** 909–915.

Reams, W. M., Shervette, R. E., and Dorman, W. H. (1968). Refractoriness of mouse dermal melanocytes to hormones. *J. Invest. Dermatol.* **50,** 338.

Reiter, R. J. (1986). Normal patterns of melatonin levels in the pineal gland and body fluids of humans and experimental animals. *J. Neural Transm.* **21S,** 35–54.

Resneik, S. (1988). Melasma induced by oral contraceptive. *J. Am. Med. Assoc.* **199,** 601–605.

Rich, P. R., Wiegand, N. K., Blum, H., Moore, A. L., and Bonner, W. D., Jr. (1978). Studies on the mechanism of inhibition of redox enzymes by substituted hydroxamic acids. *Biochem. Biophys. Acta* **525,** 325–337.

Riddiford, L. M., and Hiruma, K. (1988). Regulation of melanization in insect cuticle. *In* "Advances in Pigment Cell Research" (J. T. Bagnara, ed.), pp. 422–436. Alan Liss, New York.

Riley, P. A. (1974a). Pathological disturbances of pigmentation. *In* "The Physiology and Pathophysiology of the Skin" (A. Jarret, ed.), Vol. 3, pp. 1167–1197. Academic Press, New York.

Riley, P. A. (1974b). Biochemistry of pigment. *In* "The Physiology and Pathophysiology of the Skin" (A. Jarrett, ed.), Vol. 3, pp. 1149–1165. Academic Press, New York.

Riley, P. A. (1980). Melanins and melanogenesis. *Pathobiol. Ann.* **10,** 223–251.

Riley, P. A. (1981). Chemical leucoderma. *In* "Pigment Cell 1981: Phenotypic Expression in Pigment Cells" (M. Seiji, ed.), pp. 445–449. University of Tokyo Press, Tokyo.

Riley, P. A. (1984). Hydroxyanisole: The current status. *In* "Hydroxyanisole: Recent Advances in Antimelanoma Therapy" (P. A. Riley, ed.), pp. 1–10. IRL Press, Oxford.

Riley, P. A. (1991). Melanogenesis: A realistic target for antimelanoma therapy? *Eur. J. Cancer* **27,** 1172–1177.

Robb, D. A. (1984). Tyrosinase. *In* "Copper Proteins" (R. Loutie, ed.), Vol. 2, pp. 207–234. CRC Press, Boca Raton, Florida.

Robin, C. P. (1873). "Anatomic et Physiologie Cellulaire." Bailliers et Fils, Paris.

Rodgers, A. D., and Curzon, G. (1975). Melanin formation by human brain *in vitro*. *J. Neurochemistry* **24,** 1123–1129.

Rodighiero, G. (1985). Hyperpigmentation induced by furocoumarins. *Il Farmaco* **40,** 173–186.

Rodriguez-Lopez, J. N., Tudela, J., Varon, R., and Garcia-Canovas, F. (1991). Kinetic study on the effect of pH on the melanin biosynthesis pathway. *Biochim. Biophys. Acta* **1076,** 379–386.

Rorsman, H., and Pavel, S. (1990). Metabolic markers and melanoma. *In* "Cutaneous Melanoma Biology and Management" (N. Cascinelli, M. Santinami, and U. Veronesi, eds.), pp. 79–87. Masson. Milan.

Rorsman, H., and Tegner, E. (1988). Biochemical observations in UV-induced pigmentation. *Photodermatol.* **5,** 30–38.

Rorsman, H., Rosengren, A.-M., and Rosengren, E. (1973). Determination of 5-*S*-cysteinyldopa in melanomas with a fluorimetric method. *Yale J. Biol. Med.* **46,** 516–522.

Rorsman, H., Agrup, G., Hansson, C., Rosengren, A.-M., and Rosengren, E. (1979). Detection of phaeomelanins. *In* "Pigment Cell: Biologic Basis of Pigmentation" (S. Klaus, ed.), Vol. 4, pp. 244–252. Karger, Basel.

Rorsman, H., Agrup, G., Hansson, C., and Rosengren, E. (1983). Biochemical recorders of malignant melanoma. *In* "Pigment Cell" (R. M. MacKie, ed.), Vol. 6, pp. 93–115. Karger, Basel.

Rorsman, H., Agrup, P., Carlén, B., Hansson, C., Jonsson, N., Rosengren, E., and Tegner, E. (1986). Trichochromuria in melanisos of melanoma. *Acta Derm. Venereol. (Stockh.)* **66,** 468–473.

Rorsman, H., Albertsson, E., Edholm, L. E., Hansson, C., Ogren, L., and Rosengren, E. (1988). Thiols in the melanocyte. *Pigment Cell Res.* **1S,** 54–60.

Rosdahl, I. K. (1979). Local and systemic effect on the epidermal melanocyte population in UV-irradiated mouse skin. *J. Invest. Dermatol.* **73,** 306–309.

Rosdahl, I. K., and Rorsman, H. (1983). An estimate of the melanocyte mass in human. *J. Invest. Dermatol.* **81,** 278–281.

Rosen, C. F., Seki, Y., Farinelli, W., Stern, R. S., Fitzpatrick, T. B., Pathak, M. A., and Gange, R. W. (1987). A comparison of the melanocyte response to narrow band UVA and UVB exposure *in vivo*. *J. Invest. Dermatol.* **88,** 774–779.

Rosen, C. F., Jacques, S. L., Stuart, M. E., and Gange, R. W. (1990). Immediate pigment darkening: Visual and reflectance spectrophotometric analysis of action spectrum. *Photochem. Photobiol.* **51,** 583–588.

Rosengren, E., Linder-Eliasson, E., and Carlsson, A. (1985). Detection of 5-*S*-cysteinyldopamine in human brain. *J. Neural Transmission* **63,** 247–253.

Ruppert, S., Muller, G., Kwon, B. S., and Schultz, G. (1988). Multiple transcripts of the mouse tyrosinase gene are generated by alternative splicing. *EMBO J.* **7,** 2715–2722.

Rust, C. C., and Meyer, R. K. (1969). Hair color, molt and testis size in male, short-tailed weasels treated with melatonin. *Science* **165,** 921–922.

Rzepecki, L. M., and Waite, J. H. (1991). α,β-Dehydro-3,4-dihydroxyphenylalanine derivatives: Rate and mechanism of formation. *Arch. Biochim. Biophys.* **285,** 27–36.

Rzepecki, L. M., Nagafuchi, T., and Waite, J. H. (1991). α,β-Dehydro-3,4-dihydroxyphenylalanine derivatives: Potential sclerotization intermediates in natural composite materials. *Arch. Biochim. Biophys.* **285,** 17–26.

Sakurai, T., Ochiai, H., and Takeuchi, T. (1975). Ultrastructural change of melanosomes associated with agouti pattern formation in mouse hair. *Dev. Biol.* **47,** 466–471.

Samuelson, D. A., Ulshater, R. J., and Smith, P. (1991). Effect of low zinc diet on elemental content of the retina. *Invest. Ophthalmol. Vis. Sci.* **32,** 463.

Sarna, T. (1991). Properties and function of the ocular melanin—A photobiophysical view. *J. Photochem. Photobiol. B: Biol.* **12,** 215–258.
Sarna, T., and Sealy, R. C. (1984). Photoinduced oxygen consumption in melanin system. Action spectra and quantum yield for eumelanin and synthetic melanin. *Photochem. Photobiol.* **39,** 69–74.
Sarna, T., and Swartz, H. M. (1979). Identification and characterization of melanin in tissues and body fluids. *Folia Histochim. Cygtochem.* **16,** 275.
Sarna, T., and Swartz, H. M. (1991). Interaction of melanin with oxygen (and related species). *In* "Atmospheric Oxidation and Antioxidants" (G. Scott, ed.), Vol. 3, pp. Elsevier, Amsterdam.
Sarna, T., Hyde, J. S., and Swartz, H. M. (1976). Ion-exchange in melanin: An electron spin resonance study with lanthanide probes. *Science* **192,** 1132–1134.
Sarna, T., Duleba, A., Korytowski, W., and Swartz, H. M. (1980). Interaction of melanin with oxygen. *Arch. Biochem. Biophys.* **200,** 140–148.
Sarna, T., Korytowski, W., Pasenkiewicz-Gierula, M., and Gudowska, E. (1981). Ion-exchange studies in melanins. *In* "Pigment Cell 1981: Phenotypic Expression in Pigment Cells" (M. Seiji, ed.), pp. 23–29. University of Tokyo Press, Tokyo.
Sarna, T., Menon, I. A., and Sealy, R. C. (1984). Photoinduced oxygen consumption in melanin systems. II. Action spectra and quantum yields for pheomelanins. *Photochem. Photobiol.* **39,** 805–890.
Sato, C., Ito, S., and Takeuchi, T. (1985). Establishment of a mouse melanocyte clone which synthesizes both eumelanin and pheomelanin. *Cell Struct. Funct.* **10,** 421–425.
Sawyer, T. K., Hruby, V. J., Hadley, M. E., and Engel, M. H. (1983). Alpha-melanocyte stimulating hormone: Chemical nature and mechanism of action. *Amer. Zool.* **23,** 529–540.
Scalia, M., Geremia, E., Corsaro, C., Santoro, C., Baratta, D., and Sichel, G. (1990). Lipid peroxidation in pigmented and unpigmented liver tissues: Protective role of melanin. *Pigment Cell Res.* **3,** 115–119.
Schallreuter, K. U., and Wood, J. M. (1989). Thioredoxin reductase in control of the pigmentary system. *In* "Clinics Dermatology: Disorders of Pigmentation" (R. Mackie, ed.), Vol. 7, pp. 92–105. J. B. Lippincott, Philadelphia.
Schallreuter, K. U., Wood, J. M., and Berger, J. (1991). Low catalase levels in the epidermis of patients with vitiligo. *J. Invest. Dermatol.* **97,** 1081–1085.
Schultz, T. (1986). Chelation of 5,6-dihydroxyindole with divalent metals. *J. Invest. Dermatol.* **87,** 406.
Scislowski, P. W. D., Dlominski, A., and Bomirski, A. (1984). Biochemical characterization of three hamster melanoma variants. II. Glycolysis and oxygen consumption. *Int. J. Biochem.* **16,** 327–331.
Sealy, R. C., Felix, C. C., Hyde, J. S., and Swartz, H. M. (1980). Structure and reactivity of melanins: Influence of free radicals and metal ions. *In* "Free Radicals in Biology" (W. A. Pryor, ed.), Vol. 4, pp. 209–259. Academic Press, New York.
Sealy, R. C., Hyde, J. S., Felix, C. C., Menon, I. A., and Prota, G. (1982a). Eumelanins and pheomelanins: Characterization by electron spin resonance spectroscopy. *Science* **217,** 545–547.
Sealy, R. C., Hide, J. S., Felix, C. C., Menon, I. A., Prota, G., Swartz, H. M., Persad, S.,

and Haberman, H. F. (1982b). Novel free radical in synthetic and natural pheomelanins: Distinction between dopa melanins and cysteinyldopa melanins by ESR spectroscopy. *Proc. Natl. Acad. Sci. U.S.A.* **79,** 2885–2889.

Searle, A. G. (1968). "Comparative Genetics of Coat Colour in Mammals." Logos-Academic, London.

Searle, A. G. (1989). Comparative genetics of albinism. *Ophthalmic Pediatr. Genet.* **3,** 159–164.

Seechurn, P., and Thody, J. A. (1990). The effect of ultraviolet radiation and melanocyte-stimulating hormone on tyrosinase activity in epidermal melanocytes of the mouse. *J. Dermatol. Sci.* **1,** 283–288.

Seiji, M. (1981). Melanosomes. *In* "Pigment Cell 1981: Phenotypic Expression in Pigment Cells" (M. Seiji, ed.), pp. 3–13. University of Tokyo Press, Tokyo.

Seiji, M., Fitzpatrick, T. B., and Birbeck, M. S. C. (1961). The melanosome: A distinctive subcellular particle of mammalian tyrosinase and the site of melanogenesis. *J. Invest. Dermatol.* **36,** 243–252.

Seiji, M., Shimao, K., Birkbeck, M. S. C., and Fitzpatrick, T. B. (1963). Subcellular localisation of melanin biosynthesis. *Ann. N.Y. Acad. Sci.* **100,** 497–533.

Semple, T. U., Moore, G. E., Morgan, R. T., Woods, L. K., and Quinn, L. A. (1982). Multiple cell lines from patients with malignant melanoma: Morphology, karylogy, and biochemical analysis. *JNCI* **68,** 365–380.

Senoh, S., and Witkop, B. (1959a). Non-enzymatic conversion of dopamine to norepinephrine and trihydroxyphenethylamines. *J. Chem. Soc.* **81,** 6222–6230.

Senoh, S., and Witkop, B. (1959b). Formation and rearrangements of aminochromes from a new metabolite of dopamine and some of its derivatives. *J. Am. Chem. Soc.* **81,** 6231–6240.

Shibahara, S., Tomita, Y., Sakakura, T., Nager, C., Chaudhuri, B., and Muller, R. (1986). Cloning and expression of cDNA encoding mouse tyrosinase. *Nucleic Acids Res.* **14,** 2413–2427.

Shibahara, S., Tomita, Y., Tagami, H., Muller, R. M., and Cohen, T. (1988). Molecular basis for the heterogeneity of human tyrosinase. *J. Exp. Med.* **156,** 403–405.

Shibahara, S., Okinaga, S., Tomita, Y., Takeda, A., Yamamoto, H., Sato, M., and Takeuchi, T. (1990). A point mutation in the tyrosinase gene of BALB/c albino mouse causing the cysteine-serine substitution at position 85. *Eur. J. Biochem.* **189,** 455–461.

Shibata, T., Prota, G., and Mishima, Y. (1990). Regulatory factors of melanin monomer and polymer formation in melanogenic subcompartment of pigment cells. XIV Intl. Pigment Cell Conference, Kobe, Japan, p. 91, (Abs.).

Shono, S., Imura, M., Ota, M., Satoshi, O., and Toda, K. (1985). The relationship of skin color, UVB-induced erythema, and melanogenesis. *J. Invest. Dermatol.* **84,** 265–267.

Sichel, G. (1988). Biosynthesis and function of melanins in hepatic pigmentary system. *Pigment Cell Res.* **1,** 250–258.

Sichel, G., Corsaro, C., Scalia, M., Sciuto, S., and Geremia, E. (1987). Relationship between melanin content and superoxide dismutase (SOD) activity species of animals. *Cell Biochem. Funct.* **5,** 123–128.

Siekevitz, P., and Palade, G. E. (1958). Cytochemical study on the pancreas of the guinea pig. II. Functional variations in the enzymic activity of microsomes. *J. Biophys. Biochem. Cytol.* **4,** 309–318.

Silman, R. E., Leone, R. M., Hooper, R. J. L., and Preece, M. A. (1979). Melatonin, the pineal gland and puberty. *Nature* **282,** 301–303.

Silvers, W. K. (1979). "The Coat Colors of Mice: A Model for Mammalian Gene Action and Interaction." Springer-Verlag, New York.

Silvers, W. K., and Russell, E. S. (1955). An experimental approach to action of genes at the agouti locus in the mouse. *J. Exp. Zool.* **130,** 199–220.

Skinner, H. A. (1961). "The Origin of Medical Terms." Williams and Wilkins, Baltimore, Maryland.

Slawinska, D., and Slawinski, J. (1982). Electronically excited molecules in the formation and degradation of melanins. *Physiol. Chem. Phys.* **14,** 363–374.

Slominski, A., and Costantino, R. (1991). L-Tyrosine induces tyrosinase expression via a posttranscriptional mechanism. *Experientia* **47,** 721–724.

Slominski, A., Moellmann, G., Kuklinska, E., Bomirski, A., and Pavelek, J. (1988). Positive regulation of melanin pigmentation by two key substrates of the melanogenic pathway, L-tyrosine and L-dopa. *J. Cell Sci.* **89,** 287–296.

Smit, N. P. M., Pavel, S., Kammameyer, A., and Westerhof, W. (1990). Determination of catechol *O*-methyltransferase activity in relation to melanin metabolism using high-performance liquid chromatography with fluorimetric detection. *Analyt. Biochem.* **190,** 286–291.

Smith, P. I., and Swan, G. A. (1976). A study of the supposed hydroxylation of tyrosine catalysed by peroxidase. *J. Biochem.* **153,** 403–408.

Smyth, J. R., Jr., Porter, J. W., and Bohren, B. B. (1951). A study of pigments from red, brown, and buff feathers and hair. *Physiol. Zool.* **24,** 205–216.

Snell, R. S. (1962). Further observation on the effect of progesterone on the activity of melanocytes in the skin. *Z. Zellforsch.* **57,** 818–826.

Snell, R. S. (1967). Hormonal control of pigmentation in man and other mammals. *In* "Advances in Biology of Skin. The Pigmentary System" (W. Montagna and F. Hu, eds.), Vol. 8, pp. 447–466. Pergamon Press, New York.

Snell, R. S. (1972). Hormonal control of hair color. *In* "Pigmentation: Its Genesis and Biologic Control" (V. Riley, ed.), pp. 193–205. Appleton Century Crofts, New York.

Solano-Munoz, F., Penafiel, R., and Galindo, J. D. (1985). An electrometric method for the determination of tyrosinase activity. *Biochem. J.* **229,** 573–578.

Solomon, E. I. (1981). Binuclear copper active site: Hemocyanin, tyrosinase, and type 3 copper oxidases. *In* "Copper Proteins" (T. G. Spiro, ed.), pp. 42–107. Wiley Intersciences, New York.

Somes, R. G., Jr., and Smyth, J. R., Jr. (1965). Feather pheomelanin intensity in buff orpington, New Hampshire and Rhode Island red breeds of fowl. *Poultry Sci.* **154,** 40–46.

Song, P. S., and Tapley, K. J. (1979). Photochemistry and photobiology of psoralens. *Photochem. Photobiol.* **29,** 1177–1197.

Sorby, H. C. (1878). On the colouring matters found in human hair. *J. Anthropol. Inst. Lond.* **8,** 1–14.

Spiegel, M. A., and Henny, G. C. (1939). Studies on melanins. III. X-ray diffraction patterns of photosynthetic and genuine melanins. *J. Amer. Chem. Soc.* **61,** 2178–2179.

Sponenberg, D. P., Ito, S., Eng, L. A., and Schwink, K. (1988a). Pigment types of various color genotypes of horses. *Pigment Cell Res.* **1,** 410–413.

Sponenberg, P., Ito, S., Wakamatsu, K., and Eng, L. A. (1988b). Pigment types in sheep, goats, and llamas. *Pigment Cell Res.* **1,** 414–418.

Spritz, R. A., Strunk, K. M., Giebel, L. B., and King, R. A. (1990). Detection of mutations in the tyrosinase gene in a patient with type IA oculocutaneous albinism. *New Eng. J. Med.* **322,** 1724–1728.

Stepien, K., and Wilczok, T. (1982). Studies of the mechanism of chloroquine binding to synthetic dopa-melanin. *Biochem. Pharmacol.* **31,** 3359–3365.

Stiener, U., Rosdahl, I., Augustsson, A., and Kadegel, B. (1989). UVB irradiation induces melanocyte increase in both exposed and shielded human skin. *J. Invest. Dermatol.* **92,** 561–564.

Straub, K., Kanne, D., Hearst, J. E., and Rapoport, H. (1981). Isolation and characterization of pyrimidine–psoralen photoadducts from DNA. *J. Amer. Chem. Soc.* **103,** 2347–2355.

Strobel, M. C., Seperack, P. K., Moore, K. J., Copeland, N. G., and Jenkins, N. A. (1988). *In* "Advances in Pigment Cell Research" (J. T. Bagnara, ed.), Vol. 256, pp. 297–305. Alan Liss, New York.

Strothcamp, K. G., Jolley, R. L., and Mason, H. S. (1976). Quaternary structure of mushroom tyrosinase. *Biochem. Biophys. Res. Commun.* **70,** 519–524.

Strzelecka, T. (1982a). A hypothetical structure of melanin and its relation to biology. *Physiol. Chem. Phys.* **14,** 233–237.

Strzelecka, T. (1982b). A band model for synthetic dopa-melanin. *Physiol. Chem. Phys.* **14,** 219–222.

Sugiyama, S. (1979). Mode of redifferentiation and melanogenesis of melanocytes in mouse hair follicles: An ultrastructural and cytochemical study. *J. Ultrastruct. Res.* **67,** 40–50.

Sugiyama, S., and Kukita, A. (1976). Melanocyte reservoir in the hair follicles during the hair growth cycle: An electron microscopic study. *In* "Biology and Diseases of the Hair" (T. Kobori and W. Montagna, eds.), pp. 181–200. University of Tokyo Press, Tokyo.

Sugumaran, M. (1986). Tyrosinase catalyzes an unusual oxidative decarboxylation of 3,4-dihydroxymandelate. *Biochemistry* **25,** 4489–4492.

Sugumaran, M. (1987). Quinone methide sclerotization: A revised mechanism for β-sclerotization of insect cuticle. *Bioorg. Chem.* **15,** 194–211.

Sugumaran, M. (1990). Novel oxidation chemistry of catecholamine derivatives and related compounds. *In* "Biological Oxidation Systems" (C. C. Reddy, G. A. Hamilton, and K. M. Madyastha, eds.), Vol. 1, pp. 347–363. Academic Press, New York.

Sugumaran, M., and Semensi, V. (1990). Formation of a stable quinone methide during tyrosinase-catalyzed oxidation of alpha-methyldopa methyl ester and its implication in melanin biosynthesis. *Bioorg. Chem.* **18,** 144–153.

Sugumaran, M., and Semensi, V. (1991). Quinone methide as a new intermediate in eumelanin biosynthesis. *J. Biol. Chem.* **266,** 6073–6078.

Sugumaran, M., Semensi, V., Dali, H., and Mitchell, W. (1989a). Novel transformations of enzymatically generated carboxymethyl-*o*-benzoquinone to 2,5,6-trihydroxybenzofuran and 3,4-dihydroxymandelic acid. *Bioorg. Chem.* **17,** 86–95.

Sugumaran, M., Dali, H., Kundzicz, H., and Semensi, V. (1989b). Unusual, intramolecular cyclization and side chain desaturation of carboxyethyl-*o*-benzoquinone derivatives. *Bioorg. Chem.* **17,** 443–453.

Sundaram, C., Köster, W., and Schallreuter, K. U. (1990). The effect of UV radiation and sun blockers on free radical defence in human and guinea pig epidermis. *Arch. Dermatol. Res.* **282,** 526–531.

Swan, G. A. (1974). Structure, chemistry and biosynthesis of the melanins. *In* "Fortschritte der Chemie Organischer Naturstoffe" (W. Hertz, H. Grisenbach, and G. W. Kirby, eds.), Vol. 31, pp. 522–528. Springer-Verlag, Wien.

Swan, G. A. (1976). Studies related to the chemistry of melanins. Part XIV. The alleged formation of a *p*-quinonoid aminochrome by oxidation of 2,4,5-trihydroxyphenylethylamine. *J. Chem. Soc., Perkin. Trans.* **I,** 339–341.

Swan, G. A., and Waggott, A. (1970). Studies related to the chemistry of melanins. Part X. Quantitative assessment of different types of units present in Dopa-melanin. *J. Chem. Soc.* **(C),** 1409–1418.

Szabo, G. (1965). Current state of pigment research with special reference to macromolecular aspects. *In* "Biology of the Skin and Hair Growth" (A. G. Lyne and B. F. Short, eds.), pp. 705–726. Angus and Robertson, Sidney.

Szabo, G., Hirobe, T., Flynn, E. A., and Garcia, R. I. (1988). The biology of the melanocyte. *In* "Advances in Pigment Cell Research" (J. T. Bagnara, ed.), pp. 463–474. Alan Liss, New York.

Takahashi, H., and Fitzpatrick, T. B. (1966). Large amounts of dioxyphenylalanine in the hydrolysate of melanosomes from Harding-Passey mouse melanoma. *Nature* **209,** 888–890.

Takayama, Y., Wada, C., Kawauchi, H., and Ono, M. (1989). Structures of two genes coding for melanin-concentrating hormone of chum salmon. *Gene* **80,** 65–74.

Takeda, A., Matsunaga, J., Tomita, Y., Tagami, H., and Shibahara, S. (1989). Molecular analysis of the DNA segment cross-hybridizable to the tyrosinase gene in patients afflicted with oculocutaneous albinism. *Tohoku J. Exp. Med.* **159,** 333–340.

Takeuchi, T. (1985). Genes controlling intercellular communication during melanocyte differentiation in the mouse. *Zool. Sci.* **2,** 823–831.

Takeuchi, T. (1992). Molecular structure of the tyrosinase gene. *In* "The Pigment Cell: From the Molecular to the Clinical Level" (Y. Mishima, ed.). Munksgaard, Copenhagen. (In press).

Takeuchi, T., and Yamamoto, H. (1988). Genetic regulation of melanocyte differentiation. *Pigment Cell Res.* **1S,** 32–37.

Tamate, H. B., and Takeuchi, T. (1984). Action of the *e* locus of mice in the response of phaeomelanic hair follicles to alpha-melanocyte-stimulating hormone *in vitro*. *Science* **224,** 1241–1242.

Tamura, A., Halaban, R., Moellman, G., Cowan, J. M., Lerner, M. R., and Lerner, A. B. (1987). Normal murino melanocytes in culture. *In Vitro Cell Devel. Biol.* **23,** 519–522.

Tanaka, S., Yamamoto, H., Takeuchi, S., and Takeuchi, T. (1990). Melanization in albino mice transformed by introducing cloned mouse tyrosinase gene. *Development* **108,** 223–227.

Terao, M., Tabe, L., Garattini, S., Sartori, D., and Mintz, B. (1989). Isolation and characterization of variant cDNAs encoding mouse tyrosinase. *Biochem. Biophys. Res. Commun.* **159,** 848–853.

Thathachari, Y. T. (1976). Spatial structure of melanins. *In* "Pigment Cell: Unique Properties of Melanocytes" (V. Riley, ed.), Vol. 3, pp. 64–68. Karger, Basel.

Thody, A. J., Ridley, K., Carter, R. J., Lucas, A. M., and Shuster, S. (1984). α-MSH and coat colour changes in the mouse. *Peptides* **5,** 1031–1036.

Thody, A. J., Higgins, E. H., Wakamatzu, K., Burchill, S. A., and Marks, J. M. (1991). Pheomelanin as well as eumelanin is present in human epidermis. *J. Invest. Dermatol.* **97,** 340–344.

Thomas, M. (1955). Melanins. *In* "Modern Methods of Plant Analysis" (K. Paech and M. V. Traceye, eds.), Vol. 4, pp. 661–675. Springer-Verlag, Berlin.

Thomson, R. H. (1962). Melanins. *In* "Comparative Biochemistry" (M. Florkin and H. S. Mason, eds.), Vol. 3, pp. 727–753. Academic Press, New York.

Thomson, R. H. (1974). The pigments of reddish hair and feathers. *Angew. Chem. Int. Ed. Engl.* **13,** 305–312.

Tilders, F. J. H., and Smelik, P. G. (1975). A diurnal rhythm in melanocyte-stimulating hormone content of the rat pituitary gland and its independence from the pineal gland. *Neuroendocrinology* **17,** 296.

Tjälve, H., Nilsson, M., and Larsson, B. (1982). Thallium-201: Autoradiography in pigmented mice and melanin-binding *in vitro. Acta Pharmacol. Toxicol.* **51,** 147–153.

Toda, K., Pathak, M. A., Parrish, J. A., and Fitzpatrick, T. B. (1972). Alteration of racial differences in melanosome distribution in human epidermis after exposure to ultraviolet light. *Nature New Biol.* **236,** 143–145.

Tomita, Y., Hariu, A., Mizuno, C., and Seiji, M. (1980). Inactivation of tyrosinase by dopa. *J. Invest. Derm.* **75,** 379–382.

Tomita, Y., Fukushima, M., and Tagami, H. (1986). Stimulation of melanogenesis by cholecalciferol in cultured human melanocytes: A possible mechanism underlying pigmentation after ultraviolet radiation. *Tohuku J. Exp. Med.* **149,** 451–452.

Tomita, Y., Takeda, A., Okinaga, S., Tagami, H., and Shibahara, S. (1989). Human oculocutaneous albinism caused by single base insertion in the tyrosinase gene. *Biochem. Biophys. Res. Commun.* **164,** 990–996.

Tosti, A., and Castelli, E. (1990). The function of melanocytes. *In* "Cutaneous Melanoma Biology and Management" (N. Cascinelli, M. Santinami, and U. Veronesi, eds.), pp. 75–77. Masson, Milan.

Toussaint, O., and Lerch, K. (1987). Catalytic oxidation of 2-aminophenols and *ortho*-hydroxylation of aromatic amines by tyrosinase. *Biochemistry* **26,** 8567–8571.

Townsend, D., Oetting, W., Polman, T., and King, R. A. (1990). Purification and characterization of dopachrome tautomerase/isomerase (DT). XIV Int. Pigment Cell Conference, Kobe, Japan, p. 90, (Abs.).

Tsukamoto, K., Jackson, I. J., Urabe, K., Montague, P., and Hearing, V. J. (1992). A second tyrosinase related protein, TPR-2, is a melanogenic enzyme termed dopachrome tautomerase. *EMBO J.* **11,** 519–526.

Turner, W. A., Jr., Taylor, J. D., and Tchen, T. T. (1975). Melanosome formation in goldfish: The role of multivesicular bodies. *J. Ultrastruct. Res.* **51,** 16–31.

Van Woert, M. H. (1967). Oxidation of reduced nicotinamide adenine dinucleotide by melanin. *Life Sci.* **6,** 2605–2612.

Van Woert, M. H., Prasad, K. N., and Borg, D. C. (1967). Spectroscopic studies of *substantia nigra* pigment in human subjects. *J. Neurochem.* **14,** 707–716.

Varga, J. M., Di Pasquale, A., Pawelek, J., McGuire, J., and Lerner, A. B. (1974). Regulation of melanocyte-stimulating hormone (MSH) action at the receptor level: Discontinuous binding of MSH to synchronized mouse melanoma cells during the cell cycle. *Proc. Natl. Acad. Sci. U.S.A.* **71,** 1590–1593.

Varga, J. M., Moellmann, G., Fritsch, P., Godawska, E., and Lerner, A. B. (1976). Association of cell surface receptors for melanotropin with the Golgi region in mouse melanoma cells. *Proc. Natl. Acad. Sci. U.S.A.* **73,** 559–562.

Vaughan, J. H., Fischer, W. H., Hoeger, G., Riveri, J., and Vale, W. (1989). Characterization of melanin concentrating hormone. *Endocrinology* **125,** 1660–1665.

Vermeer, M., Schmieder, G. J., Yoshikawa, T., Van den Berg, J. W., Metzman, M. S., Taylor, J. R., and Streilein, J. W. (1991). Effects of ultraviolet B light on cutaneous immune responses of human with deeply pigmented skin. *J. Invest. Dermatol.* **97,** 729–734.

Veseley, D. L., and Hadley, M. E. (1979). Calcium requirements for melanophore-stimulating hormone action on melanophores. *Science* **173,** 923–925.

von Figura, K., Rey, M., Prinz, R., Voss, B., and Ulrich, K. (1979). Effect of tunicamycin on transport of lysosomal enzymes in cultured skin fibroblast. *Eur. J. Biochem.* **101,** 103–109.

Voorhees, M. L. (1970). Urinary excretion of dopa and metabolites by patients with melanoma. *Cancer* **26,** 146–149.

Waite, H. (1983). Evidence for a repeating 3,4-dihydroxyphenylalanine- and hydroxyproline-containing decapeptide in the adhesive protein of the mussel, *Mytilus edulis. L. J. Biol. Chem.* **258,** 2911–2915.

Wakamatsu, W., and Ito, S. (1990). Identification of ester glucuronide and sulfate conjugates of 5-hydroxy-6-methoxyindole-2-carboxylic acid and 6-hydroxy-5-methoxyindole-2-carboxylic acid in melanoma urine. *J. Dermatol. Sci.* **1,** 253–260.

Wakamatsu, W., Ito, S., and Fujita, K. (1988). Melanin-related metabolites in urine of B16 melanoma-bearing mice. *Acta Derm. Venereol. (Stockh.)* **68,** 385–389.

Wakamatsu, K., Ito, S., and Magatsu, T. (1990). Quantitative analysis of neuromelanin based on its biogenesis. XIV Int. Pigment Cell Conference, Kobe, Japan, p. 125, (Abs.).

Wakamatsu, K., Ito, S., and Nagatsu, T. (1991). Cysteinyldopamine is not incorporated into neuromelanin. *Neurosci. Lett.* **131,** 57–60.

Walker, M. J., Beattie, C. W., Patel, M. K., Ronan, S. M., and Das Gupta, T. K. (1987). Estrogen receptor in malignant melanoma. *J. Clin. Oncol.* **5,** 1256–1261.

Wasterstrom, S. A. (1984). Accumulation of drugs on inner ear melanin. *Scand. Andid.* 1–40.

Weatherhead, B. (1982). The pineal gland and pigmentation. *In* "Physiology and Pathophysiology of the Skin" (A. Jarret, ed.), pp. 2165–2179. Academic Press, New York.

Weatherhead, B. (1988). Melanotropins and melanin pigmentation in the skin of mammals. *In* "The Melanotropic Peptides" (M. E. Hadley, ed.), Vol. 2, pp. 1–19. CRC Press, Boca Raton, Florida.

Weatherhead, B., and Logan, A. (1979). Seasonal variations in the response of hair follicle melanocytes to melanocyte stimulating hormone. *J. Endocrinol.* **81,** 167–168.

Westerhof, W., Pavel, S., Hammeyer, A., and Beusenberg, F. D. (1987). Melanin-related metabolites as markers of the skin pigmentary system. *J. Invest. Dermatol.* **89,**78–81.

Westerhof, W., Estevez-Uscanga, O., Meens, J., Kammeyer, A., Durocq, M., and Cario, I. (1990). The relation between constitutional skin color and photosensitivity estimated from UV-induced erythema and pigmentation dose–response curves. *J. Invest. Dermatol.* **94,** 812–816.

Wheeler, M. H., and Bell, A. A. (1988). Melanins and their importance in pathogenic fungi. *Curr. Top. Med. Mycol.* **2,** 338–387.

Wheeler, C. E., Cawley, E. P., and Curtis, A. C. (1953). The effects of topically applied hormones on growth, pigmentation, and keratinization of the nipple and areola of immature castrated male guinea-pigs. *J. Invest. Dermatol.* **20,** 385–399.

Whimster, I. W. (1979). Compound evolution and the pigmentary system. *In* "Pigment Cell: Biologic Basis of Pigmentation" (S. N. Klaus, ed.), Vol. 4, pp. 1–12. Karger, Basel.

Whittaker, J. R. (1971). Biosynthesis of a thiouracil pheomelanin in embryonic pigment cells exposed to thiouracil. *J. Biol. Chem.* **246,** 6217–6226.

Whittaker, J. R. (1973). Tyrosinase in the presumptive pigment cells of ascidian embryos: Tyrosine accessibility may initiate melanin synthesis. *Dev. Biol.* **30,** 441–454.

Whittaker, J. R. (1979). Melanoprotein. Absence of a direct melanin–protein relationship in chick embryo melanocytes. *Biochim. Biophys. Acta* **583,** 378–387.

Wick, M. M. (1977). L-Dopa methyl ester as a new antitumor agent. *Nature* **269,** 512–513.

Wick, M. M. (1980). An experimental approach to the chemotherapy of melanoma. *J. Invest. Dermatol.* **74,** 63–65.

Wick, M. M., Byers, L., and Frei, E. (1977). L-Dopa: Selective toxicity for melanoma cells *in vitro*. *Science* **197,** 468–469.

Wick, M. M., Hearing, V. J., and Rorsman, H. (1987). Biochemistry of melanization. *In* "Dermatology in General Medicine" (T. B. Fitzpatrick, A. Z. Eisen, K. Wolff, M. Freedberg, and K. F. Austen, eds.), Vol. 1, pp. 251–258. McGraw-Hill, New York.

Wilcox, D. E., Porras, A. G., Hwang, T., Lerch, K., Winkler, M. E., and Solomon, E. I. (1985). Substrate analogue binding to the coupled binuclear copper active site in tyrosinase. *J. Am. Chem. Soc.* **107,** 4015–4027.

Wilczok, T., Bilinska, B., Buzman, E., and Kopera, M. (1984). Spectroscopic studies of chemically modified synthetic melanins. *Arch. Biochem. Biophys.* **231,** 257–262.

Wilkes, B. C., Hruby, V. J., Sherbrooke, W. C., Castrucci, A. M. L., and Hadley, M. E. (1984). Synthesis and biological actions of melanin concentrating hormone. *Biochem. Biophys. Res. Commun.* **122,** 613–619.

Wille, E., and Lüttke, W. (1971). 4,4,4′,4′-Tetramethyl-$\Delta^{2,2'}$-bipyrrolidine-3,3-dione, a compound having the basic chromophore system of indigo. *Angew. Chem. Intern. Ed. Engl.* **10,** 803–804.

Williams-Smith, D. L., and Dunne, L. J. (1976). X-ray photoelectron spectroscopy and the structure of melanins. *FEBS Lett.* **69,** 291–294.

Winkler, M. E., Lerch, K., and Solomon, E. I. (1981). Competitive inhibitor binding to the binuclear copper active site in tyrosinase. *J. Am. Chem. Soc.* **103,** 7001–7003.

Wirestrand, L. E., Hansson, C., Rosengren, E., and Rorsman, H. (1985). Melanocyte metabolites in the urine of people of different skin colour. *Acta Derm. Venerol. (Stockh.)* **65,** 354–348.

Witkop, C. J., Quevedo, W. C., Jr., and Fitzpatrick, T. B. (1978). Albinism. *In* "The Metabolic Basis of Inherited Disease" (J. B. Stanbury, J. B. Wyngaarden, and D. S. Fredickson, eds.), pp. 2983–3150. McGraw-Hill, New York.

Witkop, C. J., King, R. A., and Townsend, D. W. (1988). Human albinism and animal models of albinism. *Pigment Cell Res.* **1S,** 88–100.

Wittbjer, A., Dahlback, B., Odh, G., Rosengren, A. M., Rosengren, E., and Rorsman, H. (1989). Isolation of human tyrosinase from cultured melanoma cells. *Acta Derm. Venereol. (Stockh.)* **69,** 125–131.

Wittbjer, A., Odh, G., Rosengren, A. M., Rosengren, E., and Rorsman, H. (1990a). Isolation of tyrosinase from bovine eyes. *Pigment Cell Res.* **3,** 168–172.

Wittbjer, A., Odh, G., Rosengren, A. M., Rosengren, E., and Rorsman, H. (1990b). Isolation of soluble tyrosinase from human melanoma cells. *Acta Derm. Venereol. (Stockh.)* **70,** 291–294.

Wolbarsht, M. L., Walsh, A. W., and George, G. (1981). Melanin, a unique biological absorber. *Appl. Opt.* **20,** 2184–2186.

Wolff, K., Jimbow, K., and Fitzpatrick, T. B. (1974). Experimental pigment donation *in vivo*. *J. Ultrastruct. Res.* **47,** 400–419.

Wolfram, L. J. (1981). The reactivity of human hair. A review. *In* "Hair Research" (C. Orfanos and W. Montagna, eds.), pp. 479–500. Springer-Verlag, Berlin.

Wolfram, L., and Albrecht, L. (1987). Chemical and photo bleaching of brown and red hair. *J. Soc. Cosmet. Chem.* **82,** 179–184.

Wolfram, L. J., Hall, K., and Hui, I. (1970). The mechanism of hair bleaching. *J. Soc. Cosmet. Chem.* **21,** 875.

Wolfram, L. J., Patil, D., and Albrecht, L. (1986). Photochemical properties of pheo- and eumelanins. *J. Invest. Dermatol.* **87S,** 396.

Wood, B. J. B., and Ingraham, L. L. (1965). Labeled tyrosinase from labeled substrate. *Nature* **205,** 291–292.

Wood, J. M., and Schallreuter, K. U. (1988). Reduced thioredoxin inhibits melanin biosynthesis: Evidence for the formation of a stable bis-cysteinate complex with tyrosinase. *Inorg. Chim. Acta* **151,** 7.

Wood, J. M., and Schallreuter, K. U. (1991). Studies on the reactions between human tyrosinase superoxide anion, hydrogen peroxide and thiol. *Biochim. Biophys. Acta* **1074,** 378–385.

Woolery, G. L., Powers, L., Winkler, M., Solomon, E. I., Lerch, K., and Spiro, T. G. (1984). Extended X-ray absorption fine structure study of the binuclear copper active site of tyrosinase from *Neurospora crassa*. *Biochim. Biophys. Acta* **788,** 155–164.

Wyman, L. C. (1924). Blood and nerve as controlling agents in the movements of melanophores. *J. Exp. Zool.* **39,** 73–132.

Yaar, M., and Gilchrest, B. A. (1991). Human melanocyte growth and differentiation: A decade of new data. *J. Invest. Dermatol.* **97,** 611–617.

Yamada, I., Seki, S., Matsubara, O., Ito, S., Suzuki, S., and Kasuga, T. (1987). The cytotoxicity of cysteinylcatechols and related compounds to human melanoma cells *in vitro*. *J. Invest. Dermatol.* **88,** 538–540.

Yamada, K., Larsson, B., Roberto, A., Denker, L., and Ulberg, S. (1988). Selective incorporation of thiouracil into murine metastatic melanomas. *J. Invest. Dermatol.* **90,** 873–876.

Yamada, K., Salopek, T., Jimbow, K., and Ito, S. (1989). An extremely high content of pheomelanin in dysplastic melanocytic nevi. *J. Invest. Dermatol.* **92,** 544.

Yamada, I., Seki, S., Ito, S., Suzuki, S., Matsubara, O., and Kasuga, T. (1991). The killing effect of 4-*S*-cysteaminylphenol, a newly synthesised melanin precursor, on B16 melanoma cell lines. *Br. J. Cancer* **63,** 187–190.

Yamamoto, H., Takeuchi, S., Kudo, T., Makino, K., Nakata, A., Shinoda, T., and Takeuchi, T. (1987). Cloning and sequencing of mouse tyrosinase cDNA. *Jpn. J. Genet.* **62,** 271–274.

Yamamoto, H., Takeuchi, S., Kudo, T., Sato, C., and Takeuchi, T. (1989). Melanin production in cultured albino melanocytes transfected with mouse tyrosinase cDNA. *J. Gen.* **64,** 121–135.

Yamaura, I., Yonekura, M., Katsura, Y., Ishiguro, M., and Funatsu, M. (1980). Subunit structure of phenoloxidase purified from the larvae of housefly. *Agr. Biol. Chem.* **44,** 2437.

Yasunobu, K. T. (1959). Mode of action of tyrosinase. *In* "Pigment Cell Biology" (M. Gordon, ed.), pp. 583–608. Academic Press, New York.

Yasunobu, K. T., Peterson, E. W., and Mason, H. S. (1959). The oxidation of tyrosine-containing peptides by tyrosinase. *J. Biol. Chem.* **234,** 3291–3295.

Yohn, J. J., Norris, D. A., Yrastorza, D. G., Buno, I. J., Leff, J. A., Hake, S. S., and Repine, J. E. (1991). Disparate antioxidant enzyme activities in cultured human cutaneous fibroblast, keratinocytes and melanocytes. *J. Invest. Dermatol.* (in press).

Youdim, M. B. H., Ben-Shachar, D., and Riederer, P. (1989). Is Parkinson's disease a progressive siderosis of *substantia nigra* in iron- and melanin-induced neurodegeneration? *Acta Neurol. Scand.* **126,** 47–54.

Young, T. E., and Babbitt, B. W. (1983). Electrochemical study of the oxidation of α-methyldopamine, α-methylnoradrenaline, and dopamine. *J. Org. Chem.* **48,** 562–566.

Young, T. E., Oriswold, J. R., and Hulbert, M. H. (1974). Melanin. 1. Kinetics of oxidative cyclization of dopa to dopachrome. *J. Org. Chem.* **39,** 1980–1982.

Young, A. R., Potten, C.S., Chadwick, C. A., Murphy, G. M., Hawk, J. L. M., and Cohen, A. J. (1991). Photoprotection and 5-MOP photoprotection from UVR-induced DNA damage in humans: The role of skin type. *J. Invest. Dermatol.* **97,** 942–948.

Yurkow, E. J., and Laskin, J. D. (1989). Purification of tyrosinase to homogeneity based on its resistance to sodium dodecyl sulfate-proteinase K digestion. *Arch. Biochem. Biophys.* **1,** 122–129.

Zdarsky, E., Favor, J., and Jackson, I. J. (1990). The molecular basis of brown, an old mouse mutation, and of an induced revertant to wild type. *Genetics* **126,** 443–449.

Zecca, L., Mecacci, C., Seraglia, R., Parati, E. (1992). The chemical characterization of melanin contained in substantia nigra of human brain. *Biochim. Biophys. Acta* **1138,** 6–10.

Ziechmann, W. (1980). "Huminstoffe." Verlag Chemie, Weinheim.

Zeise, L. (1991). Bio-analytical studies of eumelanins. Ph.D. Thesis. John Hopkins University, Baltimore, Maryland.

Zimmermann, A. A., and Becker, S. W., Jr. (1959). Melanoblast and melanocytes in fetal Negro skin. *Monogr. Med. Sci.* **6,** 1–59.

Zondek, B., and Krohn, H. (1932). Hormon des Zwichenlappens der Hypophyse (Intermedin). I. Die Rotfarbung der Elritze als Testobjekt. *Klin. Worchsch.* **2,** 849–853.

Zuasti, A., Jara, J. R., Ferrer, C., and Solano, F. (1989). Occurrence of melanin granules and melanosynthesis in the kidney of *Sparus auratus. Pigment Cell Res.* **2,** 93–99.

Zviak, C., and Dawber, P. (1986). Hair structure and function. *In* "The Science of Hair Care" (C. Zviak, ed.), pp. 1–45. Marcel Dekker, New York.

Index